IFIP Advances in Information and Communication Technology 542

IFIP – The International Federation for Information Processing

IFIP was founded in 1960 under the auspices of UNESCO, following the first World Computer Congress held in Paris the previous year. A federation for societies working in information processing, IFIP's aim is two-fold: to support information processing in the countries of its members and to encourage technology transfer to developing nations. As its mission statement clearly states:

IFIP is the global non-profit federation of societies of ICT professionals that aims at achieving a worldwide professional and socially responsible development and application of information and communication technologies.

IFIP is a non-profit-making organization, run almost solely by 2500 volunteers. It operates through a number of technical committees and working groups, which organize events and publications. IFIP's events range from large international open conferences to working conferences and local seminars.

The flagship event is the IFIP World Computer Congress, at which both invited and contributed papers are presented. Contributed papers are rigorously refereed and the rejection rate is high.

As with the Congress, participation in the open conferences is open to all and papers may be invited or submitted. Again, submitted papers are stringently refereed.

The working conferences are structured differently. They are usually run by a working group and attendance is generally smaller and occasionally by invitation only. Their purpose is to create an atmosphere conducive to innovation and development. Refereeing is also rigorous and papers are subjected to extensive group discussion.

Publications arising from IFIP events vary. The papers presented at the IFIP World Computer Congress and at open conferences are published as conference proceedings, while the results of the working conferences are often published as collections of selected and edited papers.

IFIP distinguishes three types of institutional membership: Country Representative Members, Members at Large, and Associate Members. The type of organization that can apply for membership is a wide variety and includes national or international societies of individual computer scientists/ICT professionals, associations or federations of such societies, government institutions/government related organizations, national or international research institutes or consortia, universities, academies of sciences, companies, national or international associations or federations of companies.

More information about this series at http://www.springer.com/series/6102

Jason Staggs · Sujeet Shenoi (Eds.)

Critical Infrastructure Protection XII

12th IFIP WG 11.10 International Conference, ICCIP 2018
Arlington, VA, USA, March 12–14, 2018
Revised Selected Papers

Springer

Editors
Jason Staggs
Tandy School of Computer Science
University of Tulsa
Tulsa, OK, USA

Sujeet Shenoi
Tandy School of Computer Science
University of Tulsa
Tulsa, OK, USA

ISSN 1868-4238 ISSN 1868-422X (electronic)
IFIP Advances in Information and Communication Technology
ISBN 978-3-030-04536-4 ISBN 978-3-030-04537-1 (eBook)
https://doi.org/10.1007/978-3-030-04537-1

Library of Congress Control Number: 2018963824

This Springer imprint is published by the registered company Springer Nature Switzerland AG
The registered company address is: Gewerbestrasse 11, 6330 Cham, Switzerland

Contents

Contributing Authors ix

Preface xv

PART I THEMES AND ISSUES

1
A Theory of Homeland Security 3
Richard White

2
An Evidence Quality Assessment Model for Cyber Security Policymaking 23
Atif Hussain, Siraj Shaikh, Alex Chung, Sneha Dawda and Madeline Carr

3
Liability Exposure when 3D-Printed Parts Fall from the Sky 39
Lynne Graves, Mark Yampolskiy, Wayne King, Sofia Belikovetsky and Yuval Elovici

PART II INFRASTRUCTURE PROTECTION

4
Error Propagation After Reordering Attacks on Hierarchical State Estimation 67
Ammara Gul and Stephen Wolthusen

5
Securing Data in Power-Limited Sensor Networks Using Two-Channel Communications 81
Clark Wolfe, Scott Graham, Robert Mills, Scott Nykl and Paul Simon

6
Reversing a Lattice ECP3 FPGA for Bitstream Protection 91
Daniel Celebucki, Scott Graham and Sanjeev Gunawardena

7
Protecting Infrastructure Data via Enhanced Access Control, Blockchain and Differential Privacy 113
Asma Alnemari, Suchith Arodi, Valentina Rodriguez Sosa, Soni Pandey, Carol Romanowski, Rajendra Raj and Sumita Mishra

8
A New SCAP Information Model and Data Model for Content Authors 127
Joshua Lubell

PART III INFRASTRUCTURE MODELING AND SIMULATION

9
Modeling a Midstream Oil Terminal for Cyber Security Risk Evaluation 149
Rishabh Das and Thomas Morris

10
A Cyber-Physical Testbed for Measuring the Impacts of Cyber Attacks on Urban Road Networks 177
Marielba Urdaneta, Antoine Lemay, Nicolas Saunier and Jose Fernandez

11
Persistent Human Control in a Reservation-Based Autonomous Intersection Protocol 197
Karl Bentjen, Scott Graham and Scott Nykl

PART IV INDUSTRIAL CONTROL SYSTEMS SECURITY

12
A History of Cyber Incidents and Threats Involving Industrial Control Systems 215
Kevin Hemsley and Ronald Fisher

13
An Integrated Control and Intrusion Detection System for Smart Grid Security 243
Eniye Tebekaemi, Duminda Wijesekera and Paulo Costa

14
Generating Abnormal Industrial Control Network Traffic for Intrusion Detection System Testing 265
Joo-Yeop Song, Woomyo Lee, Jeong-Han Yun, Hyunjae Park, Sin-Kyu Kim and Young-June Choi

15
Variable Speed Simulation for Accelerated Industrial Control System Cyber Training 283
Luke Bradford, Barry Mullins, Stephen Dunlap and Timothy Lacey

Contributing Authors

Asma Alnemari is a Ph.D. student in Computing and Information Sciences at Rochester Institute of Technology, Rochester, New York. Her research interests include differential privacy and its application in real-world systems.

Suchith Arodi is a Software Engineer with the Office of Architecture, State Street Corporation, Raleigh, North Carolina. His research interests include blockchain, cyber security and data science.

Sofia Belikovetsky is a Ph.D. student in Information Systems Engineering at Ben-Gurion University of the Negev, Beer-Sheva, Israel. Her research focuses on the security of additive manufacturing processes and systems.

Karl Bentjen recently completed his M.S. degree in Computer Engineering at the Air Force Institute of Technology, Wright-Patterson Air Force Base, Ohio. His research interests include vehicular network security and critical infrastructure protection.

Luke Bradford is an M.S. student in Computer Science at the Air Force Institute of Technology, Wright-Patterson Air Force Base, Ohio. His research interests include cyber operations and critical infrastructure protection.

Madeline Carr is an Associate Professor of International Relations and Cyber Security, and the Director of the Research Institute for the Science of Cyber Security at University College London, London, United Kingdom. Her research interests include global cyber security, cyber norms, the Internet of Things and board/policy decision making on cyber risk.

Daniel Celebucki recently completed his M.S. degree in Cyber Operations at the Air Force Institute of Technology, Wright-Patterson Air Force Base, Ohio. His research interests include embedded systems, reconfigurable computing, reverse engineering and critical infrastructure protection.

Young-June Choi is a Professor of Computer Engineering at Ajou University, Suwon, Republic of Korea. His research interests include network security and cyber-physical systems.

Alex Chung is a Research Associate in the Department of Science, Technology, Engineering and Public Policy, University College London, London, United Kingdom. His research interests include cyber security policy, organized crime, and digital economy and society.

Paulo Costa is an Associate Professor of Systems Engineering and Operations Research at George Mason University, Fairfax, Virginia. His research interests include cyber security, transportation systems and multi-sensor data fusion.

Rishabh Das is a Ph.D. student in Computer Engineering at the University of Alabama in Huntsville, Huntsville, Alabama. His research interests include industrial control system virtualization, machine learning and embedded intrusion detection systems.

Sneha Dawda is a Research Associate in Digital Business Strategy at Forrester Research, London, United Kingdom. Her research interests include geopolitical cyber security and digital financial services security.

Stephen Dunlap is a Cyber Security Research Engineer at the Air Force Institute of Technology, Wright-Patterson Air Force Base, Ohio. His research interests include embedded systems security, cyber-physical systems security and critical infrastructure protection.

Yuval Elovici is a Professor of Information Systems Engineering, Director of the Telekom Innovation Laboratories and Head of the Cyber Security Research Center at Ben-Gurion University of the Negev, Beer-Sheva, Israel. His research interests include computer security and network security.

Jose Fernandez is an Associate Professor of Computer and Software Engineering at Ecole Polytechnique de Montreal, Montreal, Canada. His research interests include industrial control systems security, critical infrastructure security, cyber crime, cyber public health and cyber conflict.

Ronald Fisher is the Director of the Infrastructure Assurance and Analysis Division at Idaho National Laboratory, Idaho Falls, Idaho. His research interests include critical infrastructure protection and resilience, including industrial control systems.

Scott Graham is an Assistant Professor of Computer Engineering at the Air Force Institute of Technology, Wright-Patterson Air Force Base, Ohio. His research interests include vehicle cyber security, critical infrastructure protection and embedded systems security.

Lynne Graves is a Ph.D. student in Computer Science at the University of South Alabama, Mobile, Alabama. Her research focuses on additive manufacturing security.

Ammara Gul is a Ph.D. student in Information Security at Royal Holloway, University of London, Egham, United Kingdom. Her research interests include graph theory and control theory, in particular, the security of state estimation systems.

Sanjeev Gunawardena is a Research Assistant Professor of Electrical Engineering at the Air Force Institute of Technology, Wright-Patterson Air Force Base, Ohio. His research interests include satellite navigation and timing systems, embedded systems, reconfigurable computing and software-defined radio.

Kevin Hemsley is a Project Manager at Idaho National Laboratory, Idaho Falls, Idaho. His research interests include critical infrastructure protection and industrial control systems security.

Atif Hussain is a Cyber Security Researcher at the Future Transport and Cities Research Institute, Coventry University, Coventry, United Kingdom. His research interests include penetration testing, digital forensics and cyber security policymaking.

Sin-Kyu Kim is a Senior Engineering Staff Member at the National Security Research Institute, Daejeon, Republic of Korea. His research focuses on critical infrastructure protection.

Wayne King is a Project Leader at Lawrence Livermore National Laboratory, Livermore, California. His research focuses on the physics, material science, engineering and control aspects of additive manufacturing.

Timothy Lacey is an Adjunct Assistant Professor of Computer Science at the Air Force Institute of Technology, Wright-Patterson Air Force Base, Ohio. His research interests include cyber operations, critical infrastructure protection, mobile device security, and computer, network and embedded systems security.

Woomyo Lee is an Engineering Staff Member at the National Security Research Institute, Daejeon, Republic of Korea. Her research interests include applied cryptography, cyber security and cyber-physical systems.

Antoine Lemay is a Researcher in the Department of Computer and Software Engineering at Ecole Polytechnique de Montreal, Montreal, Canada. His research interests include industrial control systems security, critical infrastructure protection, cyber crime ecosystems and cyber conflict.

Joshua Lubell is a Computer Scientist in the Systems Integration Division at the National Institute of Standards and Technology, Gaithersburg, Maryland. His research interests include model-based engineering, cyber security, cyber-physical systems, information modeling and markup technologies.

Robert Mills is a Professor of Electrical Engineering at the Air Force Institute of Technology, Wright-Patterson Air Force Base, Ohio. His research interests include network security and management, cyber situational awareness and electronic warfare.

Sumita Mishra is a Professor of Computing Security at Rochester Institute of Technology, Rochester, New York. Her research interests include critical infrastructure protection, resource-constrained networking and security.

Thomas Morris is a Professor of Electrical and Computer Engineering, and the Director of the Center for Cybersecurity Research and Education at the University of Alabama in Huntsville, Huntsville, Alabama. His research interests include industrial control system virtualization, intrusion detection, machine learning and vulnerability testing of cyber-physical systems.

Barry Mullins is a Professor of Computer Engineering at the Air Force Institute of Technology, Wright-Patterson Air Force Base, Ohio. His research interests include cyber-physical systems security, cyber operations, critical infrastructure protection, computer, network and embedded systems security, wired and wireless networking, and code reverse engineering.

Scott Nykl is an Assistant Professor of Computer Science at the Air Force Institute of Technology, Wright-Patterson Air Force Base, Ohio. His research interests include visualization and the use of synthetic environments for evaluating computer vision algorithms.

Soni Pandey recently completed her M.S. degree in Computer Science at Rochester Institute of Technology, Rochester, New York. Her research interests include data management and cyber security.

Hyunjae Park is a Ph.D. candidate in Computer Engineering at Ajou University, Suwon, Republic of Korea. His research areas include cyber-physical systems and artificial intelligence.

Rajendra Raj is a Professor of Computer Science at Rochester Institute of Technology, Rochester, New York. His research interests include cyber security, data management and distributed computing.

Valentina Rodriguez Sosa is an M.S. student in Computer Science at Rochester Institute of Technology, Rochester, New York. Her research interests include enterprise system security, secure coding and cyber security education.

Carol Romanowski is a Professor of Computer Science at Rochester Institute of Technology, Rochester, New York. Her research interests include applications of data science and data mining to critical infrastructure protection, cyber security and engineering design.

Nicolas Saunier is a Professor of Transportation Engineering at Ecole Polytechnique de Montreal, Montreal, Canada. His research interests include intelligent transportation systems, road safety and information technology for transportation.

Siraj Shaikh is a Professor of Systems Security at the Future Transport and Cities Research Institute, Coventry University, Coventry, United Kingdom. His research interests include stealthy threat detection, cyber-physical systems security, especially transportation and cyber security policymaking.

Paul Simon is a Ph.D. student in Electrical Engineering at the Air Force Institute of Technology, Wright-Patterson Air Force Base, Ohio. His research interests include embedded systems security, computer communications security and critical infrastructure protection.

Joo-Yeop Song is an M.S. student in Computer Engineering at Ajou University, Suwon, Republic of Korea. His research areas include network security and artificial intelligence.

Eniye Tebekaemi is an Assistant Professor of Computer Science at Mercer University, Macon, Georgia. His research interests include cyber security, cyber-physical systems and intrusion detection systems.

Marielba Urdaneta is an M.S. student in Computer Engineering at Ecole Polytechnique de Montreal, Montreal, Canada. Her research interests include industrial control systems security and critical infrastructure protection.

Richard White is an Assistant Research Professor of Security Engineering at the University of Colorado Colorado Springs, Colorado Springs, Colorado. His research interests include risk management and critical infrastructure protection.

Duminda Wijesekera is a Professor of Computer Science at George Mason University, Fairfax, Virginia; and a Visiting Research Scientist at the National Institute of Standards and Technology, Gaithersburg, Maryland. His research interests include cyber security, digital forensics and transportation systems.

Clark Wolfe recently completed his M.S. degree in Electrical Engineering at the Air Force Institute of Technology, Wright-Patterson Air Force Base, Ohio. His research interests include computer communications security and critical infrastructure protection.

Stephen Wolthusen is a Professor of Information Security in the Faculty of Information Technology, Mathematics and Electrical Engineering at the Norwegian University of Science and Technology, Gjovik, Norway; and a Professor of Information Security at Royal Holloway, University of London, Egham, United Kingdom. His research interests include critical infrastructure protection and cyber-physical systems security.

Mark Yampolskiy is an Assistant Professor of Computer Science at the University of South Alabama, Mobile, Alabama. His research focuses on the security aspects of additive manufacturing, cyber-physical systems and the Internet of Things.

Jeong-Han Yun is a Senior Engineering Staff Member at the National Security Research Institute, Daejeon, Republic of Korea. His reseach interests include network security, cyber security and industrial control systems security.

Preface

The information infrastructure – comprising computers, embedded devices, networks and software systems – is vital to operations in every sector: chemicals, commercial facilities, communications, critical manufacturing, dams, defense industrial base, emergency services, energy, financial services, food and agriculture, government facilities, healthcare and public health, information technology, nuclear reactors, materials and waste, transportation systems, and water and wastewater systems. Global business and industry, governments, indeed society itself, cannot function if major components of the critical information infrastructure are degraded, disabled or destroyed.

This book, *Critical Infrastructure Protection XII*, is the twelfth volume in the annual series produced by IFIP Working Group 11.10 on Critical Infrastructure Protection, an active international community of scientists, engineers, practitioners and policy makers dedicated to advancing research, development and implementation efforts related to critical infrastructure protection. The book presents original research results and innovative applications in the area of infrastructure protection. Also, it highlights the importance of weaving science, technology and policy in crafting sophisticated, yet practical, solutions that will help secure information, computer and network assets in the various critical infrastructure sectors.

This volume contains fifteen revised and edited papers from the Twelfth Annual IFIP Working Group 11.10 International Conference on Critical Infrastructure Protection, held at SRI International in Arlington, Virginia, USA on March 12–14, 2018. The papers were refereed by members of IFIP Working Group 11.10 and other internationally-recognized experts in critical infrastructure protection. The post-conference manuscripts submitted by the authors were rewritten to accommodate the suggestions provided by the conference attendees. They were subsequently revised by the editors to produce the final chapters published in this volume.

The chapters are organized into four sections: (i) themes and issues; (ii) infrastructure protection; (iii) infrastructure modeling and simulation; and (iv) industrial control systems security. The coverage of topics showcases the richness and vitality of the discipline, and offers promising avenues for future research in critical infrastructure protection.

This book is the result of the combined efforts of several individuals and organizations. In particular, we thank David Balenson and Molly Keane for their tireless work on behalf of IFIP Working Group 11.10. We gratefully acknowledge the Institute for Information Infrastructure Protection (I3P), managed by George Washington University, for its sponsorship of IFIP Working Group 11.10. We also thank the U.S. Department of Homeland Security, National Security Agency and SRI International for their support of IFIP Working Group 11.10 and its activities. Finally, we wish to note that all opinions, findings, conclusions and recommendations in the chapters of this book are those of the authors and do not necessarily reflect the views of their employers or funding agencies.

JASON STAGGS AND SUJEET SHENOI

I

THEMES AND ISSUES

Chapter 1

A THEORY OF HOMELAND SECURITY

Richard White

Abstract Homeland security is a recognized practice, profession and field without a unifying theory to guide its study and application. The one previous attempt by Bellavita [2] acknowledges its own shortcomings and may be considered incomplete at best. The failure may be attributed to the lack of an underlying correlating factor. This chapter demonstrates that "domestic catastrophic destruction" is the correlating factor that unites key historical homeland security incidents and this observation is leveraged to propose a theory of homeland security that is descriptive, prescriptive and predictive. The proposed theory is descriptive because it can differentiate between what is and what is not homeland security. The theory is prescriptive because it can suggest an optimum homeland security strategy. It is predictive because it renders homeland security into a technical problem and demonstrates how its effects may eventually be blunted through the technological evolution and revolution of the critical infrastructure. Accordingly, the proposed theory embodies a set of foundational principles to guide the study and application of the practice, profession and field of homeland security.

Keywords: Homeland security, theory, foundational principles

1. Introduction

Homeland security is a recognized practice, profession and field that only recently emerged in the context of national security, which is itself a well-established practice, profession and field. Partly because of its "newness," homeland security – unlike national security – does not yet have a theory and foundational set of principles that could guide its study and application. The absence of a theory and foundational principles may also be the result of a lack of consensus on what constitutes homeland security. This chapter proposes a theory of homeland security and a set of foundational principles to help bring about consensus and guide the study and application of the practice, profession and field of homeland security.

Published by Springer Nature Switzerland AG 2018.
J. Staggs and S. Shenoi (Eds.): Critical Infrastructure Protection XII, IFIP AICT 542, pp. 3–21, 2018.
https://doi.org/10.1007/978-3-030-04537-1_1

Situation	Stage 1 Pre-Event	Stage 2 Event	Stage 3 Alarm	Stage 4 Demand	Stage 5 Difficult	Stage 6 Priorities	Stage 7 Post-Event
Simple	X				X	X	X
Complicated	X		X	X	X	X	X
Complex		X	X	X	X		
Chaotic		X					

Figure 1. Bellavita's opportunities matrix [2].

2. Previous Work

Perhaps unsurprisingly there have been few attempts to develop a theory of homeland security and the one treatment by Bellavita [2] is considered to be incomplete. Bellavita suggested that the dearth of literature may be attributed to a view by many that homeland security is a subset of existing theories and does not warrant independent status. Such a view is not unprecedented and has close parallels with computer science, which, in its early years, was considered to be a subset of mathematics and engineering. While mathematics and engineering remain integral to computer science, it eventually gained independent status due to its own distinctiveness. Bellavita felt that such distinctiveness may yet elude homeland security. However, he kickstarted the process by setting down an initial set of principles, incomplete as they may be, and letting the theory evolve from there.

Bellavita's theory of homeland security is based on an "issue-attention cycle." According to this theory, homeland security is the culmination of a series of issue-attention cycles that began with the September 11, 2001 (9/11) terrorist attacks and continued with Hurricane Katrina, the H1N1 pandemic, the merging of homeland security and national security policy by the Obama administration, and leading up to the Great Recession. Bellavita observes that each cycle proceeds in seven stages, providing an opportunity to evaluate and respond appropriately at each stage. Bellavita subsequently introduced an "opportunities matrix" for which "one could fill in the chart for a variety of decisions that have to be made during the cycle: decisions about communication, strategy, planning, technology, leadership, and so on." For example, the opportunities matrix might recommend different leadership styles during different stages depending on the type of incident. Citing the 2010 Deepwater Horizon catastrophe, Bellavita claims that an opportunities matrix could make it clear that leaders who applied complex strategies would be more effective than those who followed routine procedures. Figure 1 presents the opportunities matrix of Bellavita [2]

Bellavita's proposal satisfies two important aspects of a theory. The first is that is descriptive, offering an explanation of homeland security. The second is that it is prescriptive, offering insights on responding to homeland security incidents. However, by his own admission, Bellavita's theory fails in one important purpose – prediction. Without prediction there can be no direction and, therefore, no guide for the study and application of the practice, profession and field of homeland security. Because of the absence of the predictive characteristic, Bellavita's proposed theory must be considered incomplete at best. But, in fact, it can be proved wrong.

In his proposal, Bellavita claims that the 9/11 attacks was the initiating event for the string of issue-attention cycles that comprise homeland security. This is not the case. Homeland security did not begin in the aftermath of the 9/11 attacks. Instead, it began with the 1995 Tokyo subway attacks. On March 20, 1995, Aum Shinrikyo, a quasi-religious cult, attempted to overthrow the Japanese government and initiate an apocalypse by releasing the deadly Sarin nerve agent in the Tokyo subway system during the morning rush hour. Tragically, twelve people lost their lives, but experts believe it was sheer luck that prevented thousands more from being killed.

The 1995 Tokyo subway attacks were the first deployment of a weapon of mass destruction (WMD) by a non-state actor [8]. Before this incident, weapons of mass destruction were the exclusive domain of nation-states. The implications for national security were profound. The diplomatic, economic and military instruments of national power that kept the use of weapons of mass destruction by nation-states in check were shown to be useless against non-state actors.

Concerns about a similar attack in the United States prompted a flurry of Congressional investigations [3, 6, 7, 14–16]. In a series of reports, the Gilmore Commission, the Hart-Rudman Commission and the Bremer Commission separately agreed that the United States was unprepared for weapons of mass destruction threats involving non-state actors. Accordingly, in December 2000, the second report of the Gilmore Commission [7] recommended that the next President establish a National Office for Combating Terrorism in the Executive Office of the President. In February 2001, the third report of the Hart-Rudman Commission [16] recommended creating a new National Homeland Security Agency. In March 2001, Representative William Thornberry (R-TX) introduced House Resolution 1158 to create a National Homeland Security Agency within the Executive Branch of the U.S. Federal Government. House Resolution 1158 was still sitting in Congress when the nation was attacked six months later on September 11, 2001 [17].

Does this mean that all that is needed is to reset Bellavita's issue-attention cycle to begin with the 1995 Tokyo subway attacks? But this will not salvage the theory because it would still not have any predictive power. The reason why Bellavita's theory will not gain any predictive power – and the reason it lacks any to begin with – is that the theory does not offer any correlating factor that explains the relationship between selected events that make up homeland security. It is the absence of a correlating factor that deprives Bellavita's theory of predictive power. This does not mean there is no correlating factor that unites homeland security events. There is a correlating factor, but it just has nothing to do with issue-attention cycles. Indeed, it is the correlating factor that enables the formulation of a theory of homeland security that is descriptive, prescriptive and predictive.

3. Correlating Factor

If homeland security began with the 1995 Tokyo subway attacks, then the correlating factor that underpins homeland security must reside in some similarity between this incident and the 9/11 attacks. On September 11, 2001, nineteen hijackers gained control of four passenger jets and flew three of them into icons that represented the economic and military strength of the United States. In just two hours, the hijackers utterly destroyed the Twin Towers in New York City, and severely damaged the Pentagon outside Washington, DC. Alerted to these suicide attacks, passengers aboard the fourth aircraft rose up against their hijackers, forcing them to abort their mission against the nation's capital and crash in an empty field outside Shanksville, Pennsylvania. Altogether, the attacks left nearly 3,000 dead and caused $40 billion in direct damage. Cross-referencing the passenger manifests against CIA databases quickly revealed the hijackers to be members of Al Qaeda, a known terrorist group led by Osama bin Laden that was operating out of Afghanistan. Enraged by the presence of U.S. military forces in Saudi Arabia to protect it from aggression by Iraqi dictator Saddam Hussein, bin Laden issued an edict in 1996 that declared war on the United States. The 9/11 Commission Report [1] states that the attacks were staged to force U.S. military forces out of Saudi Arabia.

At first glance it might appear that the correlating factor is terrorism. The 1995 Tokyo subway attacks and the 9/11 attacks were terrorist attacks as defined by Title 18 Section 2331 of the United States Code [20]. Under this definition, terrorism is a crime distinguished by motive, specifically violent acts calculated to coerce government. The many commission reports stemming from the 1995 Tokyo subway attacks and the seminal 2004 9/11 Commission Report [1] clearly branded both attacks as acts of terrorism. While the Tokyo subway attacks raised the issue of homeland security in the United States, the 9/11 attack brought homeland security to the forefront of U.S. policy concerns.

Terrorism, however, is not the correlating factor underpinning the two homeland security incidents. If terrorism was, indeed, the founding principle of homeland security, then it would have become a U.S. priority policy long before the 1995 Tokyo subway attacks, because in one form or another, the United States had been the target of terrorist attacks, some would say as far back as the founding of the nation.

Hurricane Katrina provides the strongest evidence that terrorism is not the correlating factor that underpins homeland security. On August 29, 2005, Hurricane Katrina made landfall in Louisiana and crossed directly over the city of New Orleans. The wind damage was minimal, but the eight to ten inches of rain filled Lake Pontchartrain to overflowing and the canals designed to channel its waters began to fail. The levee system built to protect New Orleans breached in 53 places, rendering 80% of the city under fifteen feet of water. The extensive flooding stranded numerous residents in their homes. Many made their way to their roofs using hatchets and sledgehammers. House tops across the city were dotted with survivors; others were unable to escape and remained trapped in their homes. According to the Louisiana Department of Health, 1,464 citizens died in the storm; across the Gulf Coast, Hurricane Katrina caused nearly 1,500 deaths and $108 billion in damage [21].

Hurricane Katrina had a profound impact on the United States similar to the 9/11 attacks – both are recognized as homeland security incidents [13]. But where the 9/11 attacks was a terrorist incident, Hurricane Katrina was not. By definition, terrorism is a violent act distinguished by motive, but nature has no motive. The correlating factor between the 1995 Tokyo subway attacks, the 9/11 attacks and Hurricane Katrina in 2005 is not terrorism. The correlating factor is domestic catastrophic destruction.

Homeland security began with the 1995 Tokyo subway attacks over concerns of domestic catastrophic destruction precipitated by weapons of mass destruction in the hands of non-state actors. It was brought to the forefront of U.S. policy concerns by the 9/11 attacks, where nineteen hijackers achieved effects similar to those of weapons of mass destruction by subverting the nation's transportation infrastructure and turning passenger jets into guided missiles to inflict domestic catastrophic destruction. Hurricane Katrina was a harsh reminder that domestic catastrophic destruction can be natural as well as manmade. Although the means were different in the three incidents, the potential and the real consequences were the same for all three incidents – domestic catastrophic destruction.

4. Unique Mission

Domestic catastrophic destruction is nothing new to the United States. From its inception, the U.S. has suffered from domestic catastrophic destruction of the natural and manmade varieties. An estimated 6,000 people were killed in the 1900 Galveston Hurricane, more than twice as many as in the 9/11 attacks [22]. More than 22,000 soldiers were killed or wounded in a single day during the Battle of Antietam in the Civil War, making it the "bloodiest day in U.S. history" [24]. So what is new about domestic catastrophic destruction that makes homeland security a unique mission?

As indicated previously, the new twist in domestic catastrophic destruction is the unprecedented ability for it to be inflicted by non-state actors. The 1995 Tokyo subway attacks demonstrated the ability of a small group to acquire and deploy weapons of mass destruction. The 9/11 attacks demonstrated the ability of a small group to create weapons of mass destruction effects by subverting the critical infrastructure (CI). Because these attacks were perpetrated by non-state actors, unsanctioned by any government, the acts constituted crimes. The crimes were unprecedented in their scope – indeed, they had national and international repercussions. Because of their scope and consequences, the crimes were not ordinary and would not have been contained by traditional law enforcement alone.

As was pointed out in the many reports following the 1995 Tokyo subway attacks, the threat of domestic catastrophic destruction by a non-state actor requires an unprecedented level of coordination across all levels of government. It was also recognized that no amount of effort could ever eliminate the threat – it is impossible to always stop a determined attacker. In this regard, the threat of domestic catastrophic destruction by a non-state actor is similar to that of a natural disaster in that neither can be stopped completely. Since safety cannot be guaranteed, the best that can be accomplished is to reduce the risk of the likelihood and consequences of domestic catastrophic destruction. This requires actions across the four disaster phases – prevent, protect, respond and recover – to effectively cope with domestic catastrophic destruction.

In summary, homeland security is a unique mission because never before in human history have small groups and individuals demonstrated the ability to inflict domestic catastrophic destruction. This uniqueness makes homeland security sufficiently distinct to warrant recognition as an independent practice, profession and field.

5. Proposed Theory

Given the preceding discussion, the theory of homeland security is formulated by specifying a set of axioms that establish a firm foundation:

- **A1.0:** Domestic catastrophic destruction from natural and manmade sources is a historical threat to organized society.

- **A2.0:** Domestic catastrophic destruction perpetrated by non-state actors represents a new and unprecedented threat to organized society.

- **A3.0:** Domestic catastrophic destruction perpetrated by non-state actors is similar to that caused by natural disasters in that neither are completely stoppable.

- **A3.1:** There can be no guarantee of safety from domestic catastrophic destruction.

- **A3.2:** The best that can be accomplished is to mitigate the likelihood and consequences of domestic catastrophic destruction.

- **A3.3:** Mitigating the risks of domestic catastrophic destruction entails actions across the four disaster phases – prevent, protect, respond and recover.

- **A4.0:** It is a purpose of government to safeguard its citizens from domestic catastrophic destruction.

This set of axioms leads to the following theory of homeland security:

- **Theory:** Homeland security encompasses actions designed to safeguard a nation from domestic catastrophic destruction.

6. Descriptive Theory

The proposed theory of homeland security is descriptive because it helps identify what is and what is not homeland security. First, it tells us that homeland security is international because all nations are at risk of domestic catastrophic destruction. Consequently, any nation that engages in actions to safeguard against domestic catastrophic destruction is conducting homeland security.

The theory of homeland security thus leads to the following proposition or corollary:

- **C1.0:** Homeland security is a concern to every nation.

The theory specifies what constitutes a homeland security concern: anything that can create domestic catastrophic destruction. As stipulated by Axiom 1.0, domestic catastrophic destruction stems from two sources, natural and human (manmade). The natural sources are broadly classified as: (i) meteorological; (ii) geological; (iii) epidemiological; and (iv) astronomical. Meteorological threats encompass all types of extreme weather, including floods, heat, hurricanes and tornadoes. Geological threats cover all tectonic incidents, including earthquakes, volcanoes and tsunamis. Epidemiological threats include all forms of pandemic disease stemming from highly contagious and virulent pathogens. Astronomical threats encompass all forms of celestial phenomena, including extreme solar activity and large-body collisions. Note that large-body collisions may not necessarily include incidents such as the 1908 Tunguska event in Siberia, which experts believe was an air burst of a small asteroid or comet with the explosive equivalent of 10-15 megatons of TNT.

All these threats share the property that they may precipitate domestic catastrophic destruction in the form of a natural disaster. As noted by Axiom 3.0, they also share the property that they are unstoppable, and it is not a matter of if they will occur, but when they will occur. The inevitability of natural disasters makes it necessary to invest in emergency preparedness, actions designed to promote rapid response and recovery to catastrophic events. This presupposes two caveats: (i) the disasters are transient events of short duration; and (ii) they do not necessarily threaten human extinction. The first caveat addresses the apparent perception that threats such as climate change and cardiopulmonary disease are not immediate crises, although billions of dollars are spent every year to deal with extended droughts and floods, and cardiopulmonary disease is the leading killer of Americans. The second caveat concedes that there are no practical solutions at this time for dangers such as asteroid impacts and super volcanoes, but it also recognizes that such dangers are fortunately rare in the human time-scale.

Based on these observations, the following corollaries are derived:

- **C2.0:** Homeland security threats are transient events of a specific, short-term duration.

- **C2.1:** Emergency preparedness is a necessary investment against the inevitability of natural disasters.

With regard to manmade domestic catastrophic destruction, the threats may be broadly grouped as those committed by: (i) state actors; and (ii) non-state actors. As noted previously, manmade domestic catastrophic destruction has historically been perpetrated through warfare. Warfare is waged between sovereign nations. Like the United States, most nations have national security establishments to assert their sovereignty and defend themselves from hostile nations. National security has thus evolved to maintain a nation's sovereignty in the community of nations. However, the instruments that help maintain a nation's sovereignty are practically useless against small groups or individuals that are categorized as non-state actors. In general, non-state actors are subject to the laws of the nations in which they reside, whether or not they are citizens. Although nations use different means to enforce their laws, they were not prepared to cope with the threat of domestic catastrophic destruction posed by small groups or individuals; certainly not before the 9/11 attacks, and in some cases, not yet. This is why, according to Axiom 2.0, domestic catastrophic destruction by non-state actors constitutes a new and unprecedented threat that cannot be contained by law enforcement alone.

In the case of manmade domestic catastrophic destruction, a distinction should be made between the actions that are deliberate versus those that are accidental. While the containment of deliberate acts of manmade domestic catastrophic destruction fall in the realm of criminal justice, the containment of accidental acts of manmade domestic catastrophic destruction are the domain of safety engineering. This does not mean that an accident cannot be prosecuted as a crime. A chemical release from a pesticide plant that killed 3,787 in Bhopal, India in 1984 was ruled an accident; even so, seven ex-employees, including the former company chairman, were convicted of negligent homicide and sentenced to two years imprisonment and a fine of about $2,000 each, the maximum punishment allowed at that time under Indian law [25].

Based on these observations, the following corollaries are derived:

- **C3.0:** Homeland security and national security are related through a common objective: to safeguard a nation from manmade domestic catastrophic destruction.

- **C3.1:** National security is distinct from homeland security in that it addresses the threat of manmade domestic catastrophic destruction by recognized state actors.

- **C3.2:** Homeland security is distinct from national security in that it addresses the threat of manmade domestic catastrophic destruction by non-state actors.

- **C3.3:** Manmade domestic catastrophic destruction stemming from the actions of non-state actors may be deliberate or accidental.

- **C3.4:** Manmade domestic catastrophic destruction deliberately perpetrated by non-state actors is a crime subject to criminal justice within the jurisdiction where the act was committed.

With regard to natural and manmade disasters, as neither is completely stoppable, both require actions across the four disaster phases: prevent, protect, respond and recover. The inevitability of disasters places first responders such as police, firefighters and emergency medical services on the front-line of emergency response. By definition, since the consequences are catastrophic, local first responders are most likely to be overwhelmed. Therefore, by necessity, local first responders must have the means to quickly call for assistance and rapidly integrate capabilities from other jurisdictions to mount an efficient and effective emergency response.

Based on these observations, the following corollaries are derived:

- **C4.0:** The inevitability of disasters places first responders at the front-line of emergency response.
- **C4.1:** Efficient and effective emergency response requires the means to quickly call for assistance and rapidly integrate capabilities from other jurisdictions.

Finally, it is important to discuss what does not constitute homeland security under the proposed theory. The central property of the theory is domestic catastrophic destruction. Domestic catastrophic destruction has not been defined aside from indicating that the 9/11 attacks and Hurricane Katrina are recognized homeland security incidents. As potential benchmarks, it has been noted above that the 9/11 attacks resulted in nearly 3,000 deaths and $40 billion in damage whereas Hurricane Katrina caused about 1,500 deaths and $108 billion in damage. In March 2002, a few months after the 9/11 attacks, Williams [28] proposed a threshold of 500 deaths and/or $1 billion in property damage for catastrophic incidents. Can there be a defined threshold? Perhaps.

The more important point is that the consequences of criminal acts can far exceed those encountered previously. Title 28 §530C of the United States Code defines a mass killing as three or more killings in a single incident. In October 2017, 58 people attending a concert in Las Vegas were killed, the worst shooting incident in U.S. history [23]. Despite the horrific number of casualties, the Las Vegas shooting does not approach even the lowest threshold suggested for a catastrophic incident. The Las Vegas shooting, therefore, is not a homeland security incident; absent a motive, it cannot even be classified as a terrorist incident.

The same holds true for the 1995 Oklahoma City bombing, the worst bombing incident in United States history. The bombing killed 168 men, women and children, and inflicted $652 million in damage [26]. Still, its scope does not measure up to catastrophes such as the 9/11 attacks and Hurricane Katrina. Under the proposed theory, the Oklahoma City bombing does not constitute a homeland security incident. By the same token, the motive is inconsequential compared with the means. In fact, none of the incidents examined so far have a common motive, and nature harbors no motive at all.

Based on these observations, the following corollaries are derived:

- **C5.0:** Homeland security incidents are distinguished by catastrophic consequences.
- **C5.1:** Homeland security incidents are not distinguished by motive.
- **C5.2:** Mass killings, although tragic, are not necessarily homeland security incidents.
- **C5.3:** Terrorist incidents are not necessarily homeland security incidents.

Based on the preceding discussion, all the components constituting homeland security can be compiled into the map shown in Table 1.

7. Prescriptive Theory

The proposed theory of homeland security is prescriptive, providing a means to guide national homeland security strategy. In November 2002, the Homeland Security Act created the U.S. Department of Homeland Security to coordinate homeland security efforts across federal, state and local agencies. The department's homeland security functions were organized into critical mission areas. The original mission set was derived from the 2002 National Homeland Security Strategy and comprised the following six critical mission areas [10]:

- Intelligence and warning.
- Border and transportation security.
- Domestic counterterrorism.
- Protecting critical infrastructure.
- Defending against catastrophic terrorism.
- Emergency preparedness and response.

During the ensuing years, the mission set of the U.S. Department of Homeland Security evolved due to internal reorganizations, external events, Presidential priorities and Congressional legislation. One of the changes was instituted by the Implementing Recommendations of the 9/11 Commission Act of 2007, which mandated a systematic review of the U.S. Department of Homeland Security mission set and organization every four years starting in 2009 [18]. The first Quadrennial Homeland Security Review was released in 2010. The most recent Quadrennial Homeland Security Review, which was completed in 2014, identified the following mission set [19]:

- Prevent terrorism and enhance security.

Table 1. Homeland security map.

Theory	**Actions to Safeguard a Nation from Domestic Catastrophic Destruction**						
Threats	**Natural**				**Manmade**		
Type	**Meteorological**	**Geological**	**Epidemiological**	**Astronomical**	**State Actor**	**Non-State Actor**	
Forms	Extreme Weather	Tectonic Event	Pandemic Disease	Celestial Phenomenon	Warfare	Criminal Act	Accident
Means	Flood, Heat, Hurricane, Tornado	Earthquake, Volcano, Tsunami	Contagious Pathogen	Solar Activity, Earth Strike	Conventional, Nuclear, Asymmetric	WMD, Critical Infrastructure Subversion	WMD, Critical Infrastructure Subversion
Subclass	Emergency Preparedness	Emergency Preparedness	Emergency Preparedness	Emergency Preparedness	National Security	Criminal Justice	Safety Engineering
Prevent	Early Warning	Early Warning	Public Health	Early Warning	Deterrence Measures	Law Enforcement	Safety Design
Protect	Sheltering and Evacuation	Building Codes	Vaccinations	Sheltering and Evacuation	Defensive Measures	Security Measures	Safety Practices
Respond	Emergency Response	Emergency Response	Health Measures	Emergency Response	Emergency Response	Emergency Response	Emergency Response
Recover	Disaster Recovery	Disaster Recovery	Disaster Recovery	Disaster Recovery	Disaster Recovery	Disaster Recovery	Disaster Recovery

- Secure and manage our borders.
- Enforce and administer our immigration jaws.
- Safeguard and secure cyberspace.
- Strengthen national preparedness and resilience.

When the current U.S. Department of Homeland Security mission set is superimposed on top of the homeland security map shown in Table 1, the map shown in Table 2 is obtained. Note that the italicized items in the last five rows of Table 2 comprise the U.S. Department of Homeland Security mission set.

Based on the map in Table 2, a number of observations regarding the application of homeland security in the United States can be made:

- **Observation 1.0:** Homeland security is a team sport; the U.S. Department of Homeland Security cannot do it alone. As can be seen by the italicized items in Table 2, the U.S. Department of Homeland Security mission set does not encompass the entire mission space corresponding to the last five rows of the table. It is, therefore, incumbent upon the U.S. Department of Homeland Security to play a coordinating role across public and private agencies in what is called the "homeland security enterprise."
- **Observation 2.0:** Failure is an inevitable outcome. Nobody wants to fail. Typical strategies attempt to avoid failure at all cost. However, no amount of investment in the prevent and protect mission areas will preclude failure. Emergency preparedness, response and recovery are an inseparable part of homeland security. Accepting failure and investing in the respond and recover mission areas are essential to reducing the consequences.
- **Observation 3.0:** Unprecedented responses to unprecedented threats. Most U.S. Department of Homeland Security missions are concentrated on securing the nation from the unprecedented threats of domestic catastrophic destruction by non-state actors (i.e., security measures marked with an asterisk in Table 2). Whereas law enforcement agencies remain responsible for preventing these particularly heinous form of crimes, the U.S. Department of Homeland Security has taken the lead in protecting against the means for committing them. Aviation security, for example, keeps passenger jets from becoming guided missiles.
- **Observation 4.0:** Cyber security is essential to homeland security. Following the 1995 Tokyo subway attacks, a 1997 Presidential Commission Report examining the vulnerability of U.S. critical infrastructure to a similar attack first raised concerns about cyber security [12]. The report noted that infrastructure owners and operators were increasingly resorting to remote monitoring and control using commercial networking products to reduce costs and increase efficiency across their geographically-

Table 2. Superimposition of the U.S. Department of Homeland Security mission set on the homeland security map.

Theory	**Actions to Safeguard a Nation from Domestic Catastrophic Destruction**						
Threats	**Natural**				**Manmade**		
Type	**Meteorological**	**Geological**	**Epidemiological**	**Astronomical**	**State Actor**	**Non-State Actor**	
Forms	Extreme Weather	Tectonic Event	Pandemic Disease	Celestial Phenomenon	Warfare	Criminal Act	Accident
Means	Flood, Heat, Hurricane, Tornado	Earthquake, Volcano, Tsunami	Contagious Pathogen	Solar Activity, Earth Strike	Conventional, Nuclear, Asymmetric	WMD, Critical Infrastructure Subversion	WMD, Critical Infrastructure Subversion
	Homeland Security Enterprise						
Subclass	*Emergency Preparedness*	*Emergency Preparedness*	*Emergency Preparedness*	*Emergency Preparedness*	National Security	Criminal Justice	Safety Engineering
Prevent	Early Warning	Early Warning	Public Health	Early Warning	Deterrence Measures	Law Enforcement	Safety Design
Protect	Sheltering and Evacuation	Building Codes	Vaccinations	Sheltering and Evacuation	Defensive Measures	*Security Measures**	Safety Practices
Respond	*Emergency Response*	*Emergency Response*	Health Measures	*Emergency Response*	*Emergency Response*	*Emergency Response*	*Emergency Response*
Recover	*Disaster Recovery*	*Disaster Recovery*	*Disaster Recovery*	*Disaster Recovery*	*Disaster Recovery*	*Disaster Recovery*	*Disaster Recovery*

distributed systems. The report warned that commercial network products were making critical infrastructure increasingly vulnerable to external cyber attacks [12]. In 2007, Project Aurora demonstrated the ability to potentially destroy an electricity generator over the Internet [9]. In December 2016, the Ukrainian capital of Kiev was plunged into darkness by a cyber attack on its electric power grid [11]. If critical infrastructure provides the means for non-state actors to achieve weapons of mass destruction effects, then cyber attacks provide the opportunity.

- **Observation 5.0:** The threats from within. Keeping hostile agents and their weapons from entering the United States underpins the U.S. Department of Homeland Security's immigration and border security missions. The problem is that the weapons are already here, and the enemy need not come to the United States to set them off. The critical infrastructure, which is everywhere, is the means of destruction, and the chemical, biological, radiological and nuclear agents that comprise weapons of mass destruction are readily accessible. Cyber attacks have global reach. Physical proximity is not necessary to attack a target. Thus, an enemy can subvert the critical infrastructure or release a weapon of mass destruction by typing on a keyboard or clicking on a mouse anywhere in the world.

Based on these observations, the following prescriptive corollaries are derived:

- **C6.0:** The broad scope of the homeland security mission set exceeds the authority of the U.S. Department of Homeland Security and requires the coordinated efforts on the part of the homeland security enterprise.

- **C7.0:** Because failure is inevitable, emergency preparedness, response and recovery are also essential to homeland security.

- **C8.0:** Cyber security is essential to homeland security.

- **C8.1:** Whereas weapons of mass destruction and critical infrastructure provide the means for non-state actors to inflict domestic catastrophic destruction, cyber attacks provide the opportunity.

- **C8.2:** Cyber attacks can be launched from anywhere in the world.

8. Predictive Theory

The proposed theory of homeland security is also predictive in that it provides insights into the future of homeland security. Among its lesser predictions, Observation 2.0 indicates there will always be domestic catastrophic disasters. The case can certainly be made for natural disasters in the form of Hurricane Sandy in 2012 and Hurricane Maria in 2017. A similar case cannot be made for manmade domestic catastrophic destruction by non-state actors. But, when such a catastrophe does occur, Observations 4.0 and 5.0 make the case that it could well be the result of coordinated cyber attacks. However, the most profound prediction of the theory may be that the current concerns about homeland security will one day become irrelevant.

The worst concerns related to homeland security today are the threats of manmade domestic catastrophic destruction posed by non-state actors. The threats are predicated on the abilities of non-state actors to deploy weapons of mass destruction or to subvert the critical infrastructure. These threats provide the means and cyber attacks provide the opportunity for inflicting domestic catastrophic destruction. The means and opportunity in this case are mere technical challenges. Therefore, depriving non-state actors of the means and opportunity to inflict domestic catastrophic destruction are simply technical challenges. The word "simply" is used because technical problems are easier to solve than social problems. Technical problems take years to solve; social problems take generations to address. Eliminating the motive is a social problem. Because the proposed theory reduces homeland security to a set of technical problems, it is conceivable that the worst threats may be eliminated. The only question is how.

Can non-state actors be deprived of the opportunity to inflict domestic catastrophic destruction? Not entirely. Whereas cyber security can blunt cyber attacks, it cannot completely stop them. Like the flu, there is no cure for cyber attacks and new strains are constantly emerging. And even if cyber attacks could somehow be halted, there is still no way to halt physical attacks.

Could a non-state actor be deprived of the means to inflict domestic catastrophic destruction? Possibly. With respect to weapons of mass destruction, it is simply a matter of sequestration, keeping products and materials out of the hands of unauthorized actors. Indeed, this concept forms the foundation of the national strategy to counter weapons of mass destruction, which involves nonproliferation and counterproliferation [5]. But what about the critical infrastructure? Although most of the critical infrastructure is not designed to withstand deliberate attacks, this situation will eventually change. Through technological evolution and revolution, the critical infrastructure that sustains contemporary society will become less susceptible to deliberate attacks and less likely to incur catastrophic effects if and when failures occur.

An example of technological evolution is the U.S. telephone system. In the early decades, when human operators were replaced by computer switches, the in-band signaling system was found to be vulnerable to a form of subversion called "phreaking." So-called phreakers exploited the in-band signaling system to make free phone calls. Service providers lost millions until the phone switches were upgraded and the signaling system was taken out-of-band [27].

In a similar manner, technological evolution may eventually render cyber attacks harmless. A potential solution is the microgrid approach, which subdivides large components of the North American electric grid into much smaller, self-contained units. An attack on one unit would then be less likely to cascade across the grid and create regional outages such as the northeast blackout that affected 50 million people in 2003 [4]. Using various means, other infrastructures may similarly become immune to attacks or the consequences of their failures could be greatly reduced.

Although the need for homeland security will never be completely eliminated, the proposed theory suggests that the worst threats from non-state actors may be rendered irrelevant.

9. Implications

The proposed theory can help the practice and profession of homeland security in three ways: (i) by lending support to certain current practices; (ii) by offering justification for reducing other practices; and (iii) by providing a framework for developing a measurable strategy.

The proposed theory lends support to current practices that reinforce national emergency management. As made clear by Corollaries 4.0 and 4.1, the inevitability of natural and manmade disasters requires strong investments in first responder capabilities. One of the most significant victories that may be claimed by homeland security is the promulgation of national standards and procedures in the National Incident Management System. Before the 9/11 attacks, there was no national coordination of first responder standards. After the attacks, the U.S. Department of Homeland Security assumed the role of coordinating national standards, which has improved the ability of the nation to respond and recover to domestic catastrophic disasters.

The proposed theory justifies the reduction of practices focused on finding and apprehending potential terrorists. Corollaries 5.0 through 5.3 make it clear that homeland security is about means not motive. The current preoccupation with motive, specifically, terrorism, detracts from more productive pursuits that go after the means. In addition to terrorism, there are many potential motives for non-state actors to commit acts of domestic catastrophic destruction. However, the means for non-state actors to commit acts of domestic catastrophic destruction are limited to weapons of mass destruction and critical infrastructure subversion. Cyber attacks provide the opportunity to getting at both. This change in focus implies a greater emphasis on technical capabilities and research and development activities to cut off these avenues of attack.

Finally, the theory provides a framework for a measurable homeland security strategy. If homeland security is not a social problem but a technical problem as the theory implies, then the potential for developing a measurable strategy is within reach. As a social problem focused on terrorism, a strategy is impossible to formulate because the potential motives are unlimited and unmanageable. As a technical problem focused on weapons of mass destruction and critical infrastructure subversion, a strategy is possible because the potential means are limited and manageable. Reducing the scope of the problem to a finite set of risk factors makes a measurable risk strategy feasible. With a measurable risk strategy, it is possible to determine the current status as well as the path forward and the cost. This capability has eluded the U.S. Department of Homeland Security from its inception, but the proposed theory makes it feasible.

10. Conclusions

Developing a theory of homeland security is a daunting task, as evidenced by the dearth of literature on the topic. Bellavita [2], the only researcher who tried to do this, found it to be an overwhelming task. The resulting theory is incomplete, offering some descriptive and prescriptive analyses, but no predictive capability. Moreover, the theory could not find the correlating factor that ran through all the disparate components that claim to fall in the domain of homeland security.

The proposed theory makes the case that the correlating factor is domestic catastrophic destruction, natural and manmade. Domestic catastrophic destruction is the central concern of homeland security. Although domestic catastrophic destruction is a concern as old as civilization, the ability for it to be inflicted by non-state actors is new and unprecedented. Too large for law enforcement alone, the new threat requires a new approach that coordinates actions across the four phases of disasters – prevent, protect, respond and recover. Homeland security arose out of the 1995 Tokyo subway attacks and was brought to the forefront of U.S. policy concerns by the terrorist attacks of September 11, 2001. Correspondingly, the theory contends that homeland security encompasses actions designed to safeguard a nation from domestic catastrophic destruction.

References

[1] 9/11 Commission, The 9/11 Commission Report, Washington, DC, 2004.

[2] C. Bellavita, Waiting for homeland security theory, *Homeland Security Affairs*, vol. 8, article 16, 2012.

[3] Bremer Commission, Report of the National Commission on Terrorism, Washington, DC, 2000.

[4] Center for the Study of the Presidency and Congress, Securing the U.S. Electrical Grid, Washington, DC (`www.thepresidency.org/sites/default/files/Final%20Grid%20Report_0.pdf`), 2014.

[5] Counterproliferation Program Review Committee, Report on Activities and Programs for Countering Proliferation and NBC Terrorism, Volume I: Executive Summary, Washington, DC, 2011.

[6] Gilmore Commission, First Annual Report to The President and The Congress of the Advisory Panel to Assess Domestic Response Capabilities for Terrorism Involving Weapons of Mass Destruction, I. Assessing the Threat, Washington, DC, 1999.

[7] Gilmore Commission, Second Annual Report to The President and The Congress of the Advisory Panel to Assess Domestic Response Capabilities for Terrorism Involving Weapons of Mass Destruction, II. Towards a National Strategy for Combating Terrorism, Washington, DC, 2000.

[8] A. Neifert, Case Study: Sarin Poisoning of Subway Passengers in Tokyo, Japan in March 1995, Camber Corporation, Huntsville, Alabama, 1999.

[9] North American Electric Reliability Corporation and U.S. Department of Energy, High-Impact, Low-Frequency Event Risk to the North American Bulk Power System, Washington, DC, 2010.

[10] Office of Homeland Security, National Strategy for Homeland Security, Washington, DC, 2002.

[11] P. Polityuk, O. Vukmanovic and S. Jewkes, Ukraine's power outage was a cyber attack: Ukrenergo, *Reuters*, January 18, 2017.

[12] President's Commission on Critical Infrastructure Protection, Critical Foundations: Protecting America's Infrastructures, Washington, DC, 1997.

[13] Select Bipartisan Committee to Investigate the Preparation for and Response to Hurricane Katrina, A Failure of Initiative: Final Report of the Select Bipartisan Committee to Investigate the Preparation for and Response to Hurricane Katrina, 109th Congress, 2nd Session, U.S. House of Representatives, Washington, DC, 2006.

[14] United States Commission on National Security/21st Century, New World Coming: American Security in the 21st Century, Washington, DC, 1999.

[15] United States Commission on National Security/21st Century, Seeking a National Strategy: A Concert for Preserving Security and Promoting Freedom, Washington, DC, 2000.

[16] United States Commission on National Security/21st Century, Road Map for National Security: Imperative for Change, Washington, DC, 2001.

[17] U.S. Congress, HR 1158 – National Homeland Security Agency Act, 107th Congress, Washington, DC, 2001.

[18] U.S. Congress, Implementing Recommendations of the 9/11 Commission Act of 2007, Public Law 110-53 – August 3, 2007, Washington, DC, 2007.

[19] U.S. Department of Homeland Security, The 2014 Quadrennial Homeland Security Review, Washington, DC, 2014.

[20] U.S. Government, 18 U.S. Code §2331 – Definitions, Washington, DC (`www.law.cornell.edu/uscode/text/18/2331`), 1992.

[21] R. White, T. Bynum and S. Supinski (Eds.), *Homeland Security: Safeguarding the U.S. from Domestic Catastrophic Destruction*, CW Productions, Colorado Springs, Colorado, 2016.

[22] Wikipedia Contributors, 1900 Galveston Hurricane, *Wikipedia, The Free Encyclopedia* (`en.wikipedia.org/w/index.php?title=1900_Galveston_hurricane&oldid=817635245`), 2017.

[23] Wikipedia Contributors, 2017 Las Vegas Shooting, *Wikipedia, The Free Encyclopedia* (`en.wikipedia.org/w/index.php?title=2017_Las_Vegas_shooting&oldid=817903631`), 2017.

[24] Wikipedia Contributors, Battle of Antietam, *Wikipedia, The Free Encyclopedia* (`en.wikipedia.org/w/index.php?title=Battle_of_Antietam&oldid=816312607`), 2017.

[25] Wikipedia Contributors, Bhopal Disaster, *Wikipedia, The Free Encyclopedia* (en.wikipedia.org/w/index.php?title=Bhopal_disaster&oldid=786710332), 2017.

[26] Wikipedia Contributors, Oklahoma City Bombing, *Wikipedia, The Free Encyclopedia* (en.wikipedia.org/w/index.php?title=Oklahoma_City_bombing&oldid=785180008), 2017.

[27] Wikipedia Contributors, Phreaking, *Wikipedia, The Free Encyclopedia* (en.wikipedia.org/w/index.php?title=Phreaking&oldid=816928925), 2017.

[28] C. Williams, Prospects for macroterrorism, presented at the *Pugwash Conference on Science and World Affairs*, paper no. 25, 2002.

Chapter 2

AN EVIDENCE QUALITY ASSESSMENT MODEL FOR CYBER SECURITY POLICYMAKING

Atif Hussain, Siraj Shaikh, Alex Chung, Sneha Dawda and Madeline Carr

Abstract A key factor underpinning a state's capacity to respond to cyber security policy challenges is the quality of evidence that supports decision making. As part of this process, policy advisers, essentially a diverse group that includes everyone from civil servants to elected policy makers, are required to assess evidence from a mix of sources. In time-critical scenarios where relevant expertise is limited or not available, assessing threats, risk and proportionate response based on official briefings, academic sources and industry threat reports can be very challenging. This chapter presents a model for assessing the quality of evidence used in policymaking. The utility of the model is illustrated using a sample of evidence sources and it is demonstrated how different attributes may be used for comparing evidence quality. The ultimate goal is to help resolve potential conflicts and weigh findings and opinions in a systematic manner.

Keywords: Evidence quality assessment, cyber security, policymaking

1. Introduction

Research in cyber security tends to focus on technical factors, vulnerabilities and solutions. Some research focuses on the "human dimension," but these studies look predominantly at end-users. However, regulatory and policy frameworks also have significant implications with regard to cyber security. Policy advisers, sometimes with limited relevant expertise and often in time-critical scenarios, are asked to assess evidence from a mix of sources such as official threat intelligence, academic research and industry threat reports. The diverse evidence base is then used to make judgments about threats, risk, mitigation and consequences, and offer advice that shapes the national regulatory land-

Published by Springer Nature Switzerland AG 2018.
J. Staggs and S. Shenoi (Eds.): Critical Infrastructure Protection XII, IFIP AICT 542, pp. 23–38, 2018.
https://doi.org/10.1007/978-3-030-04537-1_2

scape, foreign and domestic security policy and/or various public and private sector initiatives. The research presented in this chapter is motivated by the need to better support decision making in the United Kingdom policy community when interpreting, evaluating and understanding evidence related to cyber security.

The decisions made by policy advisers in many ways shape the landscape and ecosystem within which other actors operate. A better understanding of the influences on such decision making is essential to identifying how the policymaking community can be supported in making sound policy decisions that foster continued innovation and mitigate current and future cyber security threats.

This research is motivated by the following key questions:

- What evidence do U.K. policymakers rely upon?

- What is the quality of the evidence?

- How effective are the judgments about threats, risks, mitigation and consequences based on the evidence?

Understanding how U.K. policymakers select evidence, why they place one source over another and how adeptly they can recognize possible weaknesses or flaws in evidence are central to addressing these research questions.

This chapter presents a simple model that supports the quality assessment of a variety of evidence sources used in cyber security policymaking. Given the diversity of the sources, some of which may be conflicting or contradictory, an evaluation of the quality of the available evidence can help resolve potential divergence. The proposed Evidence Quality Assessment Model (EQAM) is a two-dimensional map that uses a set of attributes to position evidence samples relative to each other. The attributes are derived from the literature and from a series of semi-structured interviews of policy advisers from the U.K. cyber security policy community.

2. Evidence and Policy Challenges

Policymakers use a diverse evidence base to make judgments about threats, risk, mitigation and consequences, and offer advice that shapes the national regulatory landscape, foreign and domestic security policy, and a range of public and private sector initiatives. In this context, evidence assessment for policymaking is a particular problem for three reasons:

- First, some of the evidence is contradictory and/or potentially carries within it specific agendas or goals that may impede its rigor and reliability. The "politicization" of cyber security evidence is increasingly problematic because states may trust threat intelligence based on whether the sources are located within their sovereign borders instead of the quality of the research.

- Second, it is extremely difficult to conclusively attribute cyber attacks and to quantify the costs of cyber insecurity. For policy advisers, the lack of clarity about the concrete financial implications of cyber security vulnerabilities and incidents makes it challenging to develop sound responses. Without clarity about the role of specific communities of perpetrators, policy alternatives can be disconnected from the real threats, targeting individuals or groups who may not, in fact, be the key malicious actors. These challenges mean that existing evidence often only partially supports policy advisers' evaluations of cyber security risks, threats and consequences – and the resulting recommendations.

- Third, the cyber security landscape is developing rapidly and spans many areas, including national security, human rights, commercial concerns and infrastructure vulnerabilities. Consequently, policy advisers must balance a range of possibly conflicting interests that compete for attention. Different conceptions of what "cyber security" means to different policy communities raises real impediments to a unified response. Network security, economic security, privacy and identity security, and data security all represent diverse conceptions and priorities that are commonly referred to as "cyber security."

The rise of evidence-based policy making under the Blair government prompted several studies focused on the way U.K. policy advisers engage with and interpret evidence. Early in this process, Solesbury [27] argued for careful critical analysis of what exactly constitutes "evidence," pointing out the relationship between knowledge and power, and the role that selecting and interpreting evidence plays under this approach to policymaking. This leads to several questions. What evidence do U.K. policymakers rely upon in this context? What is the quality of the evidence? How effective are the judgments about threats, risks, mitigation and consequences based on the evidence? Understanding how U.K. policymakers select evidence, why they weight one source over another and how adeptly they can recognize possible weaknesses or flaws in evidence are central to addressing these questions.

Evidence-based policymaking has been a core concept in contemporary U.K. policymaking since the 1990s. However, there is a lack of agreement in the policy community on the level of clarity and definition of evidence, and the academic or scientific standards that should be applied to the evidence. This has resulted in the popularization and politicization of evidence-based policymaking as a catch-phrase instead of a policy process that utilizes rigorous methodology and systematic analysis [6, 16, 17, 21, 34]. In addition, modern technological concerns are increasingly complex and, therefore, render an approach that solely relies on evidence-based policymaking rather simplistic compared with nuanced forms of policymaking where evidence is contextualized within the policy process and objectives. Evidence-based policymaking involves a critical approach based on replicable scientific studies. It responds to the belief that past policy decisions may have relied on the biased selection of evidence. It also seeks to address the influence of untested views of individuals or groups who represent vested interests, tradition, ideology, prejudice and/or speculation [4]. Evidence-based policymaking therefore attempts to reduce uncertainty and increase clarity in decision making by drawing on rigorous information to turn policy goals into concrete, achievable actions [26].

In recent years, the policymaking landscapes in some developed countries have led to innovative governance models for dealing with cyber security instead of relying on evidence-based policymaking or other traditional forms of policymaking such as the rational model, implied model, enlightenment model, knowledge-driven model, political model and tact model [16, 23, 32]. In the United Kingdom, newer systems take the form of adaptive (or agile) policymaking (APM). Adaptive policymaking explicitly accounts for deep uncertainties prompted by the speed with which technologies evolve [13]; this is in direct contrast to classical policymaking approaches that are ill-suited to managing the complexities associated with cyber security [16, 29, 33].

The adaptive paradigm also markedly departs from tradition by incorporating a strategic vision and framework from which policies are derived to prepare for negative eventualities; but it is also sufficiently flexible and dynamic to meet changing circumstances through short-term actions [29]. In order to facilitate this process, the proposed Evidence Quality Assessment Model seeks to validate evidence quality in a timely fashion, enabling policymakers to understand the implications of utilizing evidence and making the best judgments based on the available evidence.

3. Assessing Evidence Quality

The Strategic Policy Making Team at the U.K. Cabinet Office [28] describes evidence as expert knowledge, published research, existing statistics, stakeholder consultations, previous policy evaluations, Internet resources, costing of policy options and results from economic and statistical modeling. Davies [4] has structured different types of evidence into controlled experimental trials and studies, social surveys, econometrics, expert advisory groups, public attitudes, ethical values such as belief and aspirations, and research evidence from relevant sources that have been systematically searched, critically appraised and rigorously analyzed according to explicit and transparent criteria. However, Nutley et al. [22] note that, in practice, the U.K. public sector uses a more limited range of evidence, specifically, research and statistics, policy evaluation, economic modeling and expert knowledge.

3.1 Subject Interviews

As part of this research, sixteen policy advisers and U.K. civil servants were interviewed between November 2017 and February 2018. The subjects were employed across U.K. Government departments, including the Cabinet Office, Department for Digital, Culture, Media and Sport (DCMS), Home Office, Foreign and Commonwealth Office (FCO), Her Majesty's Revenue and Customs (HMRC) and Department of Communities and Local Government (DCLG), along with specialist agencies such as the London Mayor's Office for Policing and Crime, National Crime Agency (NCA) and National Police Chiefs' Council (NPCC).

The interviews revealed that a very wide variety of sources are used as potential evidence for policy analysis. These include research into trends from open-source material, forums, news articles, daily bulletins, media and newsletters; threat intelligence reports from academia and think tanks; intelligence reports from domestic and overseas sister agencies and restricted government information; and crime surveys for England and Wales, action fraud and general policing data from the National Crime Agency (NCA), cyber security breach surveys and Office of National Statistics (ONS) data sources and reports. Threat intelligence reports, surveys, case studies etc. are received from government sources (restricted and unrestricted), as well as from information technology giants such as BAE Systems, IBM, Microsoft, Cisco and FireEye. Policy advisers also access classified information released by law enforcement agencies and the intelligence community.

This study has not reviewed information from the various sources because the proposed model accounts for the use of such evidence. However, while one may assume that the evidence is reliable, it should be considered in the context of multiple (possibly transnational) agencies that may be trusted to varying levels.

With regard to the use of evidence in policymaking, it should be noted that decision making is often based on the best available evidence, although it may not be perfect. If one individual does not offer an informed view, then someone else who is less informed may make the decision; therefore, time is critical for a short-term response. Long-term problems are seen differently because ample time is available to institute the right approaches and gather the necessary evidence. In order to evaluate policy options and identify the options that will genuinely work, it is necessary to validate ideas and understand how to improve the process.

Two dimensions of evidence quality are proposed: (i) evidence sources; and (ii) evidence credibility.

3.2 Evidence Source

The evidence sources include data sources and human sources, both of which pose unique attributes with regard to quality.

Data Sources. Technical and survey data have been used as evidence for a variety of tasks ranging from attributing malware fragments [24] to identifying emerging trends in the technical and social spheres [30]. An artifact of evidence is subject to several considerations:

- The scope of data collection is not always perfect. As such, it may not always be complete to allow inferences. This is particularly problematic when it comes to using industry sources for threat intelligence and technological trends, which tend to increase the commercial advantage to the organizations that collect and publish the data.

- There are questions about the potential volatility of digital sources such as computers and networks [2]. The transient nature of such sources cannot be ignored because of the reliance on digital infrastructures for threat sensing. Additionally, digital forensics is subject to strict chain of custody and preservation procedures, any violation of which could cast doubt on the integrity of data.

- Analysis of data, often abstract and agnostic in nature, is open to interpretation. For example, traces of malware activity may be used to evaluate the sophistication of an attacker, which, in turn, is used as a critical criterion for attribution [7].

The subjects interviewed in the research hailed from a number of organizations. Organizations with a tradition of national data collection and statistical excellence, such as the Office of National Statistics (ONS) in the United Kingdom, are considered to be reliable sources, primarily because of their methodology and objectivity, which bolster confidence when the evidence they provide is cited in reports to ministers.

Human Sources. Human sources, either subjects of interest observed via some channel or knowledgeable experts who offer opinions, are also valuable sources of evidence. With expert knowledge and commentary comes the burden of bias and beliefs, and context and connotation. Indeed this is a substantial challenge because cyber security, as a social construct, takes various forms, including a political discourse that invokes the idea of a cyber "Pearl Harbor" [5]. Objective analysis of information from human sources is sensitive to the credibility of the entity that collects the information and the transparency of its collection method.

3.3 Evidence Credibility

This section discusses credibility in terms of the methodology and provider, both of which ultimately underpin the confidence in the presented evidence.

Methodology. The focus is on published forms of evidence to which some notion of methodology and organization could be attributed. Of course, confidential sources of threat intelligence would follow official protocols; the judgment of their quality would, therefore, be left to the relevant intelligence and policy communities.

A challenge with cyber security is the heightened interest that it attracts due to novel technological aspects. This interest lends itself to hype as well as a lack of balanced technical and broad knowledge to help policy perspectives. Indeed, the level of reporting on cyber security is routinely criticized. Lee and Rid [12] state:

> *"Cynical and overstated reports ultimately lower the quality of bureaucratic procedures and decision making. First, such reports inform decisions at both the strategic and tactical level. Intelligence reports take highly technical data, combine the information with the interpretations of analysts, and give a bottom line to fill knowledge gaps in the government and guide action ... Simply put: many of these reports are incomplete or inaccurate."*

Appropriate methodologies and analyses are key to presenting substantial claims that result from the evidentiary artifacts. These range from empirical analyses of data sets to qualitative and quantitative analyses of socio-technical information.

The legal imperative regarding cyber attacks [8] implies that several attributes are important if evidence is to be used for policy decisions related to legislation or regulation, or if a state is to respond under international norms and law. Especially important is transparency with regard to how evidence is collected, processed, stored and handled.

Provider. Over the past two decades, an entire industry dedicated to cyber threat intelligence has emerged. Cyber threat intelligence is an umbrella term that refers to the collection and analysis of threat-related activity from open-source reports, social media and dark web sources. The industry includes major information technology and telecommunications companies, such as IBM and Cisco, and niche operators, such as FireEye, that are focused on advanced threats. The industry is a major source of information for government agencies and corporations for policymaking and for making decisions about security investments.

Geopolitical affiliations have the potential to cast a shadow on providers even when their technical capabilities are acknowledged. Kaspersky Lab, headquartered in Moscow, Russia, is an example of a provider with very well regarded technical capabilities, including its efforts in detecting Stuxnet [10]. However, Kaspersky Lab software is viewed with suspicion because of the potential for its compromise by Russian Government entities. The interviews conducted in this research also revealed that threat intelligence reports from the company are discredited as a result of its reputation.

The situation in industry is paralleled by that for government agencies. An example is the National Cyber Security Centre (NCSC) in the United Kingdom, whose technical mission is to provide advice and guidance on cyber-related threats to public and private sector stakeholders. The National Cyber Security Centre provides products in various formats, from brief weekly threat reports with little transparency or detail [19] to detailed data-driven guidance with clarity on methodological approaches and data provenance, such as analysis of active cyber defense policy [14]. Indeed, the quality challenges when dealing with a complex evidence base are clearly enunciated in the threat report [19]:

> *"[It is] difficult to draw concrete conclusions – especially about causality – from our current analysis of the data. There are also some anomalies in the data that we don't understand yet. We've tried our best to be clear about our confidence in our conclusions in this paper. People will almost*

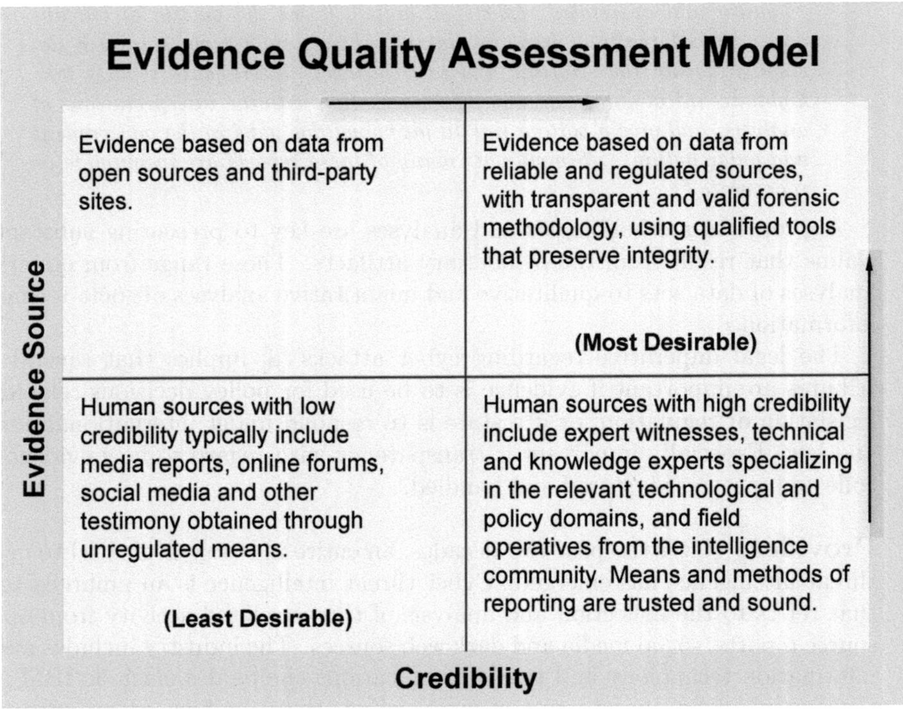

Figure 1. Evidence Quality Assessment Model.

certainly disagree with some of the conclusions we draw here. That's probably a good thing as it starts to engender an evidence-based discussion about what cyber security policy should look like going forward."

3.4 Evidence Quality Assessment Model

This section presents the Evidence Quality Assessment Model, which reflects the diverse nature of evidence sources and enables the quality of evidence to be characterized despite the diversity. The proposed model is based on the attributes discussed in Sections 3.2 and 3.3.

Figure 1 shows the proposed model. It provides a simple representation of the quality of evidence using a two-dimensional map, where the vertical axis captures the split in evidence sources between data sources and human sources, and the horizontal axis expresses credibility based on the methodology and provider. For example, the vertical axis could place the value of data sources over the value of human sources in establishing the quality of evidence. As a scale, it helps map evidence that combines both data and human sources to a quality measure. The horizontal axis, on the other hand, is a continuum, where credibility is judged on a case by case basis for each piece of evidence.

The division into four quadrants assists in mapping pieces of evidence to a relative quality metric in an intuitively appealing manner.

4. Model Analysis

This section illustrates the application of the Evidence Quality Assessment Model in a typical use case involving the analysis of a collection of evidence.

4.1 Sample Selection

The application of the Evidence Quality Assessment Model is illustrated using an evidence assessment exercise that was performed internally by a subset of the authors of this chapter. The ten pieces of evidence shown in Table 1 were chosen. The selection was deliberately broad and diverse to help understand whether the proposed model helps achieve consensus across varying levels of evidence quality. Given the current focus on the U.K. policymaking community, all the evidence items were mentioned during the interviews or in the U.K. policy discourse.

4.2 Scoring Analysis

A subset of the authors of this chapter, with expertise in technology and policy, assessed the evidence items individually. The assessors scored each item on the Evidence Quality Assessment Model vertical and horizontal scales shown in Figure 1. Similar scores were consolidated and disparate scores were discussed and a common score was negotiated by the assessors. Table 2 shows the consolidated and negotiated source and credibility scores for the ten evidence items.

Figure 2 shows the ten evidence items placed on the Evidence Quality Assessment Model map according to their consolidated and negotiated source and credibility scores listed in Table 2.

The following details pertaining to the ten evidence items provide insights into the consolidated and negotiated source and credibility scores, and their placement on the Evidence Quality Assessment Model map:

- **NCSC Weekly Threat Report (E-1):** This report is broken up into five threat bulletins. Each bulletin has distinct topics and its analysis varies. For example, the first bulletin includes facts from a survey that communicate the risk and support the claims, whereas the last bulletin only states the claims without providing details about the analysis and findings. This makes the overall threat report slightly harder to assess because the same methodology was not applied across the report. Furthermore, in some instances, the sources of evidence were not stated. For example, a Daesh (ISIL) claim was presented without any validation of its sources. Another example is that the data coverage for Android malware left some key questions unanswered: Which phone models were

Table 1. Ten evidence items used to illustrate the proposed model.

Provider	Description
NCSC	NCSC provides advice and support to the U.K. public and public sectors for addressing computer security threats. The *NCSC Weekly Threat Report* issued on December 22, 2017 contains evidence on distinct security issues [19]. *NCSC Password Security Guidance* contains advice for administrators on determining password policy; it advocates a dramatic simplification of the current approach at the system level [18].
CVE	Common Vulnerabilities and Exposures (CVE) catalogs cyber security vulnerabilities and exposures related to software and firmware in a free "dictionary" that organizations can use to to improve their security postures. *CVE-2014-0160* refers to the Heartbleed vulnerability found in the OpenSSL software library [20].
BBC	The British Broadcasting Corporation (BBC) is a British public broadcaster. *BBC 2017* highlights the main technology events that occurred in 2017 [3].
Foresight	Foresight projects, produced by the U.K. Government Office for Science, provide evidence to the policy community. The *Future of the Sea: Cyber Security* project report informs the U.K. maritime sector about cyber security response [25].
FireEye	FireEye is a cyber security company that provides products and services that protect against advanced cyber threats. *FireEye Operation Ke3chang* investigates the Ke3chang cyber espionage campaign [31]. Mandiant is a cyber security firm acquired by FireEye in 2013. The *Mandiant APT1* report implicates China in cyber espionage activities [15].
IBM	IBM X-Force Research is a security team that monitors and analyzes security issues, and provides threat intelligence content. *IBM 2017* reports IBM X-Force Research's findings for 2017 [9].
Kaspersky	Kaspersky Lab is a multinational cyber security and anti-virus provider headquartered in Moscow, Russia. The *Kaspersky Global Report* covers security events from around the globe that occurred in 2017 [11]. *Securelist* is a Kaspersky blog; an article in the blog discusses how to survive attacks that seek to access and leak passwords [1].

tested? Are all Android phones at risk? Are there any impacts on Android tablets?

Table 2. Consolidated and negotiated scores for the ten evidence items.

Quality Criteria	E-1	E-2	E-3	E-4	E-5	E-6	E-7	E-8	E-9	E-10
Source	8	15	6	12	7	13	17	6	12	2
Credibility	53	65	33	49	47	52	56	63	27	17

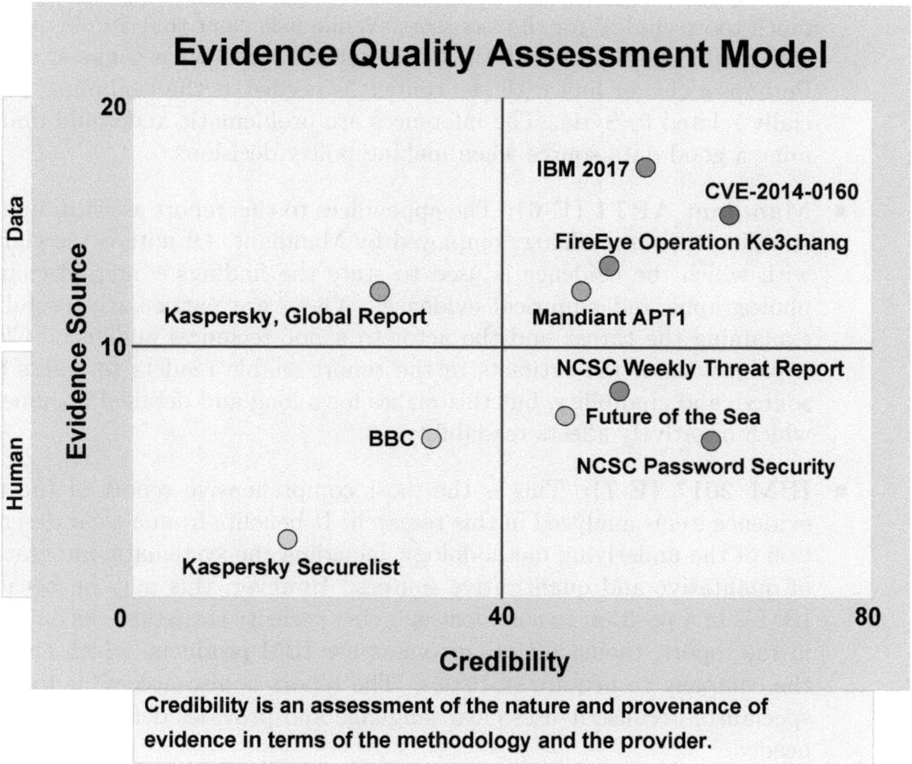

Figure 2. Placement of the ten evidence items on the EQAM map.

- **CVE-2014-0160 (E-2):** This evidence item is slightly obscure to a non-technical cyber security analyst, but the explanation of the threat and potential breadth of attacks are explained very well. A more accessible explanation would be more appropriate for non-technical consumers.

- **BBC 2017 (E-3)** This news article relies heavily on the opinions of political leaders and acknowledged experts. While the experts can be trusted to provide sound advice, individuals with strong political views may be biased.

- **Future of the Sea: Cyber Security (E-4):** This project report heavily relies on expert knowledge to provide a detailed scientific review of the topic. Such a review is subject to considerable scrutiny in terms of the scientific evidence selected and the corresponding inferences. However, the scientific evidence includes a very broad mix of research studies and technical artifacts and reports. These provide confidence in the methodology, but the evidence is drawn largely from human sources (of course, in some other cases, the evidence could be purely data-driven).

- **FireEye Operation Ke3chang (E-5):** This report was found to be much too technical for the assessors. While it is clear that ample quantitative evidence is provided, the methodology is somewhat vague at times. Perhaps a clearer link with the context is needed at the beginning, especially related to Syria. The inferences are problematic and could undermine a good data source when making policy decisions.

- **Mandiant APT1 (E-6):** The appendices to this report assist in understanding the methodology employed by Mandiant. Of note is the clarity with which the evidence is used to state the findings – myriad charts, photographs and empirical evidence. These are particularly useful in explaining the threat and the actor to a non-technical audience. Clear explanations of the artifacts in the report enable readers to assess the sources and credibility, but this makes for a long and detailed document, which negatively affects readability.

- **IBM 2017 (E-7):** This is the most comprehensive report of the ten evidence items analyzed in this research. It benefits from a clear description of the underlying methodology, including the systematic integration of qualitative and quantitative sources. However, this may be because IBM is in a position to comment on cyber security statistics – as outlined in the report, thousands of customers use IBM products, which enables the company to acquire statistics. The report is also accessible to non-specialists because it uses clear language and provides definitions where needed.

- **NCSC Password Security Guidance (E-8):** This guidance is clear in its intent: it provides readers with a visual representation of the potential threat and risks, and how to mitigate them. While there are only two instances of quantitative evidence, the qualitative advice comes from a position of authority on the topic; also, the risks are communicated very well.

- **Kaspersky Global Report (E-9):** This report is very poorly written, which distracts from the overall credibility of the report. Nevertheless, qualitative and quantitative evidence are used thoroughly, and the methodology is very clear. Kaspersky Lab suffers from a severe lack of trust as an evidence provider as far as the U.K. policymaking community is concerned. This is reflected in the low ranking of the evidence item in Figure 2.

- **Kaspersky Securelist (E-10):** This article makes sparse use of quantitative data when discussing how to survive attacks that access and leak passwords. No statistics related to prevention are presented, nor is the efficacy of prevention discussed. The data coverage is adequate to communicate the associated risk, but not enough to support the claims made in the article. For example, the guidance on using 23-character passwords is not substantiated. As before, Kaspersky Lab suffers from a severe lack of trust as an evidence provider.

5. Conclusions

It is imperative to assess the quality of the evidence base used for cyber security policymaking. The Evidence Quality Assessment Model presented in this chapter is a simple two-dimensional map that positions evidence samples relative to each other based on source and credibility. As such, it represents the first step towards a tool for assessing the fitness of evidence used in cyber security decision making. The use case involving representative items of evidence demonstrates how multiple attributes may be used to compare and contrast evidence items. The soft validation of the model also demonstrates its potential to resolve conflicts and achieve consensus when assessing evidence quality.

Future research will draw on senior members of the U.K. policymaking community who are well-versed in cyber security to help refine the evidence quality criteria and formally validate the model. The effort will leverage a repository containing a wide variety of evidence sources identified through stakeholder engagement.

Acknowledgement

This research was funded by the Engineering and Physical Science Research Council (EPSRC) as part of the project, Evaluating Cyber Security Evidence for Policy Advice: The Other Human Dimension (EP/P01156X/1), under the Human Dimensions of Cyber Security.

References

[1] D. Bestuzhev, How to survive attacks that result in password leaks? *Securelist*, Kaspersky Lab, Moscow, Russia, July 13, 2012.

[2] D. Chaikin, Network investigations of cyber attacks: The limits of digital evidence, *Crime, Law and Social Change*, vol. 46(4-5), pp. 239–256, 2006.

[3] G. Corera, If 2017 could be described as "cyber-geddon," what will 2018 bring? *BBC News*, December 30, 2017.

[4] P. Davies, Is evidence-based government possible? presented at the *Fourth Annual Campbell Collaboration Colloquium*, 2004.

[5] E. Gartzke, The myth of cyberwar: Bringing war in cyberspace back down to Earth, *International Security*, vol. 38(2), pp. 41–73, 2013.

[6] A. Glees, Evidence-based policy or policy-based evidence? Hutton and the government's use of secret intelligence, *Parliamentary Affairs*, vol. 58(1), pp. 138–155, 2005.

[7] C. Guitton and E. Korzak, The sophistication criterion for attribution: Identifying the perpetrators of cyber attacks, *The RUSI Journal*, vol. 158(4), pp. 62–68, 2013.

[8] O. Hathaway, R. Crootof, P. Levitz, H. Nix, A. Nowlan, W. Perdue and J. Spiegel, The law of cyber attack, *California Law Review*, vol. 100(4), pp. 817–886, 2012.

[9] IBM Security, IBM X-Force Threat Intelligence Index 2017, The Year of the Mega Breach, Somers, New York, 2017.

[10] E. Kaspersky, The man who found Stuxnet – Sergey Ulasen in the spotlight, *Security Matters*, Kaspersky Lab, Moscow, Russia (`www.eugene.kaspersky.com/2011/11/02/the-man-who-found-stuxnet-sergey-ulasen-in-the-spotlight`), November 2, 2011.

[11] Kaspersky Lab and Business Advantage, The State of Industrial Cybersecurity – Global Report, Woburn, Massachusetts and San Francisco, California, 2017.

[12] R. Lee and T. Rid, OMG Cyber! *The RUSI Journal*, vol. 159(5), pp. 4–12, 2014.

[13] G. Leicester, Viewpoint: The seven enemies of evidence-based policy, *Public Money and Management*, vol. 19(1), pp. 5–7, 1999.

[14] I. Levy, Active Cyber Defense – One Year On, National Cyber Security Centre, London, United Kingdom (`www.ncsc.gov.uk/information/active-cyber-defence-one-year`), 2018.

[15] Mandiant, APT1: Exposing One of China's Cyber Espionage Units, Alexandria, Virginia (`www.fireeye.com/content/dam/fireeye-www/services/pdfs/mandiant-apt1-report.pdf`), 2013.

[16] M. Monaghan, Appreciating cannabis: The paradox of evidence in evidence-based policy making, *Evidence and Policy: A Journal of Research, Debate and Practice*, vol. 4(2), pp. 209–231, 2008.

[17] G. Mulgan, Government, knowledge and the business of policy-making: The potential and limits of evidence-based policy, *Evidence and Policy: A Journal of Research, Debate and Practice*, vol. 1(2), pp. 215–226, 2005.

[18] National Cyber Security Centre, Password Guidance: Simplifying Your Approach, Guidance, London, United Kingdom (`ncsc.gov.uk/guidance/password-guidance-simplifying-your-approach`), 2016.

[19] National Cyber Security Centre, Weekly Threat Report, 22nd December 2017, Report, London, United Kingdom (`www.ncsc.gov.uk/report/weekly-threat-report-22nd-december-2017`), 2017.

[20] National Institute of Standards and Technology, CVE-2014-0160 Detail, National Vulnerability Database, Gaithersburg, Maryland (`nvd.nist.gov/vuln/detail/CVE-2014-0160`), 2014.

[21] M. Naughton, "Evidence-based policy" and the government of the criminal justice system – Only if the evidence fits! *Critical Social Policy*, vol. 25(1), pp. 47–69, 2005.

[22] S. Nutley, H. Davies and I. Walter, Evidence-Based Policy and Practice: Cross Sector Lessons from the UK, Working Paper 9, ESRC UK Centre for Evidence Based Policy and Practice, University of St. Andrews, St. Andrews, Scotland, United Kingdom, 2002.

[23] S. Nutley and J. Webb, Evidence and the policy process, in *What Works? Evidence-Based Policy and Practice in Public Services*, H. Davies, S. Nutley and P. Smith (Eds.), Policy Press, Bristol, United Kingdom, pp. 13–41, 2000.

[24] T. Rid and B. Buchanan, Attributing cyber attacks, *The Journal of Strategic Studies*, vol. 38(1-2), pp. 4–37, 2015.

[25] S. Shaikh, Future of the Sea: Cyber Security, Foresight, Government Office for Science, London, United Kingdom, 2017.

[26] L. Shaxson, Is your evidence robust enough? Questions for policy makers and practitioners, *Evidence and Policy: A Journal of Research, Debate and Practice*, vol. 1(1), pp. 101–112, 2005.

[27] W. Solesbury, Evidence Based Policy: Whence it Came and Where it's Going, Working Paper No. 1, ESRC UK Centre for Evidence Based Policy and Practice, Queen Mary, University of London, London, United Kingdom, 2001.

[28] Strategic Policy Making Team, Professional Policy Making for the Twenty-First Century, Version 2.0, Cabinet Office, London, United Kingdom, 1999.

[29] L. Tanczer, I. Brass, M. Carr, J. Blackstock and M. Elsden, The United Kingdom's emerging Internet of Things (IoT) policy landscape, to appear in *Rewired: Cybersecurity Governance*, R. Ellis and V. Mohan (Eds.), Wiley, Hoboken, New Jersey.

[30] A. Venables, S. Shaikh and J. Shuttleworth, The projection and measurement of cyberpower, *Security Journal*, vol. 30(3), pp. 1000–1011, 2017.

[31] N. Villeneuve, J. Bennett, N. Moran, T. Haq, M. Scott and K. Geers, Operation "Ke3chang:" Targeted Attacks against Ministries of Foreign Affairs, FireEye, Milpitas, California (`www.fireeye.com/content/dam/fireeye-www/global/en/current-threats/pdfs/wp-operation-ke3chang.pdf`), 2014.

[32] C. Weiss, The many meanings of research utilization, *Public Administration Review*, vol. 39(5), pp. 426–431, 1979.

[33] R. Whitt, Adaptive policy-making: Evolving and applying emergent solutions for U.S. communications policy, *Federal Communications Law Journal*, vol. 61(3), pp. 483–590, 2009.

[34] K. Young, D. Ashby, A. Boaz and L. Grayson, Social science and the evidence-based policy movement, *Social Policy and Society*, vol. 1(3), pp. 215–224, 2002.

Chapter 3

LIABILITY EXPOSURE WHEN 3D-PRINTED PARTS FALL FROM THE SKY

Lynne Graves, Mark Yampolskiy, Wayne King, Sofia Belikovetsky and Yuval Elovici

Abstract Additive manufacturing, also referred to as 3D printing, has become viable for manufacturing functional parts. For example, the U.S. Federal Aviation Administration recently approved General Electric jet engine fuel nozzles that are produced by additive manufacturing. Because additive manufacturing is integrated with cyber technology, a number of security concerns have been raised. This chapter specifically considers attacks that deliberately sabotage the mechanical properties of functional parts produced by additive manufacturing; the feasibility of these attacks has already been discussed in the literature.

Investments in security measures directly depend on cost-benefit analyses conducted by the participants involved in additive manufacturing processes. This chapter discusses the entities that can be considered to be financially liable in the event of a successful sabotage attack. The analysis employs a model that distinguishes between the levels at which the additive manufacturing process has been sabotaged. Specifically, it differentiates between the additive manufacturing service provider and the various commodity suppliers. For each possible combination of injured party and level of attack, the involved parties that may face liability exposure are identified. This is accomplished by analyzing the necessary components that establish liability. The analysis reveals that liability potential exists at all levels of the additive manufacturing process in the event of a sabotage attack. For this reason, it is imperative that the involved actors conduct or re-evaluate their cost-benefit analyses and invest in security measures.

Keywords: Additive manufacturing security, sabotage, liability

Published by Springer Nature Switzerland AG 2018.
J. Staggs and S. Shenoi (Eds.): Critical Infrastructure Protection XII, IFIP AICT 542, pp. 39–64, 2018.
https://doi.org/10.1007/978-3-030-04537-1_3

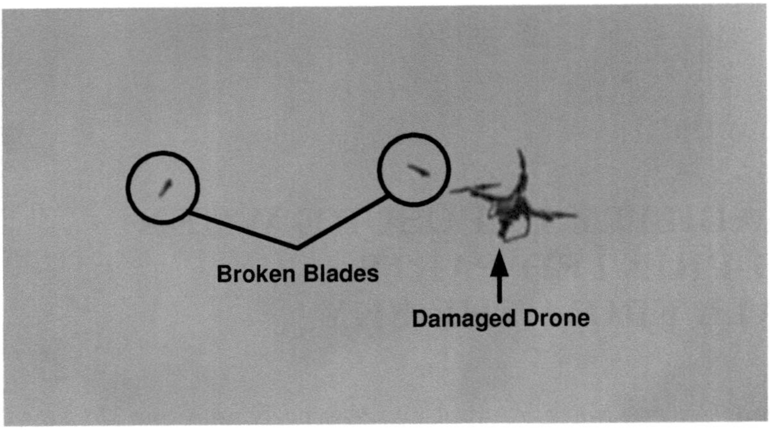

Figure 1. Failure of a sabotaged propeller in the `dr0wned` study [4].

1. Introduction

In 1947, a science fiction author envisioned 3D-printed spaceships [33]. Since then, reality has converged with vision. Additive manufacturing (AM) technology, also referred to as 3D printing, is now viable for industrial manufacturing, including the creation of functional parts for safety-critical systems. A recent example is General Electric's use of additive manufacturing to create fuel injection nozzles for the next generation LEAP jet engines [16] – a commitment of $22 billion to date [8, 20]. Meanwhile, the worldwide annual industry revenue from additive manufacturing is increasing rapidly and is expected to exceed $21 billion by 2020 [8].

The American Society for Testing and Materials (ASTM) defines seven additive manufacturing process categories [2, 46]. The shared characteristics are that they use a highly computerized process and that a 3D object is produced based on a digital model representation by depositing and fusing thin layers of source material.

Due to its reliance on computerization, additive manufacturing is susceptible to a variety of attacks. These include sabotage attacks, which deliberately degrade the mechanical properties of manufactured parts [4, 31, 51]. The `dr0wned` study [4] demonstrates the danger of sabotage attacks on functional parts. In the study, researchers compromised a benign 3D printing environment, and accessed and modified the design file of the replacement propeller of a quadcopter drone in a manner that was unique to additive manufacturing. The compromise caused the propeller to break in flight. The image in Figure 1 is taken from the video recording of the experiment. It shows the broken propeller blades and the drone falling from the sky.

Similar attacks on functional parts for safety-critical systems could result in injury and loss of life. These incidents would lead to time-consuming investigations, expensive liability litigation and reputation loss for the involved companies as well as negative public perceptions of the additive manufacturing industry. This chapter examines the various layers and avenues of liability exposure incurred by sabotage attacks on additive manufacturing.

2. Related Work

This section discusses research on additive manufacturing security and issues related to additive manufacturing liability exposure.

2.1 Additive Manufacturing Security

At the end of 2017, approximately seventy papers had been published on additive manufacturing security [47]. This section only considers research related to sabotage attacks.

Yampolskiy et al. [48] have studied the similarities and differences in security issues for additive and subtractive manufacturing (also referred to as computer numerical control (CNC) manufacturing). In their comparison, Yampolskiy and colleagues identified significant areas of overlap, including classical cyber security. However, they also identified significant and fundamental differences, including variations in possible manipulations and achievable effects.

Sturm et al. [32] have raised the possibility of attacks on large metal-alloy parts (e.g., used in jet turbines) that could cause operational failures. They identified four items that were vulnerable to attack: (i) computer-aided design (CAD) model; (ii) stereolithography (STL) file; (iii) toolpath file; and (iv) physical machine. Sturm and colleagues focused on STL files, and discussed scenarios involving corruption, scaling, indentation/protrusion, vertex movement and void attacks. They concluded that the most dangerous attacks would target structurally-strategic locations while being small enough to evade detection. They also highlighted an almost 50% decrease in failure strain for defective specimens and the inability to detect defects through mass, weight and visual inspections.

Zeltmann et al. [51] also studied similar attacks. They employed two different materials in order to embed defects. They found that the defects were undetectable with ultrasonic scans and that the defects deformed instead of cracking under stress. They also empirically investigated the impact of maliciously adjusting the printed object's orientation, an attack previously proposed by Yampolskiy et al. [49], and concluded that a 45° orientation reduced failure strain.

In their study of additive manufacturing using metals and alloys, Yampolskiy et al. [49] identified sabotage attacks that could be perpetrated by manipulating manufacturing process parameters. In the case of additive manufacturing using powder bed fusion, the alterable parameters include the scanning strategy, heat source energy and layer thickness. Another attack involves the compromise of the source material supply chain, where the source powder is substituted or mixed with a powder of different size or chemical composition, resulting in performance degradation of the manufactured parts.

Pope and Yampolskiy [26] have observed that timing disturbances in network communications (e.g., packets coming too late, too early or out of order) may impact industrial-grade additive manufacturing equipment and, by extension, the quality of the manufactured parts. Other factors include power interruptions or fluctuations to the manufacturing equipment.

Moore et al. [22] demonstrated printer firmware modification attacks. Their malicious firmware was able to substitute entire part models as well as perform less obvious modifications such as changing the extrusion rate. In earlier work, Moore et al. [21] examined the vulnerabilities of open-source software used with desktop 3D printers. They employed static source code analysis, dynamic USB communications analysis and architectural analysis to identify a number of security weaknesses.

Malicious code was key to an attack demonstrated by Belikovetsky et al. [4]. To demonstrate a complete attack chain, Belikovetsky and colleagues created a scenario in which an Internet-connected computer that controlled 3D printing was infected by malware delivered via email. The malware modified the STL file to introduce defects that would accelerate material fatigue. The modification resulted in propeller failure during flight, leading to the complete destruction of the drone and payload. A key concern brought about by the scenario is that the sabotaged propeller passed visual, weight and initial operational inspections.

2.2 Liability Exposure

Under current products liability law, parties can be held strictly liable for defective products. The concept is based on fairness, societal loss distribution and public safety [43]. To be held liable, the party must be commercially engaged in selling, must sell or distribute a product and the product must be defective [43]. Additionally, the product is expected to reach the end user without substantial change [11].

Engstrom [11] examined the liability of defective home-printed products, and identified the possible defendants as the hobbyist/inventor, digital designer and printer manufacturer. However, she argued that they are unlikely to be held liable because the hobbyist/inventor fail the commercial standard and the designer fails the product standard because code has been held not to be a product and, even if it were to change, the design code is modified significantly during the 3D printing process; for the printer manufacturer to be held liable, the printer had to be defective when it left the manufacturer's possession.

Liability can be primary or secondary. Reddy [27] discussed both types of liability when explaining the ramifications of 3D printing for intellectual property, contraband and at-home regulated item production. Primary liability evolves from the act while secondary liability can result from financial benefit and supervision or knowledge of and contribution to the act. Reddy concluded that regulations are required to address all levels of liability in additive manufacturing.

Strict liability is not the only cause of action that can be applied to 3D printers. Berkowitz [5] has analyzed the applicability of negligence and breach of warranty as well as strict liability and the related defenses. She proposed retaining strict liability for 3D printing, but creating a new affirmative defense for micro-sellers to meet the social policies of balancing protection with fairness.

Comerford and Belt [9] have also discussed strict liability, negligence and breach of warranty when they examined the exposure of scanning service providers and large-scale manufacturers. They suggested that, with definitive roles and responsibilities, the entire additive manufacturing chain can be characterized by the authorized dealer distribution chain construct, albeit virtual in nature. They also contended that contracts and insurance provide protection and indemnification in case of liability.

Supply chain categorization forms the basis of the liability analysis of Nielson [23]. Nielson examined liability in four product delivery frameworks, finding that the causes of action are difficult to pursue under all the frameworks, but more likely against a non-manufacturing seller.

Malloy [19] has proposed several avenues of recovery based on analyses of three actors: (i) printer manufacturer; (ii) computer-aided-design file creator; and (iii) object printer. He analyzed each actor with regard to design manufacturing and warnings of instruction defects, and provides strict liability grounds for each actor.

Wang [45] examined 3D printing services as a liability target. He discussed the use of risk-utility analysis to determine design defects. The analysis combines risk, utility and consumer expectations. Risk considers inherent safety and mitigability, and utility encompasses reasonable alternatives. Wang concluded that the impact of wrong materials can provide a defense to actors other than the supplier.

3. Attack Scenario

This section describes a typical additive manufacturing workflow and presents a sabotage attack scenario that targets the workflow.

3.1 Additive Manufacturing Workflow

Additive manufacturing can be used as an integral part of a manufacturer's process or it can be outsourced to external companies that provide additive manufacturing as a service. Figure 2 presents a typical additive manufacturing workflow that emphasizes the cyber and physical interactions between the various actors.

The additive manufacturing service provider infrastructure includes additive manufacturing machines, various post-processing equipment (e.g., hot isostatic pressing (HIP) equipment), non-destructive testing equipment (e.g., computer tomography system) and an information technology (IT) infrastructure. These infrastructure components are typically provided by different vendors that are often also responsible for equipment maintenance. Maintenance typically includes hardware maintenance, software and firmware updates, and equipment calibration.

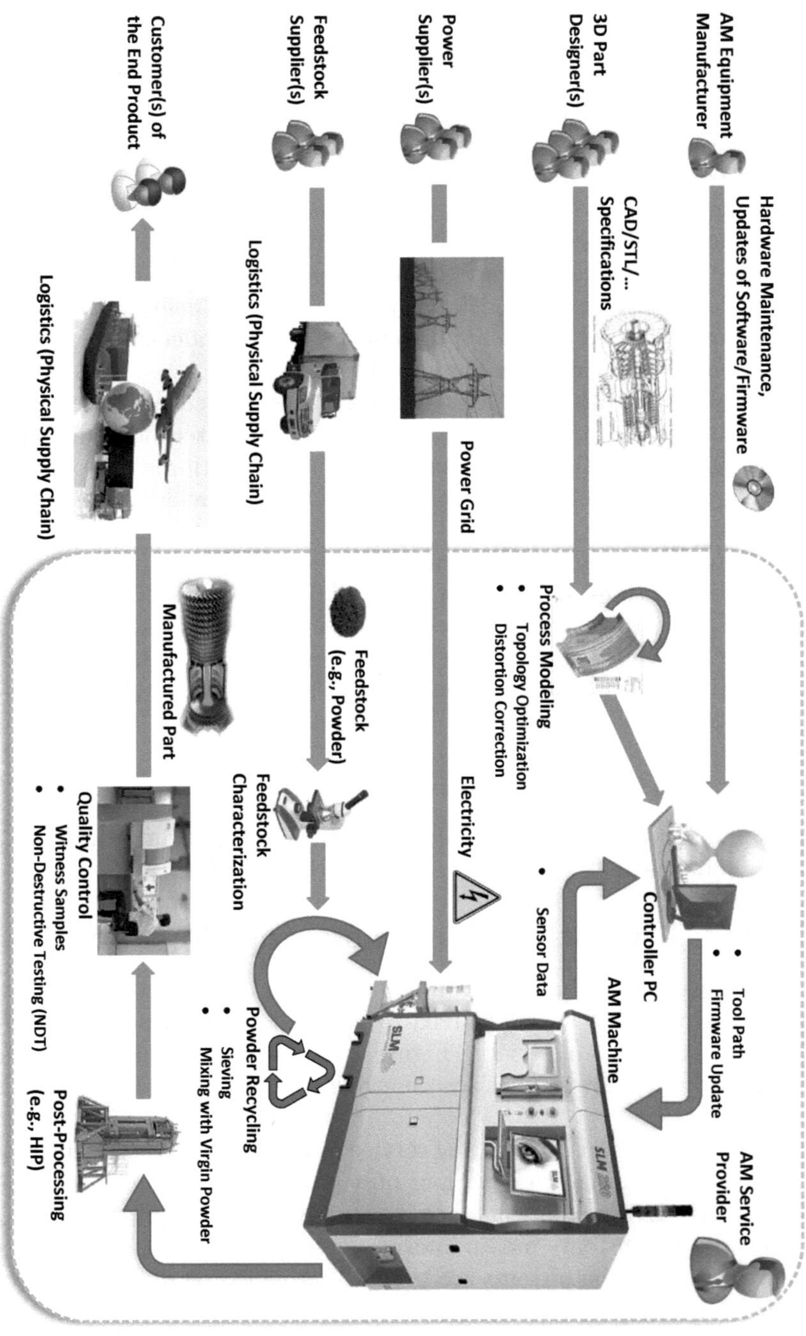

Figure 2. Additive manufacturing workflow [47].

The additive manufacturing service provider relies on digital model files (typically in the STL, AMP and 3MF file formats) and physical commodities like feedstock (i.e., source materials) and power. Depending on the expertise and scope of the manufacturer, the 3D part designer may be independent of the service provider or function as an internal entity, The physical commodities are commonly provided by external suppliers.

The physical commodities may also be involved in complex on-site processes. For example, feedstock can be characterized based on its quality. However, this is a time-consuming and expensive task. Therefore, additive manufacturing service providers often rely on characterizations provided by their suppliers. Additionally, to reduce costs and negative environmental impact, additive manufacturing processes such as powder bed fusion reclaim the unused powder, which is subsequently re-processed and reused.

The information technology infrastructure of a service provider includes computers, networks and software. In an industrial setting, software is used to optimally orient a part for a build, add support structures, lay out the build plate and slice the build into the desired layers. Process simulation software can be used to reduce geometric distortions arising from residual stress. A controller computer is used to translate a design file to equipment-specific toolpath commands that specify the 3D object to be manufactured. The toolpath commands are sent for execution to a 3D printer via a computer network. Due to the integration of *in situ* quality diagnostics in additive manufacturing machines, sensor information is commonly fed back to the controller computer via the network.

Quality assurance (QA) activities on a manufactured part may include non-destructive testing such as computer tomography and ultrasonic testing. However, while these testing methods are well-suited to subtractive manufacturing, no single technique is applicable to all types of additively-manufactured parts [1, 12, 46].

3.2 Sabotage Attack

The `dr0wned` study of Belikovetsky et al. [4] demonstrated the feasibility of sabotage attacks. Their study implemented the entire chain of a sabotage attack. They obtained backdoor access to the controller computer using a classical spear-phishing attack over an external network connection. Next, they searched the compromised computer for STL files. After locating the drone propeller STL file, they downloaded the file. Following this, they analyzed the file to determine the modifications that would accelerate fatigue; specifically, fatigue that would cause the propeller to break after a certain amount of normal operation. After they verified that the modified propeller would reliably break within three minutes, they utilized the same backdoor to replace the original STL file with the corrupt version. Subsequently, the corrupted file was used to print a replacement propeller for the quadcopter drone. During the flight test, the propeller broke in normal flight within the anticipated time frame. The drone suffered catastrophic failure and plummeted to the ground, resulting in the destruction of the drone and its payload.

The `dr0wned` scenario is a viable threat to the manufacturing industry. This assessment is supported by the fact that Belikovetsky and colleagues incorporated attack concepts that had been demonstrated in industrial settings, including a spear-phishing attack, which established a backdoor to support the exfiltration, infiltration and corruption of files. The uniqueness of the threat originates from the effects that the modifications can introduce to additive manufacturing. Indeed, the increased use of additive manufacturing to produce safety-critical parts magnifies the potential cost of failing parts beyond mere financial implications.

Although the `dr0wned` scenario involved sabotage at the service provider level, sabotage attacks are by no means restricted to direct attacks. As illustrated in the additive manufacturing workflow, other actors are indirectly involved in the manufacturing process, including electric power and feedstock suppliers. Yampolskiy et al. [49] have shown that modifications to physical commodities can also lead to the degradation of the manufactured products. Pope and Yampolskiy [26] have identified the impacts of power disturbances on the final products. Any of these methods could be leveraged in a sabotage attack.

Other exposed components in the additive manufacturing workflow are the software and firmware employed in the service provider infrastructure. Because they are frequently developed by third parties, their integrity can be compromised prior to system integration, via external network connections or pushed in by compromised updates and patches.

4. Liability Analysis Framework

Figure 3 presents the framework proposed for analyzing the liability incurred as a result of sabotage attacks on additive manufacturing.

The `dr0wned` scenario can be generalized and applied to other systems, including safety-critical systems in the automotive and aerospace industries. Failures of these systems can result in significant financial loss, serious injury and death. An injured party in such an incident could have recourse against the participants in the additive manufacturing workflow. The directly injured party could be the operator of a failed system who suffered property loss and/or physical injury. The indirectly injured party could be an innocent bystander with no connection to the additive manufacturing process or product.

An end user typically does not purchase a product directly from the manufacturer. Instead, retailers and resellers are often the final participants in the commercial distribution chain. This work does not consider the possibility of sabotage via part substitution or intentional damage at the retailer, reseller or physical carrier sites. Therefore, the retailer, reseller and physical carrier are grouped in with the end user at the consumer level. The liability analysis framework recognizes that any claim between these parties and the participants

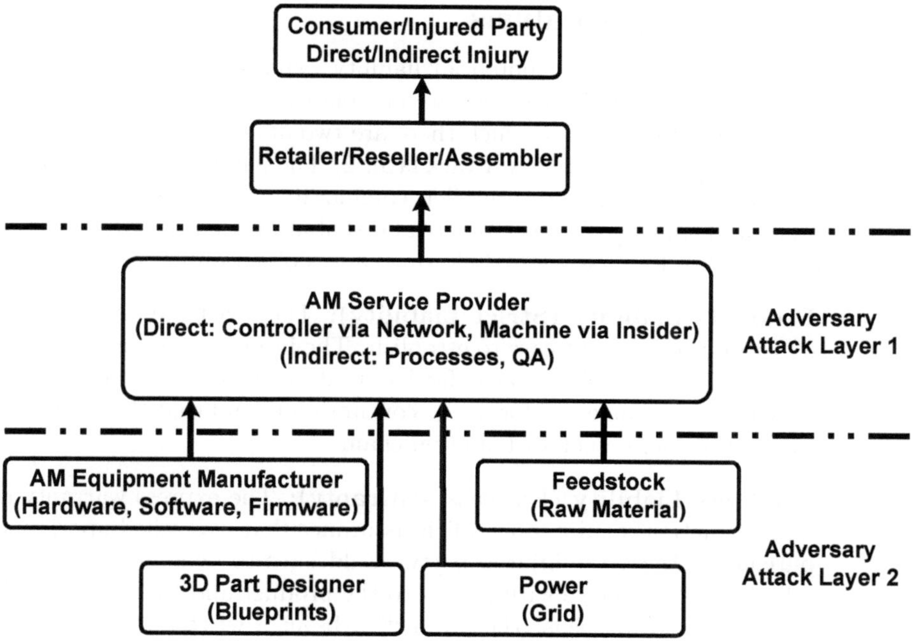

Figure 3. Liability analysis framework for sabotage attacks.

in the remainder of the additive manufacturing workflow would be governed by contractual indemnification processes, not personal injury liability.

A manufactured part is rarely an end product. Often, it is part of a multi-step assembly process that involves several business entities. Although part substitution is possible during this chain, it is not considered in this work. As with a retailer and reseller, between-party liability would be addressed as a part of the contractual relationship.

The manufacturing level is considered to be the first attack layer. An attack in this layer is similar to that perpetrated in the `dr0wned` study. What differentiates attacks in this layer is that they are the closest to the end user and can target specific end products. These attacks can be performed by modifying design files [4, 31, 51] or by compromising the additive manufacturing process [26, 30, 49, 50]. The attacks at this level are considered to operate in adversary attack layer 1.

An additive manufacturing service provider relies on a variety of physical and cyber commodities. These commodities include additive manufacturing equipment with the requisite firmware and software, object blueprints, feedstock and power supply. Any of these could be substituted or contaminated in a sabotage attack. At this level, the attack is farthest from the end user and cannot be targeted at a specific manufactured part [50]. Therefore, this layer is distinguished as the adversary attack layer 2.

5. Liability Analysis

This section analyzes the potential liability exposure in four cases – two potential litigants (i.e., parties who are eligible to sue): (i) end user victim; and (ii) bystander victim, for which there are two attack layers: (i) adversary attack layer 1; and (ii) adversary attack layer 2. The parties that could be held liable include the manufacturer, retailer, commodity supplier, service provider, merchant and members of the commercial distribution chain.

For each of the four cases, the following causes of action are considered:

- **Products Liability (Strict Liability):** The strict liability [43] cause of action is the easiest case to establish. The injured party has to demonstrate that the product was defective, that the defect made the product unreasonably dangerous for use or consumption, and that the liable party was in the commercial distribution chain.

- **Products Liability (Express Warranty):** The express warranty [37] cause of action requires an explicit assurance that was relied upon by the purchaser. Here, the injured party would have to demonstrate that the seller made a promise with regard to the product and that it factored in the decision to purchase the product. Absent a written agreement, this might be considered difficult to prove.

- **Products Liability (Implied Warranty):** The implied warranty [38] cause of action involves merchantability of average quality and ordinary purpose or a warranty that the product was fit for a particular purpose [39]. In the case of particular purpose, the injured party would have to demonstrate that the seller knew the purpose of the product and that the buyer relied on the seller to provide a suitable product. Given that additive manufacturing is increasingly used in the just-in-time and on-demand manufacturing of parts, this cause of action might be easier to prove for additive manufacturing than in the case of normal manufacturing.

- **Products Liability (Negligence):** For products liability negligence [42], the injured party has to establish a duty of care, a breach of duty and that the breach caused the injury. The focus in this situation is on the actions rather than the product, which renders the cause of action more difficult to prove.

- **Negligence in Tort:** Negligence in tort [41] differs from products liability negligence in that products liability examines the defendant's actions in terms of commercially-relevant standards as opposed to a non-commercial actor. Negligence in tort also requires an injured party to demonstrate that he or she was a foreseeable plaintiff.

The intentional torts – battery, assault, infliction of emotional distress and trespass to chattel – require an injured party to establish that there was an

Table 1. Strict liability (end user layer 1).

Strict Liability	**dr0wned Project**
Defective Product	STL file compromise
Unreasonably Dangerous	Midflight failure
Commercial Distribution Chain	Based on additive manufacturing workflow
Anyone Endangered	Flight path

intent to act. For battery [35], the intent is to harm. For assault [36], the intent is to create fear. For infliction of emotional distress [40], the intent is to cause upset. For trespass to chattel [44], the intent is to deprive someone of the use of property.

In the case of battery, the intent to harm can be the knowledge that harm is certain to occur. If the manufacturer does not take steps to protect the manufacturing environment, especially the network and software, it could be argued that the manufacturer knew that sabotage was possible and that harm would result from non-conforming parts that failed during flight. However, intent is often difficult to prove in product cases, which is why products liability is more often grounds for recovery. Products liability focuses on the product while the other causes of action focus on the defendant's actions and intent.

5.1 End User (Adversary Attack Layer 1)

This section discusses liability with regard to an end user victim in adversary attack layer 1.

Strict Liability. In the case of strict liability, the end user victim must establish that the product was defective and that the defect rendered the product unreasonably dangerous for use. Comparing the original file against the altered STL file can demonstrate the defect. Because the propeller failed, the drone crashed and injury resulted, it is possible to argue that the defective part was unreasonably dangerous to use in a drone and that the end user was endangered by the defect. To be held liable, the defendant must be in the commercial distribution chain. The additive manufacturing workflow establishes that the retailer and service provider are in the commercial distribution chain. Table 1 summarizes the products liability strict liability elements.

Express Warranty. In the case of express warranty, the injured end user has to prove the terms of the warranty and that the propeller failure demonstrated a breach of the warranty. Depending on the terms of an express warranty, part failure may not be sufficient to prove the breach. The defendant must be a seller, demonstrated by the additive manufacturing workflow and commercial transaction, while the plaintiff could be the buyer, a household

Table 2. Express warranty (end user layer 1).

Express Warranty	**dr0wned Project**
Terms	Transaction specific
Breach	Midflight failure
Seller	Based on additive manufacturing workflow
Buyer/Expected User	Transaction specific

member, guest or someone else expected to use, consume or be affected by the part. Table 2 summarizes the products liability express warranty elements.

Table 3. Implied warranty (end user layer 1).

Implied Warranty	**dr0wned Project**
Average Quality	Derived from design requirements
Fit for Use	STL file comparison
Seller	Based on additive manufacturing workflow
Buyer/Expected User	Transaction specific

Implied Warranty. Implied warranty involves merchantability or fit for a particular use. Merchantability requires that the part be of average quality and fit for ordinary purposes. The propeller failed the average quality and fit for particular use requirements due to premature fatigue. The quality and fit requirements were arguably captured in the design files; the failure to meet the requirements can be confirmed by comparing the executed files against the design files. Under implied warranty, the buyer relies on the seller to produce a conforming part and to protect the marketplace. The plaintiff can be any buyer, household member, guest or someone else expected to use, consume or be affected by the product. In the case of implied warranty, the defendant is a merchant in goods of that kind. Table 3 summarizes the products liability implied warranty elements.

Negligence. Key to products liability negligence is demonstrating a duty of care. Although a defendant might argue that standards are not established in the additive manufacturing industry, duty of care in the industry could be expected to combine manufacturing care with cyber security standards for the information technology infrastructure. The question would be whether the additive manufacturing service provider implemented available protections and defenses or those comparable with other cyber-physical systems, especially with regard to open-source software and network connectivity, as well as man-

Table 4. Negligence (end user layer 1).

Negligence	dr0wned Project
Reasonable Person	Analysis of security decisions
Breach	File compromise with failure
Manufacturer	Based on additive manufacturing workflow
Foreseeably Endangered	Flight path

ufacturing quality assurance for the purpose of detecting problems. Table 4 summarizes the products liability negligence elements.

Table 5. Negligence in tort (end user layer 1).

Negligence in Tort	dr0wned Project
Reasonable Person	Analysis of security decisions
Breach	File compromise with failure
Actual Cause	Sabotaged part failure
Legal Cause	Based on additive manufacturing workflow
Foreseeable	Flight path

Negligence in Tort. Negligence in tort has more components to establish for recovery. The reasonable person standard of care is owed to a foreseeable plaintiff. Negligence also requires demonstrating a breach of the duty of care and that the defendant's actions caused the injury. In the `dr0wned` attack, the modified file along with the destruction and injury would demonstrate the breach. Note that the cause must be actual and legal. Actual cause dictates that the injury would not have occurred, but for the retailer's or service provider's action in furnishing the sabotaged part. Legal cause requires a direct injury with no intervening cause or an indirect injury that was a foreseeable result. Table 5 summarizes the negligence elements.

5.2 Bystander (Adversary Attack Layer 1)

This section discusses liability with regard to a bystander victim in adversary attack layer 1.

Strict Liability. In the case of strict liability, the plaintiff is someone who was endangered by a defect. Thus, the bystander victim would use the same arguments as the end user victim to establish liability. Table 6 summarizes the products liability strict liability elements.

Table 6. Strict liability (bystander layer 1).

Strict Liability	**dr0wned Project**
Defective Product	STL file compromise
Unreasonably Dangerous	Midflight failure
Commercial Distribution Chain	Based on additive manufacturing workflow
Anyone Endangered	Flight path

Table 7. Express warranty (bystander layer 1).

Express Warranty	**dr0wned Project**
Terms	Transaction specific
Breach	Midflight failure
Seller	Based on additive manufacturing workflow
Expected to be Affected	Transaction and flight path

Express Warranty. In the case of express warranty, the plaintiff may be a guest or someone who is expected to be affected by a product. Thus, the injured bystander would use the same arguments as an end user victim to establish liability. However, the bystander may have greater difficulty in establishing the fact of a warranty depending on his or her relationship to the buyer. Table 7 summarizes the products liability express warranty elements.

Table 8. Implied warranty (bystander layer 1).

Implied Warranty	**dr0wned Project**
Average Quality	Derived from design requirements
Fit for Use	STL file comparison
Seller	Based on additive manufacturing workflow
Expected to be Affected	Transaction and flight path

Implied Warranty. As in the case of express warranty, the plaintiff can be a guest or someone who is expected to be affected by the product. Thus, the injured bystander would use the same arguments as an end user victim to establish liability. Table 8 summarizes the products liability implied warranty elements.

Negligence. The plaintiff in a products liability negligence case can be someone who has been foreseeably endangered. Therefore, the injured by-

Table 9. Negligence (bystander layer 1).

Negligence	**dr0wned Project**
Reasonable Person	Analysis of security decisions
Breach	File compromise with failure
Manufacturer	Based on additive manufacturing workflow
Foreseeably Endangered	Flight path

stander would need to establish that the additive manufacturing defendants could have foreseen injury to the bystander in addition to the actual user. It is arguable that the manufacturer could have foreseen that people other than the operator would be injured by a drone falling from the sky due to a sabotaged part, although other circumstances such as the relationship to the operator and operating location would be considered. After being established as a foreseeable plaintiff, the injured bystander could use the same products liability negligence arguments as the end user plaintiff. Table 9 summarizes the products liability negligence elements.

Table 10. Negligence in tort (bystander layer 1).

Negligence in Tort	**dr0wned Project**
Reasonable Person	Analysis of security decisions
Breach	File compromise with failure
Actual Cause	Sabotaged part failure
Legal Cause	Based on additive manufacturing workflow
Foreseeable	Flight path

Negligence in Tort. In the case of negligence in tort, the reasonable standard of care is owed to the foreseeable plaintiff. In the `dr0wned` sabotage attack, it is arguable that a machine that fell from the sky and resulted in injury to an innocent bystander would violate the reasonable person standard, especially since the sabotage occurred under the control of the manufacturer. It is also arguable that the bystander is a foreseeable plaintiff. The additive manufacturer produced a propeller used in a flying machine that could cause indiscriminate harm if it fell from the sky upon failure. Demonstrating the breach could include showing a failure to use available means to prevent and detect the sabotage, along with a comparison of the original and actual files. In the event of a compromised jet nozzle resulting in potentially more loss of life and property damage, competing concerns of social utility and societal loss distribution would have to be balanced. Table 10 summarizes the negligence elements.

Table 11. Strict liability (end user layer 2).

Strict Liability	drOwned Project
Defective Product	Comparison of design specifications against compromised commodities
Unreasonably Dangerous	Midflight failure
Commercial Distribution Chain	Based on additive manufacturing workflow
Anyone Endangered	Untargeted attack and flight path

5.3 End User (Adversary Attack Layer 2)

This section discusses liability with regard to an end user victim in adversary attack layer 2.

Strict Liability. In attack layer 2, the defect is introduced via one of the cyber or physical commodities. However, under strict liability, the focus is on the product and whether its defect rendered it unreasonably dangerous for use rather than the source of the defect. In the `drOwned` scenario, it would be the same for end user recovery whether the fatigue was introduced in layer 1 by the altered STL file or in layer 2 by contaminated feedstock, power fluctuations or firmware updates. As such, liability against the manufacturer would be established as with the layer 1 attack. The layer 2 attack introduces an additional liable party, the commodity supplier, which the injured party could argue is part of the commercial distribution chain. Table 11 summarizes the products liability strict liability elements.

Express Warranty. In the case of express warranty, the focus is on the warranty and the breach. For a layer 2 attack, the terms of the warranty would determine whether the source of the defect was relevant to the cause of action. Depending on the warranty, a layer 2 attack might not necessarily excuse the manufacturer from liability while also exposing the commodity supplier. However, the greater the distance of the end user from the source of the defect, the more complicated it would be to establish the necessary relationship or that the commodity supplier is a liable party. Table 12 summarizes the products liability express warranty elements.

Implied Warranty. As in the case of strict liability, the cause of action in implied warranty focuses on the product and the defect instead of the source of the defect. The failure of the propeller to meet average quality or fit for a particular use standard is independent of the defect's origin. The buyer's reliance on the manufacturer to produce a conforming part and to protect the marketplace has not changed. Rather, the manufacturer's placement between the end user and the source of the sabotage underscore its role in protecting

Table 12. Express warranty (end user layer 2).

Express Warranty	drOwned Project
Terms	Transaction specific
Breach	Midflight failure
Seller	Based on additive manufacturing workflow
Buyer/Expected User	Transactional distance

Table 13. Implied warranty (end user layer 2).

Implied Warranty	drOwned
Average Quality	Derived from design specifications
Fit for Use	Specified commodity quality
Seller	Based on additive manufacturing workflow
Buyer/Expected User	Transactional distance

the marketplace. In addition to not excusing the manufacturer, the layer 2 attack exposes the commodity supplier to liability because the end user can be categorized as affected by the defect regardless of origin. For example, if the `drOwned` defect was created when the contaminated material did not fuse, then the end user was affected by the contaminated feedstock. Table 13 summarizes the products liability implied warranty elements.

Negligence. The duty of care to the end user in a layer 2 attack might arguably include screening activities at the commodity supplier and manufacturer levels. In exercising due care, the commodity supplier could be expected to screen cyber and physical commodities for flaws, bugs and other compromises prior to shipping. The manufacturer could be expected to conduct screening at intake to detect layer 2 compromises. In the case of feedstock, it is common for the manufacturer to rely on the supplier's characterization. In the `drOwned` sabotage scenario, this would enable contaminated feedstock to compromise the propeller leading to the drone failure and injury to the end user. Table 14 summarizes the products liability negligence elements.

Negligence in Tort. In the case of negligence in tort, the injured end user would have to show standing as a foreseeable plaintiff and that the liable parties violated a reasonable person standard of care. In a layer 2 attack, it is arguably foreseeable that harm would reach the end user because sabotage at the commodity supplier level cannot be targeted, but could impact anyone along the manufacturing process chain up to and including the end user. For the reasonable person standard, the injured user could include prevention

Table 14. Negligence (end user layer 2).

Negligence	dr0wned Project
Reasonable Person	Analysis of screening/security decisions
Breach	Commodity compromise with failure
Commercial Seller	Based on additive manufacturing workflow
Foreseeably Endangered	Untargeted attack and flight path

Table 15. Negligence in tort (end user layer 2).

Negligence in Tort	dr0wned Project
Reasonable Person	Analysis of screening/security decisions
Breach	Commodity compromise with failure
Actual Cause	Sabotaged part failure
Legal Cause	Based on additive manufacturing workflow
Foreseeable	Untargeted attack and flight path

and detection measures employed in other similar industries to demonstrate protections against and attempts to detect compromised cyber and physical supplies. The measures could also be used to demonstrate the reasonableness of deployment at the manufacturing level given the susceptibility of cyber and manufacturing systems to attack. The cause must be actual and legal. Actual cause dictates that the injury would not have occurred but for the sabotage, which can be established with a showing that the parts do not fail under the same circumstances and that the injury results from the part failure. Legal cause requires a direct injury with no intervening cause or an indirect injury that was a foreseeable result. The injured party would argue that, although the various steps of the process chain might appear to be intervening causes, the part failure is a foreseeable result when a compromise disrupts the manufacturing process. Table 15 summarizes the negligence elements.

5.4 Bystander (Adversary Attack Layer 2)

This section discusses liability with regard to a bystander victim in adversary attack layer 2.

Strict Liability. For the bystander victim of a layer 2 sabotage attack, the focus is still on the product and whether the bystander victim was endangered by the defect. The bystander victim would use the same arguments as the end user victim of a layer 1 attack to establish liability. With the focus on the product, the source of the defect would be irrelevant to the manufacturer's exposure to bystander liability. The commodity supplier would also be exposed

Table 16. Strict liability (bystander layer 2).

Strict Liability	**dr0wned Project**
Defective Product	Comparison of design specifications against compromised commodities
Unreasonably Dangerous	Midflight failure
Commercial Distribution Chain	Based on additive manufacturing workflow
Anyone Endangered	Untargeted attack and flight path

to strict liability recovery, although the supplier could attempt to argue that it was not part of the commercial distribution chain or that its sabotaged contribution to the product was not the source of the defect that injured the bystander. Table 16 summarizes the products liability strict liability elements.

Table 17. Express warranty (bystander layer 2).

Express Warranty	**dr0wned Project**
Terms	Transaction specific
Breach	Midflight failure
Seller	Based on additive manufacturing workflow
Expected to be Affected	Transactional distance and flight path

Express Warranty. As in the case of a layer 1 attack, the bystander victim could use the same arguments as the end user because the express warranty extends to anyone who is expected to be affected by a product. The layer 2 attack would pose the same challenges to the bystander as it does to an end user with regard to the warranty terms and the ability to include or extend the warranty to the commodity supplier. Table 17 summarizes the products liability express warranty elements.

Implied Warranty. Since the bystander victim could be someone who is expected to be affected by the product, the same layer 2 arguments for the end user with regard to implied warranty could be applied by the bystander. The commodity supplier could argue that the product was the sabotaged commodity instead of the compromised propeller and, as such, the bystander was not in the expected class of user. However, the commodity supplier has a role in protecting the marketplace, as does the manufacturer, which is the underlying social policy for implied warranty liability. Thus, the manufacturer and the commodity supplier arguably would be exposed to implied warranty liability because the bystander was injured as a result of the layer 2 sabotage attack. Table 18 summarizes the products liability implied warranty elements.

Table 18. Implied warranty (bystander layer 2).

Implied Warranty	dr0wned Project
Average Quality	Derived from design specifications
Fit for Use	Specified commodity quality
Seller	Based on additive manufacturing workflow
Expected to be Affected	Transactional distance and flight path

Table 19. Negligence (bystander layer 2).

Negligence	dr0wned
Reasonable Person	Analysis of screening/security decisions
Breach	Commodity compromise with failure
Commercial Seller	Based on additive manufacturing workflow
Foreseeably Endangered	Untargeted attack and flight path

Negligence. The foreseeability of a bystander victim as someone endangered by the sabotaged product is again an issue with this cause of action. As in the case of a layer 1 attack, it is arguable that the manufacturer could have foreseen that individuals other than the operator could be injured by a drone falling from the sky due to a sabotaged part. It is also arguable that the commodity supplier could have foreseen that anyone up the workflow, including bystanders, could be injured by the sabotage of items under its control. Due to the untargeted nature of a layer 2 attack, it is perhaps even more arguable that an unsuspecting bystander would be endangered. After a bystander victim is established as a foreseeable plaintiff, the bystander could use the same products liability negligence arguments as the end user plaintiff to hold the manufacturer and commodity supplier liable. Table 19 summarizes the products liability negligence elements.

Negligence in Tort. In the case of negligence in tort, the liability argument for a layer 2 sabotage would resemble that of an end user victim because the attack was indiscriminate and could foreseeably have injured anyone after the point of the compromise. If the defendant claims that the sheer indiscriminate nature contradicts any foreseeability, the bystander could argue that it is exactly why he/she is a foreseeable plaintiff and why the reasonable person standard would examine what measures could and should have been deployed to prevent indiscriminate injury. The nature of the control of the manufacturer and commodity supplier of the component and the preventative measures, along with the indiscriminate nature and extent of the harm, combine to form the basis for meeting the foreseeable plaintiff standard and the breach of a reason-

Table 20. Negligence in tort (bystander layer 2).

Negligence in Tort	dr0wned Project
Reasonable Person	Analysis of screening/security decisions
Breach	Commodity compromise with failure
Actual Cause	Sabotaged part failure
Legal Cause	Based on additive manufacturing workflow
Foreseeable	Untargeted attack and flight path

able person standard of care. As in the case of the bystander in layer 1 and end user in layer 2, the defendants could argue that intervening events and actions affected the actual and legal cause elements. However, traceability from the introduction of sabotage (by power fluctuations, compromised firmware or contaminated feedstock) to the end product would establish actual cause. The fact that injury resulted from a failed compromised part that caused the drone to fall from the sky would establish the legal cause. If the compromise was not traceable to the original sabotage, then the injured bystander could argue that it was further indication that the reasonable person standard was violated because the commodity supplier did not sufficiently audit its processes and materials and the service provider did not sufficiently audit its supplies. Table 20 summarizes the negligence elements.

6. Discussion

This chapter has discussed the financial liability of the entire additive manufacturing supply chain in the event of a sabotage attack. However, there are some topics that are out of scope, but still bear mentioning. This section briefly discusses the financial liability between participants in the manufacturing process, corporate criminal liability and nation-state actors.

6.1 Liability between Process Chain Elements

Three areas should be considered when making decisions about security investments to combat sabotage attacks: (i) liability to external parties; (ii) liability between parties; and (iii) shifting risk through insurance. Liability to external parties has been covered in detail. This section briefly discusses the remaining two areas.

Liability between the participants in the additive manufacturing chain, from supplier to manufacturer, can be considered to be a contractual situation. It is anticipated that workflow component liability would be governed by the contracts between the parties [9, 24]. Insurance adds another factor to liability between the participants in the additive manufacturing workflow because it shifts the risk outside the workflow [17, 34]. Liability between parties and insurance are both considerations for additive manufacturing components with regard to liability exposure and the detection and prevention of 3D printer sabotage attacks.

6.2 Corporate Criminal Liability

Criminal liability is not likely for corporate behavior. An exception was the 2010 Deepwater Horizon explosion that killed eleven people and spilled millions of gallons of oil into the Gulf of Mexico [28]. The company (BP) plead guilty to fourteen criminal charges and paid $1.256 billion in fines [18]. By comparison, BP was levied $18.7 billion in fines for environmental and economic damage [29]. Company employees were also charged, but the harshest sentence was probation [14].

If a corporation is to be held criminally liable for an act by an employee, then the act must be in the scope of employment, it must benefit the company and there must be intent that can be imputed to the company [10, 13]. An act can be a decision to omit quality control. It can also be a decision not to implement security measures (e.g., based on risk analysis). For a corporation to be held liable in a sabotage attack, intent would again be an issue as in the civil liability analysis presented in this chapter. Additionally, the act of sabotage would not normally be in the corporation's interest.

6.3 Nation-State Actors

If a nation-state actor were to launch a sabotage attack, the Foreign Sovereign Immunities Act would make it difficult to pursue liability. There is, however, a commercial activity exemption that could be invoked for a civil cause of action [3, 15]. In this case, attribution is required. Based on the prior cases, tracing an attack on a cyber system has proven to be difficult [3, 6, 7]. The additive manufacturing workflow adds complexity due to the number of participants and the avenues of attack. Given the distributed nature of cyber systems and additive manufacturing environments, there is a strong likelihood that the saboteur would have launched a remote attack, which would raise jurisdictional issues. Trans-jurisdictional investigation and prosecution could be considered to be insurmountable problems [3, 7, 25]. Beyond the technical limitations related to attribution and jurisdiction, political considerations impose additional restrictions because governments generally avoid exposing their investigative capabilities.

7. Conclusions

The `dr0wned` study [4] demonstrated the feasibility and impact of a sabotage attack on additive manufacturing. The question now is not if, but when such attacks will occur.

This chapter has analyzed liability exposure arising from sabotage attacks on additively-manufactured functional parts. It established the sabotage attack layers, developed a framework for analyzing liability for sabotage attacks on functional parts and analyzed the civil liability exposure of the additive service provider and commodity suppliers in the event of an attack that results in injury to an end user and/or bystander. The analysis reveals that the parties are exposed to potential liability that would result in expensive investigations and defense costs regardless of whether or not they are ultimately held responsible and incur financial penalties. Additionally, additive manufacturing service providers and the nascent industry would suffer reputation loss as a result of injury-causing accidents. This would be especially true if the additive manufacturing industry is viewed as being more susceptible to sabotage attacks compared with the traditional manufacturing industry or is portrayed as failing to implement prevention and detection techniques in pursuit of profit. It is, therefore, important that all the additive manufacturing actors conduct or re-evaluate their cost-benefit analyses and invest in security measures.

References

[1] M. Albakri, L. Sturm, C. Williams and P. Tarazaga, Non-destructive evaluation of additively-manufactured parts via impedance-based monitoring, *Proceedings of the Twenty-Sixth International Solid Freeform Fabrication Symposium*, pp. 1475–1490, 2015.

[2] American Society for Testing and Materials, Standard Terminology for Additive Manufacturing Technologies, ASTM F2792-12a, West Conshohocken, Pennsylvania, 2012.

[3] P. Anderson, Cyber attack exception to the Foreign Sovereign Immunities Act, *Cornell Law Review*, vol. 102(4), pp. 1087–1114, 2017.

[4] S. Belikovetsky, M. Yampolskiy, J. Toh, J. Gatlin and Y. Elovici, `dr0wned` – Cyber-physical attack with additive manufacturing, *Proceedings of the Eleventh USENIX Workshop on Offensive Technologies*, 2017.

[5] N. Berkowitz, Strict liability for individuals? The impact of 3-D printing on products liability law, *Washington University Law Review*, vol. 92(4), pp. 1019–1053, 2015.

[6] S. Brenner, At light speed: Attribution and response to cybercrime/terrorism/warfare, *Journal of Criminal Law and Criminology*, vol. 97(2), pp. 379–476, 2007.

[7] C. Brown, Investigating and prosecuting cyber crime: Forensic dependencies and barriers to justice, *International Journal of Cyber Criminology*, vol. 9(1), pp. 55–119, 2015.

[8] L. Columbus, 2015 roundup of 3D printing market forecasts and estimates, *Forbes*, March 31, 2015.

[9] P. Comerford and E. Belt, 3DP, AM, 3DS and products liability, *Santa Clara Law Review*, vol. 55(4), pp. 821–836, 2015.

[10] C. Doyle, Corporate Criminal Liability: An Overview of Federal Law, Congressional Research Service, Washington, DC, 2013.

[11] N. Engstrom, 3-D printing and products liability: Identifying the obstacles, *University of Pennsylvania Law Review Online*, vol. 162, pp. 35–41, 2013.

[12] W. Frazier, Metal additive manufacturing: A review, *Journal of Materials Engineering and Performance*, vol. 23(6), pp. 1917–1928, 2014.

[13] A. Geraghty, Criminal Corporate Liability, Seventeenth Survey of White Collar Crime, *American Criminal Law Review*, vol. 39, pp. 327–354, 2002.

[14] J. Gill, Disaster prosecution is, well, a disaster, *The New Orleans Advocate*, March 12, 2016.

[15] S. Gilmore, Suing the surveillance states: The (cyber) tort exception to the Foreign Sovereign Immunities Act, *Columbia Human Rights Law Review*, vol. 46(3), pp. 227–287, 2014.

[16] T. Kellner, An epiphany of disruption: GE additive chief explains how 3D printing will upend manufacturing, *GE Reports*, November 13, 2017.

[17] M. Koch and B. Stansbury, 3-D printing: Innovation, opportunities and risk, *Law360*, February 24, 2016.

[18] C. Krauss and J. Schwartz, BP will plead guilty and pay over $4 billion, *The New York Times*, November 15, 2012.

[19] E. Malloy, Three-dimensional printing and a laissez-faire attitude towards the evolution of the products liability doctrine, *Florida Law Review*, vol. 68(4), pp. 1199–1226, 2016.

[20] Markets and Reports, 3D Printing Market Trends: Global Market Growth and Forecasting 2015–2020, DART Consulting, Bangalore, India, October 10, 2015.

[21] S. Moore, P. Armstrong, T. McDonald and M. Yampolskiy, Vulnerability analysis of desktop 3D printer software, *Proceedings of the IEEE Resilience Week*, pp. 46–51, 2016.

[22] S. Moore, W. Glisson and M. Yampolskiy, Implications of malicious 3D printer firmware, *Proceedings of the Fiftieth Hawaii International Conference on System Sciences*, pp. 6089–6098, 2017.

[23] H. Nielson, Manufacturing consumer protection for 3-D printed products, *Arizona Law Review*, vol. 57(2), pp. 609–622, 2015.

[24] L. Osborn, Regulating three-dimensional printing: The converging worlds of bits and atoms, *San Diego Law Review*, vol. 51, pp. 553–621, 2014.

[25] E. Podgor, Cybercrime: National, transnational or international? *Wayne Law Review*, vol. 50, pp. 97–108, 2004.

[26] G. Pope and M. Yampolskiy, A hazard analysis technique for additive manufacturing, presented at the *Better Software East Conference*, 2016.

[27] P. Reddy, The legal dimension of 3D printing: Analyzing secondary liability in additive layer manufacturing, *Columbia Science and Technology Law Review*, vol. XVI, pp. 222–247, 2014.

[28] C. Robertson and C. Krauss, Gulf spill is the largest of its kind, scientists say, *The New York Times*, August 2, 2010.

[29] C. Robertson, J. Schwartz and R. Perez-Pena, BP to pay $18.7 billion for Deepwater Horizon oil spill, *The New York Times*, July 2, 2015.

[30] A. Slaughter, M. Yampolskiy, M. Matthews, W. King, G. Guss and Y. Elovici, How to ensure bad quality in metal additive manufacturing: In-situ infrared thermography from the security perspective, *Proceedings of the Twelfth International Conference on Availability, Reliability and Security*, article no. 78, 2017.

[31] L. Sturm, C. Williams, J. Camelio, J. White and R. Parker, Cyber-physical vulnerabilities in additive manufacturing systems, *Proceedings of the Twenty-Fifth International Solid Freeform Fabrication Symposium*, pp. 951–963, 2014.

[32] L. Sturm, C. Williams, J. Camelio, J. White and R. Parker, Cyber-physical vulnerabilities in additive manufacturing systems: A case study attack on the .STL file with human subjects, *Journal of Manufacturing Systems*, vol. 44(1), pp. 154–164, 2017.

[33] Technovelgy, Plastic Constructor (3D Printer) (`www.technovelgy.com/ct/content.asp?Bnum=2445`), 2017.

[34] A. Thierer and A. Marcus, Guns, limbs and toys: What future for 3D printing? *Minnesota Journal of Law, Science and Technology*, vol. 17(2), pp. 805–854, 2016.

[35] Thomson Reuters Editorial Staff, § 87 Battery, *American Jurisprudence Second*, vol. 6, 2017.

[36] Thomson Reuters Editorial Staff, § 90 Causing apprehension, *American Jurisprudence Second*, vol. 6, 2017.

[37] Thomson Reuters Editorial Staff, § 631 Express warranties, *American Jurisprudence Second*, vol. 63, 2017.

[38] Thomson Reuters Editorial Staff, § 676 Implied warranties, *American Jurisprudence Second*, vol. 63, 2017.

[39] Thomson Reuters Editorial Staff, § 676 Implied warranty of fitness for particular purpose, *American Jurisprudence Second*, vol. 63, 2017.

[40] Thomson Reuters Editorial Staff, § 37 Intentional infliction of emotional distress, *American Jurisprudence Second*, vol. 74, 2017.

[41] Thomson Reuters Editorial Staff, § 1 Negligence, *American Jurisprudence Second*, vol. 57A, 2017.

[42] Thomson Reuters Editorial Staff, § 207 Negligence liability, *American Jurisprudence Second*, vol. 63, 2017.

[43] Thomson Reuters Editorial Staff, § 508 Strict liability in tort, *American Jurisprudence Second*, vol. 63, 2017.

[44] Thomson Reuters Editorial Staff, § 11 Trespass to chattel, *American Jurisprudence Second*, vol. 75, 2017.

[45] S. Wang, When classical doctrines of products liability encounter 3D printing: New challenges in the new landscape, *Houston Business and Tax Law Journal*, vol. 16, pp. 104–126, 2016.

[46] Wohlers Associates, Wohlers Report 2017: 3D Printing and Additive Manufacturing State of the Industry, Annual Worldwide Progress Report, Fort Collins, Colorado, 2017.

[47] M. Yampolskiy, W. King, J. Gatlin, S. Belikovetsky, A. Brown, A. Skjellum and Y. Elovici, Security of additive manufacturing: Attack taxonomy and survey, *Additive Manufacturing*, vol. 21, pp. 431–457, 2018.

[48] M. Yampolskiy, W. King, G. Pope, S. Belikovetsky and Y. Elovici, Evaluation of additive and subtractive manufacturing from the security perspective, in *Critical Infrastructure Protection XI*, M. Rice and S. Shenoi (Eds.), Springer, Cham, Switzerland, pp. 23–44, 2017.

[49] M. Yampolskiy, L. Schutzle, U. Vaidya and A. Yasinsac, Security challenges of additive manufacturing with metals and alloys, in *Critical Infrastructure Protection IX*, M. Rice and S. Shenoi (Eds.), Springer, Cham, Switzerland, pp. 169–183, 2015.

[50] M. Yampolskiy, A. Skjellum, M. Kretzschmar, R. Overfelt, K. Sloan and A. Yasinsac, Using 3D printers as weapons, *International Journal of Critical Infrastructure Protection*, vol. 14, pp. 58–71, 2016.

[51] S. Zeltmann, N. Gupta, N. Tsoutsos, M. Maniatakos, J. Rajendran and R. Karri, Manufacturing and security challenges in 3D printing, *Journal of the Minerals, Metals and Materials Society*, vol. 68(7), pp. 1872–1881, 2016.

II

INFRASTRUCTURE PROTECTION

Chapter 4

ERROR PROPAGATION AFTER REORDERING ATTACKS ON HIERARCHICAL STATE ESTIMATION

Ammara Gul and Stephen Wolthusen

Abstract State estimation is vital to the stability of control systems, especially in power systems, which rely heavily on measurement devices installed throughout wide-area power networks. Several researchers have analyzed the problems arising from bad data injection and topology errors, and have proposed protection and mitigation schemes. This chapter employs hierarchical state estimation based on the common weighted-least-squares formulation to study the propagation of faults in intermediate and top-level state estimates as a result of measurement reordering attacks on a single region in the bottom level. Although power grids are equipped with modern defense mechanisms such as those recommended by the ISO/IEC 62351 standard, reordering attacks are still possible. This chapter concentrates on how an inexpensive data swapping attack in one region in the lower level can influence the accuracy of other regions in the same level and upper levels, and force the system towards undesirable states. The results are validated using the IEEE 118-bus test case.

Keywords: Power systems, hierarchical state estimation, reordering attacks

1. Introduction

Efficient and reliable supervisory control and data acquisition (SCADA) systems along with energy management systems (EMSs) contribute to the safe and efficient operation of power grids. A SCADA system located at a control center collects data from remote substations in order to manage the power grid. An energy management system at the control center processes the collected data using an on-line application called state estimation. State estimation enables an operator to obtain accurate estimates of the system state despite noisy or faulty measurement data using a steady state flow model of the physical system [1, 13].

Published by Springer Nature Switzerland AG 2018.
J. Staggs and S. Shenoi (Eds.): Critical Infrastructure Protection XII, IFIP AICT 542, pp. 67–79, 2018.
https://doi.org/10.1007/978-3-030-04537-1_4

Many energy management system applications (e.g., for contingency analysis) use the estimated system state, which makes accurate state estimation vital to safe and efficient power grid operations.

Modern power systems are becoming more interconnected and less likely to be dependent on a single control center for operations. Positioning operators throughout the system in a hierarchical or distributed structure improves operational efficiency. Each operator located at his/her own control center uses SCADA and emergency management systems to manage a certain region of the overall system. Examples of such interconnected systems are the ENTSO-E in Europe and Western Interconnect (WECC) in the United States. Future power systems are expected to be even more interconnected than before and, thus, systems without any central coordinators should be anticipated. The timely exchange of accurate information between regional operators is essential to maintaining the safety of a large interconnected power network. At the same time, data exchange is limited for reasons of sensitivity. This complicates the tasks of operators who use local state estimates for command and control in their regions, which, in turn, contribute to the estimated state of the entire system.

Hierarchical state estimation requires control centers at each level to exchange data regularly. The Inter-Control Center Communications Protocol (ICCP) is widely used to transmit information from one level to another during hierarchical state estimation. This protocol supports access control, but it does not provide key-based authentication for the exchanged data. Therefore standard protocols such as TLS as mandated by IEC 62351 are used to implement authentication for ICCP associations [6]. As a result, ICCP messages may be passed in the clear to the protocol stack to provide authentication. An adversary who installs a Trojan could compromise all incoming and outgoing ICCP messages [19]. The vulnerability of control systems to such attacks is exacerbated by the fact that ICCP relations are often formed between hosts in the various regions.

This chapter examines the conditions under which a compromised region in a lower level can have undesirable impacts on other regions in the same hierarchical level as a result of the propagation of faults to the top level and then back down to each level. Although an attacker can impact other regions by manipulating a single region, in reality, the magnitudes of the changes that can be induced are limited. This chapter determines a necessary condition that enables the formulation of a minimum cost attack to realize a maximum (negative) impact.

2. Related Work

The effects of bad data on state estimation in power systems have been studied extensively [14–16]. Typically, a bad data detection algorithm is executed during state estimation; this algorithm removes outliers based on simple statistical thresholds.

When the measurement data collected by a SCADA system is compromised, the resulting incorrect state estimation can force the system into an undesirable state. Without further constraints on data and data correlations, Liu et al. [12] have relied on DC power flows. Other studies have attempted to determine the minimal undetectable attacks that require the least manipulation of data [5, 10].

Van Cutsem and Ribbens-Pavella [18] were among the earliest researchers to focus on hierarchical state estimation; their seminal survey paper is still used to construct models. Lakshminarasimhan and Girgis [11] have proposed a two-level hierarchical state estimation for wide-area power systems that assumes a highly reliable phasor measurement unit (PMU) at every boundary bus. Vukovic and Dan [19] have described several types of data attacks on decentralized state estimation, but they do not provide details about the computational complexity. Moreover, their proposed mitigation scheme involving an outlier approach can detect errors only after hundreds of iterations and, even then, the attack may not be identified.

False data injection attacks, which were initially studied in the context of conventional state estimation, have been shown to be possible in hierarchical topologies as well [7]. Baiocco and Wolthusen [3] have employed automated (graph) partitioning to support robust hierarchical state estimation during unexpected failures of single or multiple lines, or attacks. Shepard et al. [17] have described GPS spoofing attacks on phasor measurement units, which can result in ill-conditioned Jacobian matrices and divergence by introducing jitter in the communications channels during hierarchical state estimation [2]. A number of state estimators have been proposed, but studies of robustness to attacks have focused on centralized topologies. However, Baiocco et al. [4] have discussed the hierarchical case in the context of smart grid and microgrid environments.

Gul and Wolthusen [9] have highlighted the vulnerability of a communications infrastructure to an attack that reorders measurement vectors, resulting in incorrect estimates and potentially undesirable system states. It is worth noting that Gul and Wolthusen assume that the preceding and present measurement vectors are known to the attacker. In the two distinct scenarios they analyzed, the system diverged as a result of an ill-conditioned Jacobian matrix.

3. Power System State Estimation

A power system is denoted by a graph $\mathcal{G}$ with a set of buses $\mathcal{V}$ and a set of transmission lines $\mathcal{E}$. An AC power flow model is assumed. This is expressed as:

$$\mathbf{z} = h(\mathbf{x}) + \mathbf{e} \tag{1}$$

where $\mathbf{z} \in R^m$ is the measurement vector; $\mathbf{x} \in R^n$ is the state vector ($m > n$); h is the measurement function relating $\mathbf{z}$ to $\mathbf{x}$; and $\mathbf{e}$ is the noise vector with a mean of zero and known co-variance $\mathbf{R}$. The errors are assumed to be independent; therefore, $\mathbf{R} = diag\{\sigma_1^2, \sigma_2^2, \cdots, \sigma_m^2\}$ is a diagonal matrix.

The states $\hat{\mathbf{x}}$ are estimated by solving the following normal equations:

$$[F^T R^{-1} F]\Delta\hat{\mathbf{x}} = F^T R^{-1}[\mathbf{x} - f(\mathbf{x})] \tag{2}$$

Following this, bad data analysis is performed based on the residual values:

$$\mathbf{r} = \mathbf{z} - h(\hat{\mathbf{x}}) \tag{3}$$

Residual values that are larger than a statistical threshold τ are identified and the corresponding measurements are flagged as bad. After the bad measurements are removed, state estimation is re-run until the system converges. Unfortunately, bad data detection is difficult when there are multiple bad measurements. In practice, bad data goes undetected due to the presence of other bad data, or good measurements are flagged as bad for other reasons such as a change to the topology. Interested readers are referred to [1] for more details about state estimation.

4. Hierarchical State Estimation

Conventional or centralized state estimation can be followed by a multi-region hierarchical procedure in which local state estimators process all the raw measurements that are available locally; thus, only manageable amounts of data are sent to the immediate higher level. This process continues upward until the highest level is able to compute the state of the entire system, which is then conveyed to the lower levels for crucial tasks such as bad data processing [8].

The multi-region hierarchical structure can be symmetric or asymmetric. A symmetric hierarchy has a balanced division of bus-bars/tie-lines over all the regions whereas an asymmetric hierarchy has an unbalanced distribution of bus-bars/tie-lines. While symmetric hierarchical state estimation is trivial, asymmetric hierarchical state estimation models real-world power systems, but is more complex. Only asymmetric hierarchical state estimation is considered in this work. The formulation is taken from [2, 4].

Baiocco et al. [4] have introduced a tree structure for multi-region hierarchical state estimation with the tree root (level k) denoting the highest level state estimation. A lower level may have child nodes; a lower level without child nodes is a leaf node and resides in the lowest level (level 1) of the hierarchy. Each node performs its own state estimation using measurements of the estimated states from lower nodes; for level 1, the measurements are obtained by computing power flows. It is assumed that robust partitioning is already performed and that there are no overlaps between regions, except for common tie-lines that connect neighboring regions.

When a node estimates its state vector, it sends this vector (including the gain matrix) to all its children or to the parent node. This type of multi-region hierarchical state estimation involves two-way transmission of information from the lower levels to the higher levels until the root node is reached, upon which the overall state estimate is sent downwards towards the leaf nodes so that the state estimate is passed to all the tie-line branches.

A general k-level multi-region hierarchical state estimation is expressed as:

$$\begin{aligned} y_{0,j_1} &= f_{1,j_1}(y_{1,j_1}) + e_{1,j_1}, \quad j_1 = 1, \cdots, r_1 \\ y_{0,b_1} &= f_{1,b_1}(y_1) + e_{1,b_1} \\ y_{1,j_2} &= f_{2,j_2}(y_{2,j_2}) + e_{2,j_2}, \quad j_2 = 1, \cdots, r_2 \\ y_{1,b_2} &= f_{2,b_2}(y_2) + e_{2,b_2} \\ &\vdots \\ y_{0,b_1} &= f_{1,b_1}(y_1) + e_{1,b_1} \end{aligned} \tag{4}$$

where y_{0,j_1} is the local measurement vector in S_{j1} in level 1; y_{0,b_1} is the border measurement vector in level 1; y_{1,j_2} is the local measurement vector in S_{j2} in level 2; y_{1,b_2} is the border measurement vector in level 2; y_k is the state vector of the overall system; f_l is the corresponding non-linear measurement function for each level l; and e_l is the corresponding Gaussian measurement noise vector.

Level 1 Multi-Region State Estimation. For level 1, each region S_j estimates its own state $\tilde{y}_{1j}$ by solving the following normal equations iteratively:

$$\begin{aligned} [F^T_{1,j_1} R^{-1}_{1,j_1} F_{1,j_1}]\Delta\tilde{y}_{1,j_1} &= F^T_{1,j_1} R^{-1}_{1,j_1}[y_{0,j_1} - f_{1,j_1}(y_{1,j_1}(k))] \\ [F^T_{1,b_1} R^{-1}_{1,b_1} F_{1,b_1}]\Delta\tilde{y}_{1,j_1} &= F^T_{1,b_1} R^{-1}_{1,b_1}[y_{0,b_1} - f_{1,b_1}(y_{1,j_1}(k))] \end{aligned} \tag{5}$$

where the inputs at this level include the measurement vectors y_{0,j_1} and y_{0,b_1}; Jacobian matrices $F_{1,j1}$ and $F_{1,b1}$; and gain matrices R_{1,j_1} and R_{1,b_1}. Note that the Jacobian matrices are updated at every iteration.

Level *i* Multi-Region State Estimation. The following equations must be solved for each intermediate level hierarchically from the lower levels:

$$\begin{aligned} &[F^T_{i,j_{i-1}} G_{i-1,j_{i-1}} F_{i,j_{i-1}}]\Delta\tilde{y}_{i-1,j_{i-1}}(k) = \\ &\qquad\qquad F^T_{i,j_{i-1}} G_{i-1,j_{i-1}}[\tilde{y}_{i-1,j_{i-1}} - f_{i,j_{i-1}}(y_i(k))] \\ &[F^T_{i,b_i} G_{i-1,b_{i-1}} F_{i,b_i}]\Delta\tilde{y}_{i-1}(k) = F^T_{1,b_1} G_{i-1,b_{i-1}}[\tilde{y}_{i-1} - f_i(y_i(k))] \end{aligned} \tag{6}$$

Using the estimate $\tilde{y}_{i-1,j_{i-1}}$ from level $l-1$ as the measurements in a distributed approach, $\tilde{y}_{i,j_i}$ can be obtained as described in [7]. The Jacobian matrices are revised based on the estimates from levels i and $i+1$.

Level l Multi-Region State Estimation. Using the vector $\tilde{y}_{l_1}$ supplied by the lower level $l-1$ as the measurement vector, the system state can be estimated by iteratively solving the following equations:

$$\begin{aligned} &[F_{l,j_{l-1}}^T G_{l-1,j_{l-1}} F_{l,j_{l-1}}]\Delta\tilde{y}_{l-1,j_{l-1}}(k) = F_{l,j_{l-1}}^T G_{l-1,j_{l-1}}[\tilde{y}_{l-1,j_{l-1}} - f_{l,j_{l-1}}(y_l(k))] \\ &[F_{l,b_l}^T G_{l-1,b_{l-1}} F_{l,b_l}]\Delta\tilde{y}_{l-1}(k) = F_{1,b_1}^T G_{l-1,b_{l-1}}[\tilde{y}_{l-1} - f_l(y_l(k))] \end{aligned} \tag{7}$$

Note that the hierarchical state estimation process outlined above requires two-way exchange of data between local state estimators in each layer of the hierarchy [2].

5. Three-Level Simplification

This section presents a simplification of the multilevel model as a three-level model. The three-level model is given by:

$$\begin{aligned} y_{0,j_1} &= f_{1,j_1}(y_{1,j_1}) + e_{1,j_1}, \quad j_1 = 1,2 \\ y_{0,b} &= f_{1,b}(y_{1,b}) + e_{1,b} \\ y_{1,j_2} &= f_{2,j_2}(y_{2,j_2}) + e_{2,j_2}, \quad j_2 = 1,2 \\ y_{1,b} &= f_{2,b}(y_{2,b}) + e_{2,b} \\ y_2 &= f_3(x) + e_3 \end{aligned} \tag{8}$$

where the measurement vectors y_{0,j_1}, y_{1,j_1} and $y_{0,b}$, $y_{1,b}$; state vectors y_{1,j_1}, y_{2,j_2} and y_{b,j_1}, y_{b,j_2}; and non-linear measurement functions f_{1,j_1}, f_{2,j_2} and $f_{1,b}$, $f_{2,b}$ are as described above.

In order to simplify the process, it is assumed that there are no border variables and that the measurement functions are linear. The resulting three-level model is given by:

$$\begin{aligned} y_{0j} &= F_{1j} y_{1j} + e_{1j}, \quad j = 1,2 \\ y_{1j} &= F_{2j} y_{2j} + e_{2j}, \quad j = 1,2 \\ y_2 &= F_3 x + e_3 \end{aligned} \tag{9}$$

where F_{1j}, F_{2j} and F_3 are the Jacobian matrices of the corresponding measurement functions.

For each region, state estimation employs an iterative algorithm that determines the local state vector along with another iterative process involving the two levels [7]:

- **Level 1:** The inputs to the first level are y_{1j} for regions $j = 1,2$ (assuming two regions) and the weighting matrix R_{1j}^{-1}. The output, which corresponds to the local state vector $\hat{y}_{1j}$ for each region, is obtained by solving the following normal equation iteratively for each region:

$$[F_{1j}^T R_{1j}^{-1} F_{1j}^T]\hat{y}_{1j} = F_{1j}^T R_{1j}^{-1} y_{0j} \tag{10}$$

- **Level 2:** The inputs to the second level are y_{1j} for regions $j = 1, 2$ (assuming two regions) and the weighting matrix R_{1j}^{-1}. The output, which corresponds to the local state vector $\hat{y}_{1j}$ for each region, is obtained by solving the following normal equation iteratively for each region:

$$[F_{2j}^T R_{2j}^{-1} F_{2j}^T]\hat{y}_{2j} = F_{2j}^T R_{2j}^{-1} y_{1j} \tag{11}$$

- **Level 3:** The inputs to the third level are the state vectors of the second level $\hat{y}_2$ and the gain matrices $G_2 = F_{1j}^T R_{2j}^{-1} F_{2j}^T$ (corresponding to the weighting matrix). The output $\hat{x}$, which is the state of the entire system, is obtained by solving the following normal equation for the third level:

$$[F_3^T G_2^{-1} F_3^T]\hat{x} = F_3^T G_2^{-1} \hat{y}_2 \tag{12}$$

  where y_2 and G_2 are obtained by juxtaposing the corresponding y_{2j} and G_{2j}, respectively.

6. Attack Model

The attacker's goal is to disrupt hierarchical state estimation. It is assumed that the attacker can reorder the measurement set $\mathbf{y}^0$ of only one partition $S^0 \in S$ in the lowest level l_1 of the hierarchy, where S is the set of partitions. As a result, incorrect state variables are transmitted to the partitions in the upper levels at the beginning of each hierarchical state estimation iteration.

The structured reordering attack leverages internal knowledge of the partitions in order to maximize its impact. The knowledge required for the success of the reordering attack includes some previous plausible measurement set $\mathbf{y}_{old}$ of the targeted partition. The principal goal of the attack is to have a false local state estimate that propagates to the higher levels to produce an incorrect estimate $\mathbf{x}$.

The following constraints are imposed on an attack on the three-level hierarchical structure:

- After the attack is launched on a single partition in level l_1, the data exchange between the upper two levels (i.e., l_2 and l_3) remains normal. This means that there is no further attack on the upper levels.

- The network configurations (i.e., sub-region partitioning) in levels l_2 and l_3 are not permitted to change over the course of a complete top-down synchro-upgrade. Note that this constraint is usually not imposed on hierarchical state estimation [2].

After the attack, the flow equation for the first level l_1 is:

$$[F_{1j}^T R_{1j}^{-1} F_{1j}^T]\hat{y}_{1j}^* = F_{1j}^T R_{1j}^{-1} y_{0j}^* \tag{13}$$

where y_{0j}^* is the swapped measurement vector of one of the sub-regions in level l_1. The inputs to the second level y_{1j}^* for regions $j = 1, 2$ are the false estimates from the first level l_1:

$$[F_{2j}^T R_{2j}^{-1} F_{2j}^T]\hat{y}_{2j}^* = F_{2j}^T R_{2j}^{-1} y_{1j}^* \quad (14)$$

Finally, the output $\hat{x}^*$, which is the state of the entire system, is obtained by solving the following normal equation for the third level l_3:

$$[F_3^T G_2^{-1} F_3^T]\hat{x}^* = F_3^T G_2^{-1} \hat{y}_2^* \quad (15)$$

where y_2^* and G_2 are as defined above.

In the case of a false data injection attack, the symbol **a** denotes the attack vector that expresses the amount of change to the original measurement vector [12]:

$$\mathbf{a} = \mathbf{Fc} \quad (16)$$

where the vector **c** denotes the magnitude of change and is bounded by some stealthy condition.

Jamming or delay attacks can be seen as a sub-class of reordering attacks because they resend the previous data after a time interval. Also, attacks that replay or block measurement vectors can be considered to be a special case of reordering attacks with time constraints. The common aspect of all of these attacks is that no attack vector has to be added. Instead, the attacker simply drops/blocks a measurement or injects jitter in the measurement regardless of whether or not it is secure/protected by hacking the communications infrastructure. Therefore, the general term, "reordering of the measurement vector" is introduced to convey that the attacker replaces the true measurement vector with a previous plausible (true) vector.

In this case, the time horizon is critical to the attacker because it determines the strength of the attack. It is assumed that the attacker has measurement information from the present back to some point in time. From among these measurements, the attacker chooses the measurement vector to be swapped with the present measurement vector while continuing to maintain stealth. The term "stealth" implies that the attack is successful in forcing the system state without being detected by the model-based bad data detection algorithm. Sophisticated detection criteria certainly exist, but they are mainly used to determine which measurement devices (vector entries) are compromised, and, therefore, are not relevant to the case at hand. Other models rely on message redundancy to determine compromise, but this approach is not feasible for network-based attacks.

7. Reordering Attack Cost and Impact

The minimum attack cost Γ_y corresponds to the situation where the attacker expends the least effort to obtain the maximum mean square error (MSE). Power grid regions can be secured in one of the three ways: (i) non-tamperproof authentication ($S_{ntp} \subseteq S_m$); (ii) tamperproof authentication ($S_{tp} \subseteq S_m$); or (iii) other protection. Non-tamperproof authentication is implemented by a bump-in-the-wire device or a remote terminal unit (RTU) with a non-tamperproof authentication module; regions with this type of authentication are only susceptible to attacks that involve physical access to the region from where the measurements originate. In contrast, tamperproof authentication is not susceptible to attacks. Other protection mechanisms include security guards and video surveillance systems that are generally not vulnerable to attacks. However, to be realistic, all the regions of a power grid cannot be protected and there will be at least one vulnerable region $S_{m'}$. If the region where the measurement vector is to be attacked is protected and uses non-tamperproof or tamperproof authentication, then the measurement is not vulnerable and it is assumed that $\Gamma_y = \infty$.

Otherwise, for a measurement y, Γ_y is defined as:

$$\begin{aligned} &\Gamma_y = \min \ \|a\| \ \ s.t. \ \ a = Fc = \hat{y}^{new} - \hat{y}^{old} \ \ \text{and} \\ &a(y) \neq 0 \Longrightarrow |S(m')| \neq 0, \ \ s.t. \ \ S = S(m) \cup S(m') \end{aligned} \tag{17}$$

where S_m denotes the authenticated regions; and $S_{m'}$ denotes the vulnerable regions such that $S = S(m) \cup S(m')$.

In addition, it is assumed that the attacker is free to choose the set of plausible measurements in a particular time frame to be used in a reordering attack. As a result of this freedom and the attack cost Γ_y mentioned above, the maximum attack impact is taken to correspond to the attacker's outcome $\mathcal{I}_y$, which is given by:

$$\begin{aligned} &\mathcal{I}_y = \max \ I = \sqrt{\sum (\tilde{y}^{new} - \tilde{y}^{old})^2} \\ &\text{s.t.} \ \ t^{new} - t^{old} \geqslant \epsilon \end{aligned} \tag{18}$$

where t is the time slot from among the time frames available to the attacker; and ϵ is a pre-defined threshold that limits the attacker's choice. The superscripts "old" and "new" denote the original measurement and the measurement to be inserted in its place, respectively.

8. Experimental Results

Before discussing the experimental results, it is important to recall that, in order to perform a reordering attack, the attacker must have knowledge of the system topology. It is assumed that the topology does not change or the topology is static for the duration of the attack.

This section evaluates the proposed model by considering reordering attacks on hierarchical state estimation involving regions of the standard IEEE 118-bus system. The IEEE 118-bus system is divided into six regions, and an intermediate level exists between the top and bottom levels (Figure 1).

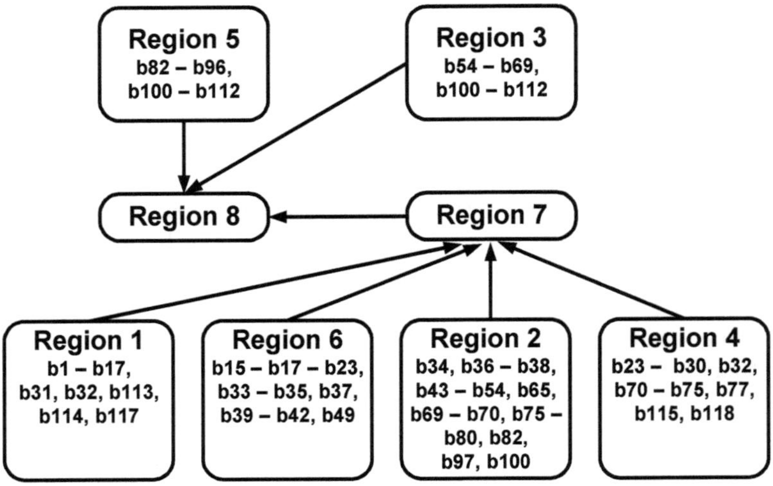

Figure 1. Bus-bar distribution in the IEEE 118-bus system.

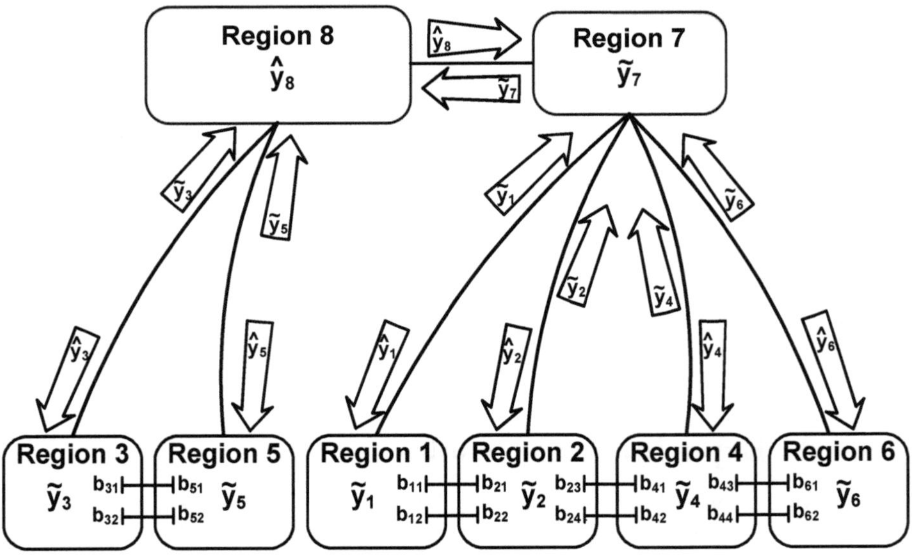

Figure 2. Information flow during hierarchical state estimation.

As shown in Figure 2, since the hierarchical model involves two-way synchro-upgrades (i.e., from the lower levels to the upper levels and subsequently from the topmost level down to the lower levels), it is particularly interesting to observe the error propagation after an attack. The attacker is free to choose data from a certain time frame (i.e., the attacker has a limited amount of knowl-

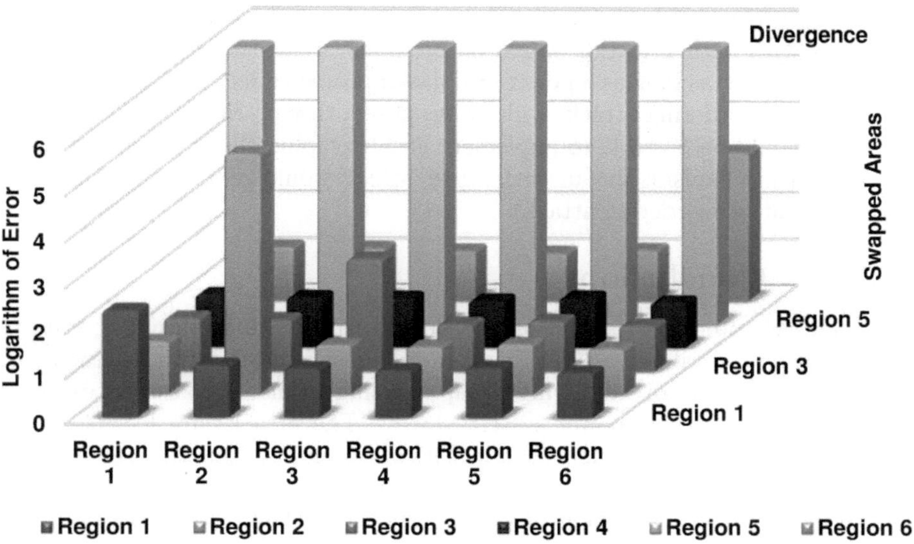

Figure 3. Impact of regionwise reordering in the lower level.

edge about the previous data). The weighted-least-squares (WLS) technique was used to estimate the state and the open-source MATPOWER package was used to load the data associated with the IEEE 118-bus system.

Figure 3 shows the mean squared error after performing least cost reordering attacks on the IEEE 118-bus system. The figure shows the logarithm (base 10) of the mean squared error for a complete round of the weighted-least-squares state estimation – from the lower layer to the top layer and all the way down, detailing how the error propagates up from the lowest level to the top level and back down.

It is clear that, at the end of a complete round after a reordering attack, all the regions are affected regardless of the intensity and the reordering of the individual regions.

A key observation is the epidemic characteristic of the attack, where the error propagates from an infected region in the lower level to all the regions in the lower level. The plot also illustrates how a single region in a lower level influences all the regions in the same level, implying that the attacker can choose the cheapest and most vulnerable region to launch the attack. Clearly, the error is maximum for the region where the attack originates. In the specific partitioning of the IEEE 118-bus system, Region 5 appears to be the most vulnerable because the system diverges when the input data is reordered. However, it is worth noting that the partitioning of the IEEE 118-bus system for the reordering attack is a particular case and other cases may exist.

The measurement reordering attack as described above works when some portions of the power system have integrity protection mechanisms. This is not an unreasonable assumption because implementing timestamped measurements with authentication would be prohibitively expensive for current power grids. Indeed, as long as a power grid has unprotected legacy components, measurement reordering attacks will always pose a threat. However, in a decade or so, it should be possible to implement cryptographically timestamped authentication mechanisms for an entire grid, which would reduce, if not eliminate, the threat of reordering attacks.

9. Conclusions

This chapter has focused on reordering attacks on hierarchical state estimation as described in [9], where an adversary reorders measurement data without injecting or modifying data, resulting in incorrect estimates and potentially undesirable power system states. The attacks are feasible because it is not possible to implement authentication mechanisms throughout a large power grid. Therefore, this chapter has studied targeted reordering attacks on the most vulnerable region of a power system, which cause errors to propagate all over the system, and not just the attacked region. The results also demonstrate that an attacker can force incorrect estimates in a protected (i.e., authenticated) region of a power system by launching a clever attack on a less protected region.

Future research will attempt to develop protection and mitigation techniques for hierarchical or fully-distributed state estimation as employed in a smart grid. Research will also investigate the number of measurements and the specific measurements that would be swapped by an attacker to achieve maximal impact.

References

[1] A. Abur and A. Gomez-Exposito, *Power System State Estimation: Theory and Implementation*, CRC Press, Boca Raton, Florida, 2004.

[2] A. Baiocco, C. Foglietta and S. Wolthusen, Delay and jitter attacks on hierarchical state estimation, *Proceedings of the IEEE International Conference on Smart Grid Communications*, pp. 485–490, 2015.

[3] A. Baiocco and S. Wolthusen, Dynamic forced partitioning of robust hierarchical state estimators for power networks, *Proceedings of the Power and Energy Society Innovative Smart Grid Technologies Conference*, 2014.

[4] A. Baiocco, S. Wolthusen, C. Foglietta and S. Panzieri, A model for robust distributed hierarchical electric power grid state estimation, *Proceedings of the Power and Energy Society Innovative Smart Grid Technologies Conference*, 2014.

[5] S. Cui, Z. Han, S. Kar, T. Kim, H. Poor and A. Tajer, Coordinated data-injection attack and detection in the smart grid: A detailed look at enriching detection solutions, *IEEE Signal Processing*, vol. 29(5), pp. 106–115, 2012.

[6] T. Dierks and E. Rescorla, The Transport Layer Security (TLS) Protocol, Version 1.2, RFC 5246, 2008.

[7] Y. Feng, C. Foglietta, A. Baiocco, S. Panzieri and S. Wolthusen, Malicious false data injection in hierarchical electric power grid state estimation systems, *Proceedings of the Fourth International Conference on Future Energy Systems*, pp. 183–192, 2013.

[8] A. Gomez-Exposito, A. Abur, A. de la Villa Jaen and C. Gomez-Quiles, A multilevel state estimation paradigm for smart grids, *Proceedings of the IEEE*, vol. 99(6), pp. 952–976, 2011.

[9] A. Gul and S. Wolthusen, Measurement reordering attacks on power system state estimation, *Proceedings of the IEEE Power and Energy Society Innovative Smart Grid Technologies Conference Europe*, 2017.

[10] O. Kosut, L. Jia, R. Thomas and L. Tong, On malicious data attacks on power system state estimation, *Proceedings of the Forty-Fifth International Universities Power Engineering Conference*, 2010.

[11] S. Lakshminarasimhan and A. Girgis, Hierarchical state estimation applied to wide-area power systems, *Proceedings of the IEEE Power Engineering Society General Meeting*, 2007.

[12] Y. Liu, P. Ning and M. Reiter, False data injection attacks against state estimation in electric power grids, *Proceedings of the Sixteenth ACM Conference on Computer and Communications Security*, pp. 21–32, 2009.

[13] A. Monticelli, *State Estimation in Electric Power Systems: A Generalized Approach*, Springer, New York, 1999.

[14] F. Schweppe and D. Rom, Power system static-state estimation, Part II: Approximate model, *IEEE Transactions on Power Apparatus and Systems*, vol. PAS-89(1), pp. 125–130, 1970.

[15] F. Schweppe and J. Wildes, Power system static-state estimation, Part I: Exact model, *IEEE Transactions on Power Apparatus and Systems*, vol. PAS-89(1), pp. 120–125, 1970.

[16] F. Schweppe and J. Wildes, Power system static-state estimation, Part III: Implementation, *IEEE Transactions on Power Apparatus and Systems*, vol. PAS-89(1), pp. 130–135, 1970.

[17] D. Shepard, T. Humphreys and A. Fansler, Evaluation of the vulnerability of phasor measurement units to GPS spoofing attacks, *International Journal of Critical Infrastructure Protection*, vol. 5(3-4), pp. 146–153, 2012.

[18] T. van Cutsem and M. Ribbens-Pavella, Critical survey of hierarchical methods for state estimation of electric power systems, *IEEE Transactions on Power Apparatus and Systems*, vol. PAS-102(10), pp. 3415–3424, 1983.

[19] O. Vukovic and G. Dan, On the security of distributed power system state estimation under targeted attacks, *Proceedings of the Twenty-Eighth Annual ACM Symposium on Applied Computing*, pp. 666–672, 2013.

Chapter 5

SECURING DATA IN POWER-LIMITED SENSOR NETWORKS USING TWO-CHANNEL COMMUNICATIONS

Clark Wolfe, Scott Graham, Robert Mills, Scott Nykl and Paul Simon

Abstract Confidentiality and integrity of wireless data transmissions are vital for sensor networks used in critical infrastructure assets. While the challenges could be addressed using standard encryption techniques, the sensors are often power-limited, bandwidth-constrained or too rudimentary to accommodate the power and latency overhead of robust encryption and decryption implementations. To address this gap, this chapter proposes a novel methodology in which data is split between two distinct wireless channels to achieve acceptable levels of data confidentiality and/or integrity. Threat scenarios are discussed in which an attacker gains access to one of the two communications channels to either eavesdrop on or modify data in transit. Given these threats, five data splitting methods are presented that employ the two-channel communications concept to detect and adapt to the attacks, and provide varying levels of data security. Additionally, a simple proof-of-concept packet structure is introduced that facilitates data transmission over the two channels in accordance with the data-splitting methods.

Keywords: Wireless sensor networks, data security, two-channel communications

1. Introduction

Data security includes the challenge of protecting data in transit from eavesdropping and unauthorized tampering such as data modification. Normally, this challenge is met by applying encryption in the form of industry-standard symmetric-key algorithms such as the advanced encryption standard (AES). However, for small, low-powered devices, such as those used in remote sensor networks, the additional computational resources required for robust encryption may consume more power and time than are acceptable [4]. This chapter presents a proof-of-concept methodology that partially mitigates eavesdrop-

J. Staggs and S. Shenoi (Eds.): Critical Infrastructure Protection XII, IFIP AICT 542, pp. 81–90, 2018.
https://doi.org/10.1007/978-3-030-04537-1_5

ping and data modification threats using two-channel communications while reducing the encryption overhead.

2. Background

This section briefly discusses the threats to data in transit, the overhead imposed by encryption and the concept of two-channel communications.

2.1 Data Threats

A sensor network is an interconnected system of small sensors, each containing computing and communications elements. Sensor networks are used in numerous industries to monitor conditions or control equipment in remote locations. They often comprise large numbers of low-powered devices that are designed to conserve battery life while communicating critical information over wireless links [2].

Because of their wireless nature, sensor networks face a multitude of attacks. This work focuses on two types of man-in-the-middle (MiTM) attacks: (i) eavesdropping; and (ii) data modification. Eavesdropping is the unauthorized interception of confidential data. In the case of a wireless sensor network, eavesdropping could occur by placing an unauthorized receiver within signal range of the sensor network to collect transmitted data [6]. In a data modification attack, a network intruder modifies the data after it is sent, but before it reaches the intended recipient [5].

2.2 Encryption Overhead

Encryption provides confidentiality at the cost of computational resources such as memory, power and time. Wireless sensor networks typically have limited computational and power resources and, therefore, modern encryption standards such as 128-bit AES can impose significant burden on individual nodes. According to one study [7], using 128-bit AES to encrypt just one 128-bit block of data required 946 bytes of random access memory (RAM), 23.57 μJ and 1.1 ms on an IEEE/ZigBee 802.15.4 board commonly used in low-power wireless sensor networks. These resources add up quickly as increasing amounts of data are transmitted over the lifespan of the sensor. For example, according to the following equation:

$$1GB \times \frac{8\,\text{bits}}{1\,\text{byte}} \times \frac{23.57\,\mu\text{J}}{128\,\text{bits}} \times \frac{1\text{Wh}}{3600J} = 0.41\text{Wh} \tag{1}$$

a sensor encrypting one GB of data would expend 0.41 Wh of energy just to encrypt the transmitted data. Such power consumption would significantly affect battery life in a device that may have a few watt-hours of energy.

Lightweight encryption schemes, such as the SIMON and SPECK encryption ciphers, attempt to address this issue in low-powered devices by offering more efficient, but less robust encryption options [1]. However, no encryption

Figure 1. Policy development process.

scheme can eliminate the overhead completely. Fortunately, the two-channel communications concept presented in this chapter can achieve adequate levels of data confidentiality and integrity without introducing significant encryption overhead.

2.3 Two-Channel Communications

As its name suggests, the two-channel communications technique transmits data over two channels in order to increase the security profile of data in transit. A simple example is a wireless network operating over the 2.4 GHz and 5 GHz industrial, scientific and medical (ISM) radio bands. The proposed methodology for a wireless sensor network requires each sensor to be equipped with full-duplex communications over two data links with distinct frequencies. The two-channel data splitting occurs at the physical layer, which enables industry-standard data transmission protocols to ride on top of the two-channel implementation.

The methods utilized to split data between the two channels operate under the assumption that the attacker has gained access to only one of the two channels. This is because the situation where an attacker successfully targets both communications paths reduces to the single-channel man-in-the-middle attack scenario. For simplicity of analysis, it is assumed that the two channels have the same bandwidth. Finally, while the methodology could be applied to any number of channels, the focus is on two channels for reasons of simplicity.

3. Proposed Methodology

This section describes the proposed two-channel methodology in which data is split between two distinct wireless channels to achieve acceptable levels of data confidentiality and/or integrity.

3.1 Threat Scenario Development

The first step in developing the two-channel solution for combating eavesdropping and data modification attacks is to model the threat scenarios. Following this, the techniques for mitigating the attacks are developed. Finally, the mitigation techniques are specified in terms of two-channel policies that leverage both channels to reduce or eliminate the threats. Figure 1 summarizes the policy development process.

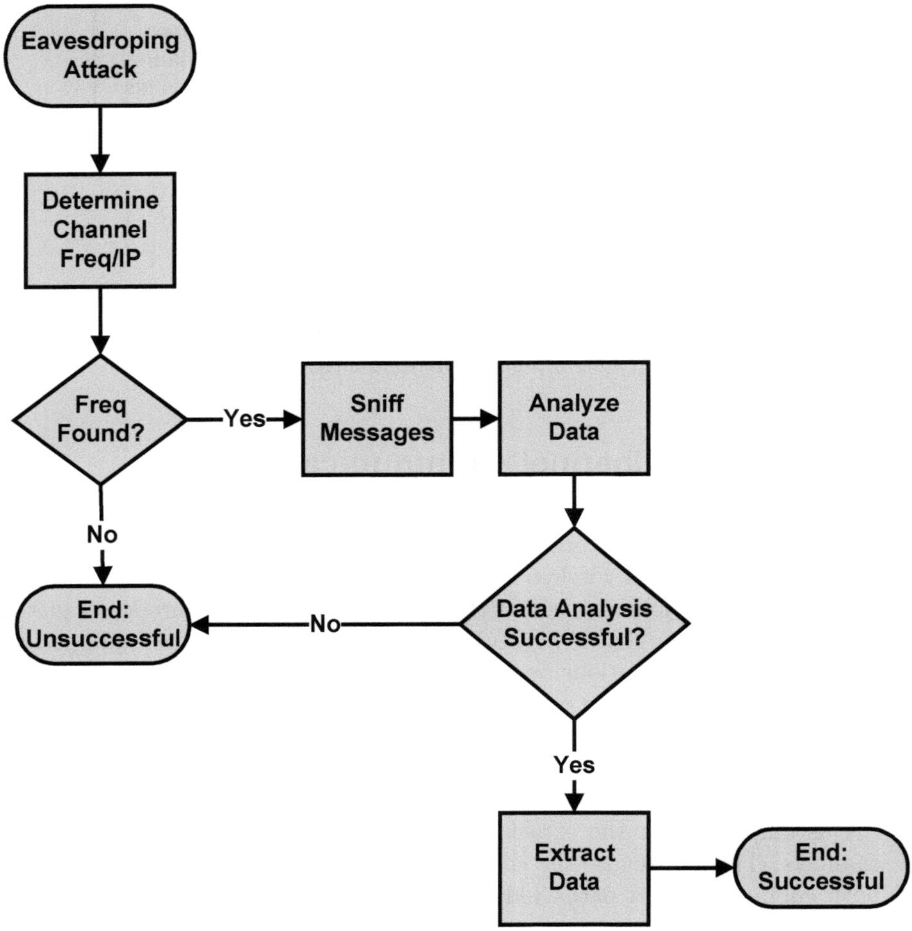

Figure 2. Eavesdropping attack.

3.2 Eavesdropping Scenario

The eavesdropping threat scenario involves an attacker compromising the confidentiality of data in transit. The threat model assumes that the attacker is able to gain access to one of the two channels used for communications.

Figure 2 shows a flowchart that models the attacker's possible courses of action. Note that, in order to be successful, the attacker must locate one of the two channels and properly analyze the data.

Table 1 presents three mitigation strategies based on the threat model along with their outcomes. The first mitigation strategy enables the attacker to obtain only the portion of the data that is sent over the compromised channel. Whether or not the data accessed is adequate to accomplish the attacker's

Table 1. Eavesdropping attack mitigation strategies.

Strategy 1	**Mitigation**	Split the data between the two channels.
	Outcome	Eavesdropper is limited to the collection of partial data (only the data sent over the compromised channel).
Strategy 2	**Mitigation**	Send no data over the compromised channel.
	Outcome	Eavesdropper has no data, but the eavesdropper may become suspicious and search for other frequencies because no data is being sent. The data transfer rate is cut in half.
Strategy 3	**Mitigation**	Send the data over the uncompromised channel. Send faux data over the compromised channel.
	Outcome	Eavesdropper only has worthless or misleading data. The data transfer rate is cut in half.

goal depends on the type of data being sent and the percentage of data that traverses the compromised path. This mitigation strategy enables the sender and receiver to tailor the volume of revealed data to meet their security posture. For example, if confidentiality is not a priority, the communicating entities may choose to send half the data over the compromised channel in order to obtain the best data transfer rate.

The second mitigation strategy sends no data over the compromised channel. It is appropriate when confidentiality is of upmost importance. The strategy defeats the attacker by sending no data via the compromised channel, but the absence of data flow in the compromised channel could alert the attacker to the mitigation strategy. Additionally, the data transfer rate is cut in half.

The third strategy sends faux data over the compromised channel. The attacker does not know about the mitigation and is misled; however, the data transfer rate is cut in half.

The three eavesdropping mitigation strategies are formalized as the two-channel data transmission policies shown in Table 2. In the example, Channel A is assumed to be secure whereas Channel B is assumed to be compromised.

3.3 Data Modification Scenario

The second threat scenario involves data modification, where the attacker changes a portion of the data in transit. This attack compromises data integrity.

Figure 3 shows a flowchart that models the attacker's possible courses of action. The attacker has to modify the data successfully and ensure that the recipient does not discover that the data has been modified. If the recipient notices that the data has been changed, the sender could be requested to retransmit the data over the known secure channel.

Leveraging this fact, a mitigation strategy is formulated that enables the receiver to detect data modification. This is accomplished by computing a

Table 2. Eavesdropping attack policies (Channel B is compromised).

Policy	Channel	Data Sent
1	A	50% of data per packet
	B	50% of data per packet
2	A	100%
	B	0%
3	A	100% of actual data
	B	Random data (0% real data)

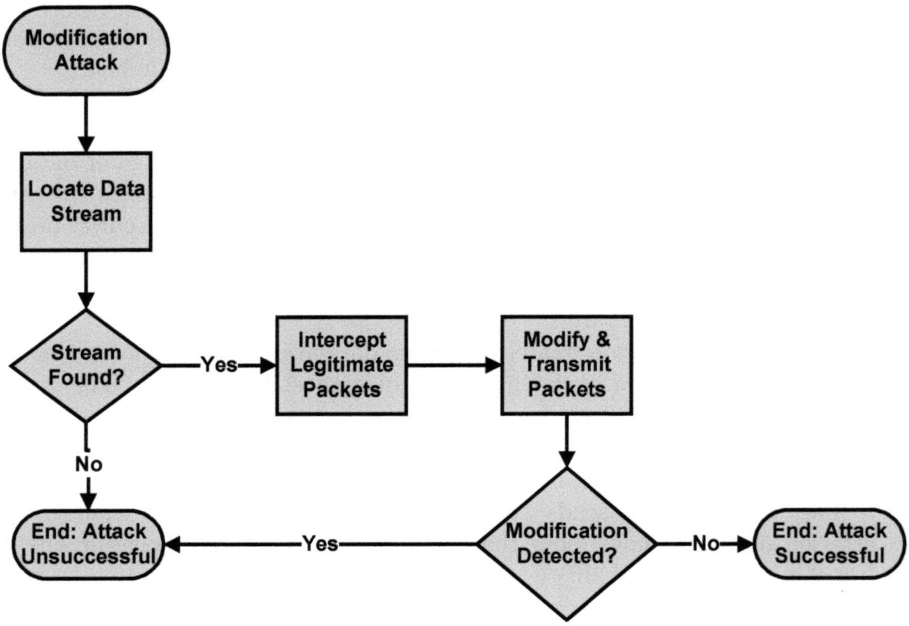

Figure 3. Data modification attack.

cyclic redundancy code (CRC) at the sender and verifying the code at the receiver to ensure that the data has not been modified. A sufficiently strong CRC could enable any amount of data modification to be detected. While this may not be the most efficient method, it is suitable to demonstrate the concept [3].

Consider the case where 50% of the data is sent over each channel and a CRC is computed for each packet of data before it is split between the two channels. The CRC itself is split into two parts with each part sent over a different channel. Note that an additional CRC would be computed on the data sent

Table 3. Data modification attack mitigation strategies.

Strategy 1	**Mitigation**	Compute a CRC for the data on one channel to detect if it has been modified. If the attack is detected, cease data transfer over the compromised channel. Transfer all the data over the secure channel.
	Outcome	Attacker is unable to modify the data without being detected. However, the attacker may become suspicious if data transfer is ceased. The data transfer rate is reduced because only one channel is used.
Strategy 2	**Mitigation**	Compute a CRC for the data on one channel to detect if it has been modified. If the attack is detected, transfer only faux data over the compromised channel. Transfer all the data over the secure channel.
	Outcome	Attacker is unable to modify the data without being detected and is unaware that the attack has been detected if data continues to be sent over the compromised channel. The data transfer rate is reduced because only one channel is used.

over each channel to protect the data being transmitted. Also, the attacker who has access to only one channel cannot generate the correct CRC for the modified data sent over the compromised channel. This is because the other half of the data is unknown to the attacker. As a result, any data modification would be detected when the receiver combines the data and CRC halves and checks the combined CRC. Table 3 presents the two mitigation strategies along with their outcomes.

The two mitigation strategies for data modification are formalized as the two-channel data transmission policies shown in Table 4.

3.4 Packet Structure Development

In order to implement the five two-channel policies introduced above, it is necessary to design a packet structure that incorporates the data splitting and CRC schemes. Table 5 shows a proof-of-concept two-channel packet structure.

The 26-bit packet structure incorporates the following five fields:

- **Policy:** This three-bit field specifies the policy used to send the packet. The policy numbers (1 through 5) correspond to the five policies presented in Sections 3.2 and 3.3. For example, a `101` in the field denotes Policy 5. The receiver uses this field to ensure that the packets sent over the two channels have matching policy numbers before processing them.

Table 4. Data modification attack policies (Channel B is compromised).

	Channel	Data Sent
Policy 4	A	50% of the data per packet + CRC → Switch to 100% of the data after unauthorized data modification is detected.
	B	50% of the data per packet + CRC → Switch to 0% of the data after unauthorized data modification is detected.
Policy 5	A	50% of the data per packet + CRC → Switch to 100% of the data after unauthorized data modification is detected.
	B	50% of the data per packet + CRC → Switch to 100% faux data after unauthorized data modification is detected.

Table 5. Proof-of-concept two-channel packet structure.

	Bits 0-2	Bits 3-10	Bits 11-14	Bits 15-17	Bits 18-25
Channel A	Policy	MessageA	CRC-DataA	Packet#	CRC-Msg1
Channel B	Policy	MessageB	CRC-DataB	Packet#	CRC-Msg2

- **MessageX:** This eight-bit field contains the data bits. The first eight bits are loaded into the MessageA field while the second eight bits are loaded into the MessageB field.

- **CRC-DataX:** This four-bit field contains the data CRC required by Policy 4 and Policy 5 in order to detect data modification attacks. It is formed by generating an eight-bit code from the sixteen bits of data (MessageA + MessageB). Then, the first four-bits of the eight-bit data CRC are loaded into the CRC-DataA field and the second four-bits are loaded into the CRC-DataB field.

- **Packet#:** This three-bit field records the packet number. Packets sent over one channel have a matching packet with the identical packet number sent over the other channel. The packet numbers help ensure that the correct packets are processed together by the receiver.

- **CRC-Msg#:** This eight-bit field is used for error detection during message transmission. The eight-bit CRC for a message is generated using the entire eighteen bits of the message, which is verified by the receiver.

4. Conclusions

Maintaining confidentiality and trust for wireless data transmissions are vital to sensor networks used in critical infrastructure assets. However, remote sensors are often power-limited, bandwidth-constrained or too rudimentary to accommodate the power and latency overhead of robust encryption and decryption operations. The two-channel communications methodology presented in this chapter splits the transmitted data over two wireless channels to provide acceptable levels of data confidentiality and/or integrity for non-encrypted remote sensor networks.

The threat scenarios considered involve an attacker gaining man-in-the-middle access to one of the two communications channels to eavesdrop on or modify data in transit. To combat these threats, five data splitting policies are presented that detect and adapt to the attacks while providing varying levels of data security.

Future research will attempt to create additional two-channel policies that can combat other threat scenarios such as denial-of-service attacks and spoofing attacks [2]. These policies will be simulated in software or implemented in hardware to evaluate their effectiveness in real-time applications. Additionally, a measurement and comparison framework will be constructed to gauge the effectiveness of the policies and corresponding packet structures in combating data threats.

Note that the views expressed in this chapter are those of the authors and do not reflect the official policy or position of the U.S. Air Force, U.S. Department of Defense or U.S. Government.

References

[1] R. Beaulieu, S. Treatman-Clark, D. Shors, B. Weeks, J. Smith and L. Wingers, The SIMON and SPECK lightweight block ciphers, *Proceedings of the Fifty-Second ACM/EDAC/IEEE Design Automation Conference*, 2015.

[2] H. Kalita and A. Kar, Wireless sensor network security analysis, *International Journal of Next-Generation Networks*, vol. 1(1), pp. 1–10, 2009.

[3] P. Koopman and T. Chakravarty, Cyclic redundancy code (CRC) polynomial selection for embedded networks, *Proceedings of the International Conference on Dependable Systems and Networks*, pp. 145–154, 2004.

[4] T. Nie, L. Zhou and Z. Lu, Power evaluation methods for data encryption algorithms, *IET Software*, vol. 8(1), pp. 12–18, 2014.

[5] G. Padmavathi and D. Shanmugapriya, A survey of attacks, security mechanisms and challenges in wireless sensor networks, *International Journal of Computer Science and Information Security*, vol. 4(1-2), paper no. 20070913, 2009.

[6] Y. Shiu, S. Chang, H. Wu, S. Huang and H. Chen, Physical layer security in wireless networks: A tutorial, *IEEE Wireless Communications*, vol. 18(2), pp. 66–74, 2011.

[7] F. Zhang, R. Dojen and T. Coffey, Comparative performance and energy consumption analysis of different AES implementations on a wireless sensor network node, *International Journal of Sensor Networks*, vol. 10(4), pp. 192–201, 2011.

Chapter 6

REVERSING A LATTICE ECP3 FPGA FOR BITSTREAM PROTECTION

Daniel Celebucki, Scott Graham and Sanjeev Gunawardena

Abstract Field programmable gate arrays are used in nearly every industry, including consumer electronics, automotive, military and aerospace, and the critical infrastructure. The reprogrammability of field programmable gate arrays, their computational power and relatively low price make them a good fit for low-volume applications that cannot justify the non-recurring engineering costs of application-specific integrated circuits. However, field programmable gate arrays have security issues that stem from the fact that their configuration files are not protected in a satisfactory manner. Although major vendors offer some sort of encryption, researchers have demonstrated that the encryption can be overcome. The security problems are a concern because field programmable gate arrays are widely used in industrial control systems across the critical infrastructure. This chapter explores the reverse engineering process of a Lattice Semiconductor ECP3 field programmable gate array configuration file in order to assist infrastructure owners and operators in recognizing and mitigating potential threats.

Keywords: Field programmable gate arrays, threats, reverse engineering

1. Introduction

As field programmable gate arrays (FPGAs) become more powerful and less expensive, they are increasingly being adopted in industry. Key applications areas of FPGAs are industrial control systems used for managing critical infrastructure assets and hardware-in-the-loop simulations used for industrial process system design and training [15]. Low latency, high computational power and an abundance of embedded resources enable FPGAs to implement complex control algorithms with an excellent performance-to-cost ratio. However, FPGAs have security issues that stem from the fact that their configuration files are not protected in a satisfactory manner. A number of attacks targeting FPGAs and FPGA-based systems have been devised. These include hardware Trojans,

J. Staggs and S. Shenoi (Eds.): Critical Infrastructure Protection XII, IFIP AICT 542, pp. 91–111, 2018.
https://doi.org/10.1007/978-3-030-04537-1_6

crippling attacks and fault injection attacks, as well as attacks that reveal sensitive information for subsequent exploitation, such as side-channels, reverse engineering, readback and counterfeiting [2, 4, 9, 16].

This chapter explores the reverse engineering process of a Lattice Semiconductor ECP3 FPGA. The focus is on two key FPGA building blocks – the input/output block and look-up tables. The reverse engineering efforts have resulted in a proof-of-concept parser that analyzes FPGA bitstreams (circuit configuration files) for errors and malicious modifications without revealing any sensitive intellectual property.

2. Background

This section discusses FPGAs, bitstream synthesis, the applications of FPGAs in the critical infrastructure and FPGA threats.

2.1 Field Programmable Gate Arrays

FPGAs were first introduced in 1984 by Xilinx and have since increased in capacity and speed by factors of 10,000 and 100, respectively [17]. Unlike traditional application-specific integrated circuits (ASICs) that are customized for a particular use, FPGAs are reprogrammable. This is accomplished using a combination of configurable logic blocks (CLBs), an input/output block and a series of configurable interconnects. The interconnects are sometimes referred to as the switching matrix.

Figure 1 shows an example FPGA architecture with configurable logic blocks, an input/output block and interconnects. Configurable logic blocks, which comprise digital circuits such as look-up tables, multiplexers and flip-flops, can be configured to perform various combinational functions. These functions can also be registered within a configurable logic block to implement synchronous logic. An input/output block provides connections to external stimuli. The interconnects link the configurable logic blocks and input/output block to complete the desired circuit.

The penalties incurred for FPGA reconfigurability include larger chip area, slower speed and higher power consumption compared with an ASIC that implements the same circuit [6]. This is primarily due to the additional area and propagation delays introduced by the programming circuitry in an FPGA.

The initial steps in designing a digital system are largely identical for FPGAs and ASICs; they involve design capture and simulation using a hardware description language (HDL). After the correct functionality is verified via simulation, the hardware-description-language-based design is synthesized into a form that represents logic elements and registers, which is referred to as the register-transfer level. At this point, the logic elements and registers are mapped to implementable components contained in a target technology library. In the case of ASICs, this is usually a standard cell library. However, for FPGAs, the design is mapped to functional primitives comprising look-up tables and registers. Following the placement and routing, the final design is converted

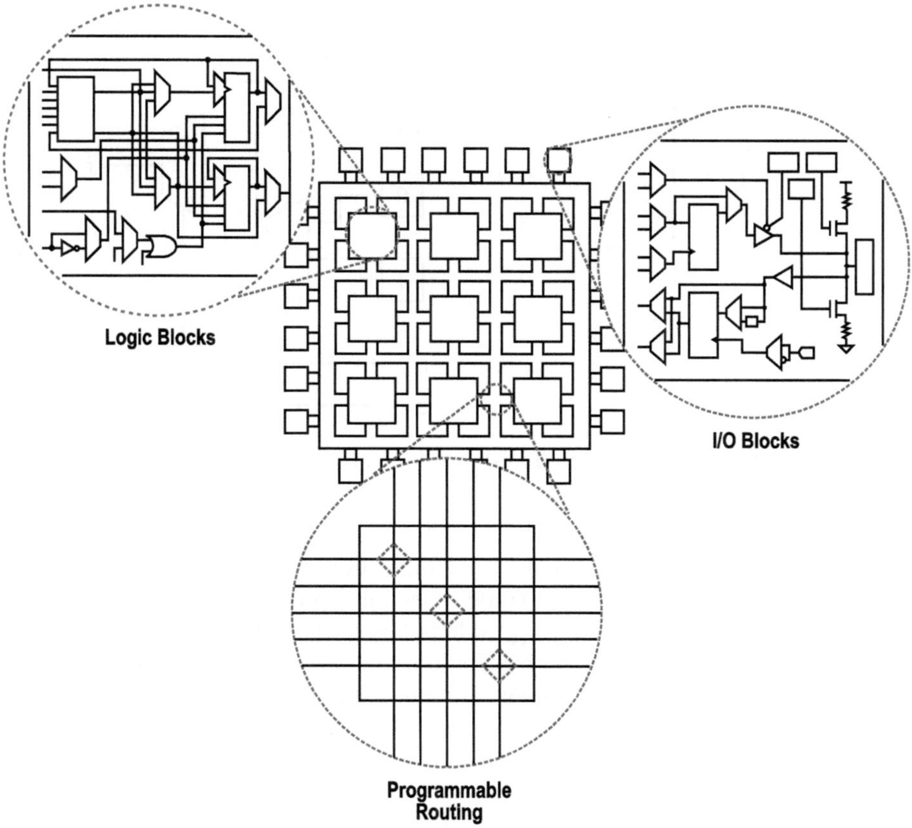

Figure 1. FPGA architecture [12]

to a "bitstream," a series of zeros and ones that specifies the configuration options of the configurable logic blocks, input/output block and interconnects in order to implement a given circuit. FPGA vendors have their own proprietary bitstream formats whose details are rarely released to the public.

Configurable Logic Blocks. Configurable logic blocks enable an FPGA to implement logic. Although there are differences in vendor implementations of configurable logic blocks, they commonly include look-up tables, multiplexers and flip-flops. Unlike traditional ASICs that use hardware logic gates to implement digital logic for the desired circuits, FPGAs employ look-up tables. Figure 2 shows a two-input look-up table that uses multiplexers to implement digital logic. Inputs `a` and `b` are selectors for the multiplexers and the inputs to the multiplexers are the output values of the desired truth table. The look-up table implements a circuit that is logically equivalent to an AND gate by setting the inputs to the multiplexers as `0001`. Different input values are provided to

a	b	f
0	0	0
0	1	0
1	0	0
1	1	1

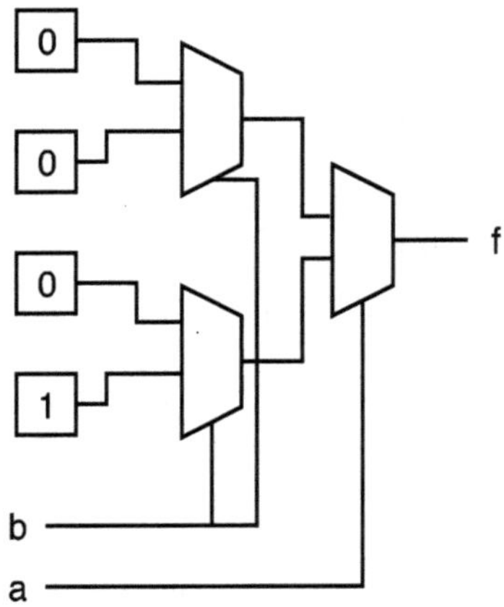

Figure 2. Two-input look-up table example.

the multiplexers to implement a new digital circuit without having to change the hardware. Look-up tables trade space for reprogrammability; the hardware needed to implement a look-up table is larger than that needed to implement the digital circuit replicated by the look-up table. However, a look-up table can be reprogrammed to implement any logic function that can be modeled by a truth table. Designs that require more inputs are implemented by daisy chaining look-up tables.

Input/Output Block. An input/output block connects the internal logic of an FPGA to external components. Since an input/output block usually allows inputs and outputs on the same physical pad, the choice of whether a certain pin is an input or output is decided at configuration time. Other configuration options determine the physical characteristics of the signal at a pin such as pullmode, slew rate and drive level; these may vary from FPGA to FPGA. Figure 3 shows an example input/output block that is configured as an output.

Switching Matrix. A switching matrix connects the configurable logic blocks and the input/output block to produce the desired digital logic circuit [5]. The large number of routes that have to be accommodated make the switching matrix the largest portion of an FPGA in terms of silicon area.

2.2 Bitstream Synthesis

Before a circuit design can be implemented on an FPGA it must be transformed into a configuration file – called a bitstream – that can be loaded on the FPGA. Figure 4 shows how a design proceeds from a hardware description language file to the final bitstream for a specific FPGA. Hardware description language code is first synthesized into a netlist that contains the list of com-

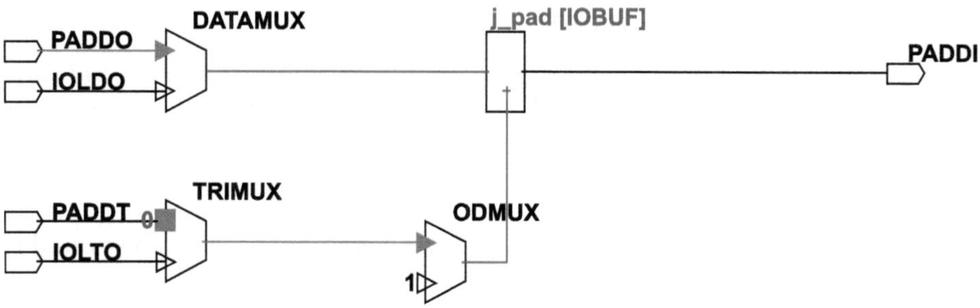

Figure 3. Example input-output block.

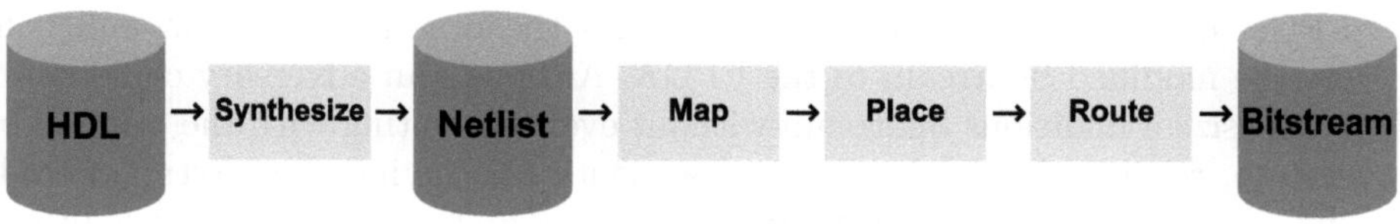

Figure 4. Process for generating a bitstream from HDL [8]

ponents in the circuit and the nodes to which they are connected. The map function maps the components in the netlist to the components on the FPGA. The place function then selects the locations of the components on the FPGA. Since the FPGA typically has numerous instances of the same components, the place function determines which components will actually be part of the circuit. The route function then makes the connections between all the placed components on the board. After the circuit has been placed and routed, it is converted to a bitstream file that configures the correct components and connections on the board to create the circuit.

2.3 Critical Infrastructure Applications

Industrial control systems rely heavily on FPGAs. These systems must be powerful and have low latency to ensure high performance, flexibility and reliability [10]. FPGAs are well suited for this purpose. They provide a higher performance-to-cost ratio than ASICs due to continual advances in FPGA computing power and high per-unit costs for low-volume ASIC designs [14]. The reconfigurability of FPGAs supports rapid prototyping as well as control algorithm upgrades throughout the lifespan of an industrial control system using the same deployed hardware. Additionally, system-on-a-chip (SoC) platforms can implement advanced control techniques [15]. Finally, the availability of third-party intellectual property cores that can be licensed or purchased enables infrastructure owners to implement portions of, or complete, FPGA systems by outsourcing the work.

2.4 FPGA Threats

FPGA complexity increases the potential cyber attack surface and, hence, the risk of cyber attacks [1]. These threats can be particularly dangerous to FPGAs used in the critical infrastructure due to the potential impacts on industry, the economy and society.

Bitstream Modification. Chakraborty et al. [2] have demonstrated that a bitstream can be modified to introduce hardware Trojans without knowing the hardware description language (source) code used to create the bitstream. In one instance, they inserted ring oscillators to elevate the temperature of an FPGA, which increased the probability of failure. This attack could be implemented by an insider who is responsible for loading the bitstream on the FPGA or by an adversary who intercepts the original bitstream and delivers the modified bitstream to the FPGA. Although an adversary could simply synthesize a malicious bitstream without ever interacting with the original bitstream, reverse engineering and subsequently modifying the bitstream enable the adversary to implement a hard-to-detect attack that maintains the original functionality of the FPGA design.

Covert Channels. Covert channels allow for the transmission of information between components that are not supposed to be communicating. In an industrial control system setting, this could involve the exfiltration of the control algorithm or sensitive data while the industrial control system is performing its intended functions. The exfiltration of a proprietary control algorithm and sensitive data could give competitors an advantage.

Intellectual Property Theft. An industrial control system vendor that develops its own FPGAs should be wary of adversaries potentially reverse engineering its designs. This is important because bitstreams are not inherently protected and, given enough time, an adversary could reverse engineer them and obtain valuable intellectual property [13]. Malicious entities also would be interested in reverse engineering bitstreams and creating exploits that could be used in future attacks on critical infrastructure assets.

In theory, encryption can be used to protect a bitstream, but this feature is usually offered by expensive FPGA models. In any case, encryption has been shown to be breakable through side-channel analysis [11]. Additionally, encryption requires an energy source, usually in the form of a battery, to keep the key from being cleared if the board loses power. Also, in some cases, an FPGA cannot be accessed after it is deployed or its battery cannot be replaced without significant effort [3].

3. Reverse Engineering Methodology

This section discusses the reverse engineering methodology for a Lattice Semiconductor ECP3 FPGA configuration file.

3.1 Target System

Numerous articles have been published about reverse engineering efforts directed at Xilinx and Altera (now part of Intel) FPGAs. However, Lattice FPGAs have received much less attention apart from the very small iCE40 FPGAs [18]. The target system chosen for this research was the Lattice ECP3 LFE3-35EA-8FN484C FPGA with the Lattice ECP3 Versa Development Kit. Comparisons are made between the reverse engineering process for Lattice FPGAs and the reverse engineering processes for Xilinx and Altera FPGAs.

All the bitstreams were designed using Lattice Diamond software version 3.9.1 and the Lattice Synthesis Engine. Additionally, Tool Command Language scripts were used to generate design variations to explore the effects on the bitstreams.

3.2 Input/Output Block Reversal

This section describes the process of mapping the relationship between a bitstream file and the configuration of the input/output block; this was easily set using the spreadsheet view provided by the Lattice Diamond software. Because changes made to the configuration options in the spreadsheet view were present in the Lattice preference file (LPF) when the changes were saved, modifying the Lattice preference file directly enabled the automated generation of a bitstream.

The reverse engineering of the input/output block has three goals: (i) map a number of the configuration options for each pin to their respective indices and values in the bitstream file; (ii) determine whether a pin is an input or output based on the bitstream file; and (iii) determine whether a pin is connected to logic blocks in the design based on the bitstream file.

Pullmode. Although input/output blocks have a variety of configuration options, the reverse engineering process of the different configuration options is very similar. Therefore, only the process for reverse engineering the pullmode attribute is described here.

The pullmode is responsible for describing how a signal is interpreted at a pin. The pullmode can be set to the following four modes:

- **Up:** The input is attached to a pull-up resistor, i.e., the pin is tied to a logical 1.

- **Down:** The input is attached to a pull-down resistor, i.e., the pin is tied to ground or a logical 0.

- **Keeper:** This mode is neither pull-up nor pull-down. It drives a weak 0 or 1 level to match the level of the last logic state present on the pad to prevent the pad from floating.

- **None:** The input is not set to any of the above three modes.

Algorithm 1 : Pullmode bitstream generation.

```
for Pin p in all I/O Pins do
    for val in UP, DOWN, KEEPER, NONE do
        Replace IOBUF line in LPF with "IOBUF PORT "a" PULLMODE=val"
    end for
    Replace Location line in LPF with "LOCATE COMP "<input/output pin
  name>" SITE p"
end for
```

In this chapter, an index refers to the byte address where the contents of a bitstream have been changed due to a design modification. A change to the pullmode of a pin resulted in three to six change indices in the bitstream. This was much more manageable compared with the 70 to 100 indices when the location of a configurable logic block was moved slightly. The reason for the varying change indices is because the bitstream did not abide by the byte boundaries; this is discussed later in this chapter.

After the configuration option was sufficiently isolated, a Tool Command Language script synthesized bitstreams for every pullmode option for every pin; every pin set was first used as an input and subsequently every pin set was used as an output. Algorithm 1 shows the pseudocode of the script. The objectives were to determine which indices were responsible for the pullmode option for each pin and whether a common pattern could be used for every pin to identify the pullmode. The hypothesis was that each pin would have a different location in the bitstream where its configuration options were stored, and the values at each location would follow the same pattern in terms of representing the pullmode in the bitstream.

The 2,296 bitstreams generated by the script were compared to find the indices responsible for the pullmode configuration option for each pin. Table 1 shows the bitstream indices responsible for the pullmode configuration option for six pins. The indices for each pin were generated by comparing the four bitstreams (pull-up, pull-down, bus keeper and none) for the pin and listing all the indices in the bitstreams that were different from any of the other bitstreams. For each pin, the first column with hex values (i.e., second column overall) refers to the values in the bitstream associated with pullmode pull-up, the second refers to pull-down, the third refers to bus keeper and the fourth column refers to none. The numbers in the leftmost (i.e., first) column are the bitstream indices where the changes occurred. For example, when comparing the four bitstreams generated for pin A2 and set as an input, the only differences between the four bitstreams were at bytes 429, 476 and 477. In fact, Table 1 reveals that relatively few indices were changed.

Note that indices 476 and 477 appear for almost every pin. However, for the generated bitstreams, it was impossible to know whether all the indices listed for each pin were necessary to configure the various pullmode options or if only a subset of the indices for each pin was necessary.

Table 1. Indices for the pullmode configuration option for various pins.

Pin A2					**Pin A3**				
429	00	06	02	04	416	00	06	02	04
476	f3	9e	57	3a	476	20	5e	0a	74
477	dc	b4	07	6f					
Pin A4					**Pin A6**				
412	00	18	08	10	395	01	61	21	41
476	82	e2	a2	c2	476	d4	cd	5c	45
477	ba	ea	8a	da	477	bf	30	39	b6
Pin A7					**Pin A8**				
326	01	61	21	41	308	00	01	00	01
476	c5	63	27	81	309	04	84	84	04
477	02	ab	66	cf	476	47	e3	a4	00
					477	7c	44	97	af

To solve this problem the bitstream generation script was executed again with different logic designs mapped to different portions of the FPGA. The reasoning was to isolate the indices responsible for the pullmode configuration option. If the same generation script was executed with different logic designs and the logic mapped to different portions of the FPGA, then the indices responsible for the pullmode configuration would have the same values across all the runs while the indices affected by the switching matrix or configurable logic blocks would change. The following gates and placements were employed:

- One-input NOT gate at R2C73D.
- One-input NOT gate at R23C53A.
- Two-input AND gate at R3C70B.
- 1553 encoder placed by the compiler.

When exploring the designs and placements required to isolate the indices, it became clear that the configurable logic blocks had to be varied in diverse ways. This was achieved using a NOT gate, an AND gate and the intellectual property core of a MIL-STD-1553 encoder. The simple gates represented small designs whereas the encoder represented a large design. Each gate was then placed in a different slice within the configurable logic blocks on different corners of the FPGA and the 1553 encoder was placed by the tool. This variation proved to be enough initially. If none of the indices expressed different values, then additional variation could be introduced before considering that all the indices listed were necessary to represent the pullmode configuration.

The assumption that all the indices listed were responsible for the pullmode was excluded for a few reasons. First, the number of indices that changed for each pin when comparing bitstreams was not constant. Some pins only had three indices change whereas other pins had six indices change. It is unlikely that a designer would use extra indices to represent the same change in different pins. Additionally, each pin had indices that were different, but some pins also had indices that were the same. A designer would likely not have the same information located in two locations, especially since a larger file would increase the FPGA configuration time and complexity of the process used to parse the bitstream. In fact, it is more likely that some indices were being changed due to some other variation that was occurring as a result of changing the pullmode. The difference in the numbers of changed indices and shared indices was later attributed to the bitstream not abiding by the byte boundaries and some indices acting as internal checksums.

After analyzing which indices were changing for each pin, the pins were organized into six groups based on the indices responsible for their changes. Pins that shared the same indices were grouped together as well as pins that shared a pattern in the offset of their indices. The binary values located at all the indices in each of the six groups were then printed for each pullmode for each pin in a group, facilitating the visual inspection of all the indices simultaneously in order to discern changes. If the hypothesis was correct, there would be a regular pattern of 1s and 0s cascading down the created file. To facilitate visual inspection, 1s were replaced with black spaces and 0s with white spaces.

Figure 5 shows a selection of the pins in the first group after the 1s were replaced with black spaces and 0s with white spaces. A single bitstream runs horizontally from left to right and each group of four bitstreams relates to the same pin. For example, the first four bitstreams in Figure 5 are the bitstreams related to the pullmode configuration of D19. The first is pull-up, the second pull-down, the third bus-keeper and the fourth none. The next four bitstreams follow the same pullmode pattern, but for pin D18, the next four for pin B20, and so on.

Additionally, the first bitstream for each pin is always pullmode up, followed by down, keeper and none. The figure reveals significant information about how the bitstream is organized with respect to the pins. First, the bitstreams do not adhere to strict byte boundaries. The columns of black squares running vertically through the picture represent 1s that are the boundaries for where information about a certain pin appears. The 1s between the columns correspond to different configuration options. The pullmode configuration option can be observed at each of the pins as the only change in a pin's space in the boxes. Pullmode up is represented as `00`, down as `11`, keeper as `01` and none as `10`. This pattern was observed in all six pin groups investigated in the research.

3.3 Configurable Logic Block Reversal

Configurable logic blocks were more difficult to reverse engineer than the input/output block because there are many more configurable logic blocks, and the Lattice preference file modification cannot be used to change the configuration options in a straightforward manner. This is because the look-up tables in the configurable logic blocks are configured based on the hardware description

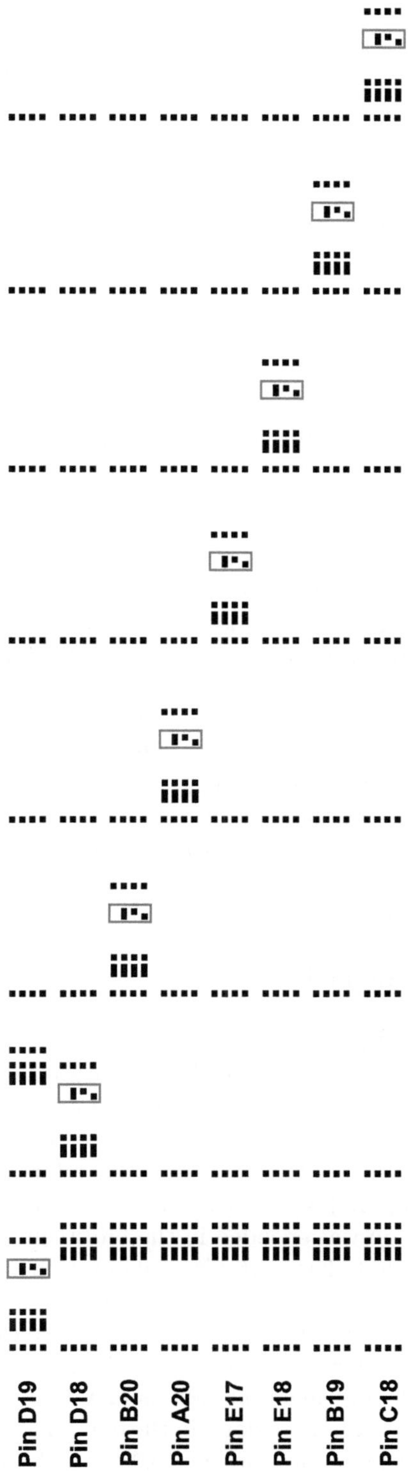

Figure 5. Selection of bitstreams.

Table 2. Derived truth table.

A	0	0	0	0	0	0	0	0	1	1	1	1	1	1	1	1
B	0	0	0	0	1	1	1	1	0	0	0	0	1	1	1	1
C	0	0	1	1	0	0	1	1	0	0	1	1	0	0	1	1
D	0	1	0	1	0	1	0	1	0	1	0	1	0	1	0	1
F	0	0	1	0	0	0	1	0	0	0	1	0	1	1	1	1

language when they were synthesized. The Lattice preference file is not used until the map, place and route steps in the bitstream generation process; therefore, it was not even considered until after the look-up tables were configured.

In order to overcome this issue, Lattice primitives were used along with hardware description language attributes. Each Lattice FPGA has a library of primitives supported by the device. In the case of the ECP3, the LUT4 primitive was used so that the look-up table could be directly initialized to the desired configuration value. The hardware description language attributes were used to set other attributes such as location instead of modifying the Lattice preference file simply to avoid having to change two different files. The look-up tables were initialized based on the outputs of their desired truth tables shown in Table 2. An initialization value of `0xF444` yielded a look-up table that produced the outputs in the truth table. When the synthesis process translates hardware description language code to the bitstream, it replaces the logic in the design with the look-up tables that are initialized to produce the same outputs. Based on this information, it was hypothesized that the initialization information appears somewhere in the bitstream. Therefore, in order to understand the digital logic implemented in the configurable logic block, it is only necessary to determine how the look-up tables were initialized and then recreate the truth table.

Single Look-Up Table Reversal. The process for reverse engineering the configuration of a look-up table involved the creation of a set of bitstreams using Tool Command Language scripts that had a variety of configuration values for the same look-up table. The bitstreams were compared to identify the indices that were responsible for the configuration information. The bitstreams were then visually compared with each other at the indices to reveal how the configuration information was encoded in the bitstream.

Since each look-up table has a 16-bit configuration value, the 16-bit value was assumed to be stored somewhere in the bitstream. Therefore, at least sixteen bitstreams had to be generated for each look-up table in order to locate the indices. However, if other indices were also changed as a result of modifying the configuration value, additional bitstreams may be necessary to identify the correct configuration indices. Therefore, in the experiments, 61 bitstreams were initially generated for the look-up table – `0x0` through `0xF` for each symbol in

```
0000  0000111111110000  0010100111100110  0000111111110000  0010110110111101
0001  0000111111110000  0010100111100110  0000111111100000  1111011111110011    1's place
0002  0000111111110000  0010100111100110  0000111111010000  0001100100100100    2's place
0003  0000111111110000  0010100111100110  0000111111000000  1100001101101010
0004  0000111111110000  0010100111100110  0000111011110000  0000100101110000    4's place
0005  0000111111110000  0010100111100110  0000111011100000  1101001100111110
0006  0000111111110000  0010100111100110  0000111011010000  0011110111101001
0007  0000111111110000  0010100111100110  0000111011000000  1110011110100111
0008  0000111111110000  0010100111100110  0000110111110000  0110010000100111    8's place
0009  0000111111110000  0010100111100110  0000110111100000  1011111001101001
000A  0000111111110000  0010100111100110  0000110111010000  0101000010111110
000B  0000111111110000  0010100111100110  0000110111000000  1000101011110000
000C  0000111111110000  0010100111100110  0000110011110000  0100000011101010
000D  0000111111110000  0010100111100110  0000110011100000  1001101010100100
000E  0000111111110000  0010100111100110  0000110011010000  0111010001110011
000F  0000111111110000  0010100111100110  0000110011000000  1010111000111101
```

Figure 6. Bitstreams with different configuration values.

the four-digit hex number. This provided adequate information to determine how the configuration was stored in the bitstream.

When comparing a set of bitstreams with different configuration options for the same look-up table, between six to eight bytes were observed to change in the bitstream. The variation in the number of changed bytes has to do with the bitstream not abiding by the byte boundaries and the presence of some checksum-like bits that also change. Sixteen indices correspond to the 16-bit initialization value used for look-up table configuration and 32 bits serve as a checksum. However, the configuration information is encoded. If the initialization value of the look-up table is considered to a 16-bit binary number, then the indices are negated in that 1s are replaced with 0s, and vice versa. Additionally, the 16-bits are not placed next to each other, but are spread across two to four bytes that can be hundreds of indices apart in the bitstream. The bits responsible for encoding the configuration information were discerned by analyzing the differences between the individual bitstreams.

Figure 6 illustrates this process. Each line corresponds to one of the first sixteen bitstreams from look-up table 1 in R2C40D, each with a different configuration value. The four values on the left show the hex representation of the 16-bit value used to initialize the look-up table and the indices enclosed by boxes correspond to the 1-, 2-, 4- and 8-place locations of the corresponding hex value. This is observed by comparing which locations change from line to line. For example, the 1-place for the first hex value was confirmed by comparing the `0x0002` and `0x0003` initialization value lines. All the indices in the last sixteen indices were not changed in a regular manner, so they can be ignored. However, the change from `0x0002` to `0x0003` has a 0 in in the same location where the `0x0001` line has a 0. This was also confirmed by comparing the `0x0004` line and the `0x0005` line or any other line where the binary representation was changed in the 1-place. This process was repeated for the remaining configuration values.

After the process was completed, many of the remaining bitstreams were removed to reveal the mask shown in Figure 7. The mask is the set of locations

```
0001 00001111 11110000 00101001 11100110 00001111 11100000 11110111 11110011  1 place
0002 00001111 11110000 00101001 11100110 00001111 11010000 00011001 00100100  2 place
0004 00001111 11110000 00101001 11100110 00001110 11110000 00001001 01110000  4 place
0008 00001111 11110000 00101001 11100110 00001101 11110000 01100100 00100111  8 place
0010 00001111 11110000 00101001 11100110 00001111 10110000 01000100 10001111  16 place
0020 00001111 11110000 00101001 11100110 00001111 01110000 11111111 11011001  32 place
0040 00001111 11110000 00101001 11100110 00001011 11110000 10111110 10001001  64 place
0080 00001111 11110000 00101001 11100110 00000111 11110000 10001011 11010000  128 place
0100 00001111 11100000 11110011 10101000 00001111 11110000 00101101 10111101  256 place
0200 00001111 11010000 00011101 01111111 00001111 11110000 00101101 10111101  512 place
0400 00001110 11110000 00001101 00101011 00001111 11110000 00101101 10111101  1024 place
0800 00001101 11110000 01100000 01111100 00001111 11110000 00101101 10111101  2048 place
1000 00001111 10110000 01000000 11010100 00001111 11110000 00101101 10111101  4096 place
2000 00001111 01110000 11111011 10000010 00001111 11110000 00101101 10111101  8192 place
4000 00001011 11110000 10111010 11010010 00001111 11110000 00101101 10111101  16384 place
8000 00000111 11110000 10001111 10001011 00001111 11110000 00101101 10111101  32768 place
```

Figure 7. Mask for R2C40D look-up table 1.

that encode the 16-bit initialization value for the look-up table. The same process was then performed for the remaining look-up tables on the board to obtain their masks.

Table 3. HDL used to generate a three-input AND gate.

```
module four_and_gate (a, b, c, d, i) /* synthesis LOC="R2C40D" */;
   input a /* synthesis LOC="E18" */;
   input b /* synthesis LOC="B20" */;
   input c /* synthesis LOC="A20" */;
   input d /* synthesis LOC="D18" */;
   output i /* synthesis LOC="D19" */;
assign i = a&b&c;
endmodule
```

Mask Correctness Confirmation. After the mask for a specific look-up table was fully reversed, it was necessary to confirm that the information was correct and useful for understanding a bitstream. To accomplish this, bitstreams for a three-input AND gate and a more complicated design of $AB + C\bar{D}$ were synthesized at look-up table 1 in the R2C40D configurable logic block. As shown in Tables 3 and 4, both were designed using hardware description language operators instead of initializing the primitives directly to ensure the initialization value in the bitstream was generated by the synthesis engine when translating the design. The bitstreams were then inspected at the locations corresponding to the look-up table configuration value and the truth tables were recovered.

Table 5 shows the bitstream values (underlined) at the R2C40D look-up table 1 indices for a three-input AND gate. Compared with the mask for the look-up table shown in Figure 7, there are 0s in the 16,384-place and 32,768-place, resulting in a hex value of `0xC000`. When this value was used to derive

Table 4. HDL used to generate a more complicated logic function.

```
module four_logic_gate (a, b, c, d, i) /* synthesis LOC="R2C40D" */;
   input a /* synthesis LOC="E18" */;
   input b /* synthesis LOC="B20" */;
   input c /* synthesis LOC="A20" */;
   input d /* synthesis LOC="D18" */;
   output i /* synthesis LOC="D19" */;
assign i = (a&b)|(c&~d);
endmodule
```

Table 5. Bitstream values for a three-input AND gate.

```
00000011 11110000 00011100 10111111 00001111 11110000 00101101 10111101
```

the truth table, the inputs `1110` and `1111` yielded an output of `1`. This matches the logic for a three-input AND gate, implying that the mask is correct.

Table 6. Bitstream values for a more complicated logic function.

```
00000100 11000000 00001100 00001011 00000101 11110000 11000010 01001010
```

The logic for the second design is more difficult to reconstruct. Table 6 shows the bitstream values (underlined) at the R2C40D look-up table 1 indices.

Table 7 shows the reconstructed truth table. This truth table was used to recover the digital logic function:

$$\bar{A}\bar{B}CD + \bar{A}BCD + A\bar{B}\bar{C}\bar{D} + A\bar{B}\bar{C}D + A\bar{B}CD + ABCD$$

which can be further reduced to:

$$A\bar{B} + CD$$

Although this function is not in the same form as the hardware description language, it is important to note that the synthesis process has control over how the inputs are routed to the look-up table. This is logically equivalent to what was specified in the hardware description language and that the mask can be used to correctly predict the logic in a look-up table. Although the routing cannot be inferred, it is still possible to understand the logic function embodied in a look-up table by analyzing the bitstream. Thus, the logic embodied in every look-up table on the board can be analyzed although the connections between the look-up tables are not fully reverse engineered.

Table 7. Recovered truth table.

W	X	Y	Z	F
0	0	0	0	0
0	0	0	1	0
0	0	1	0	0
0	0	1	1	1
0	1	0	0	0
0	1	0	1	0
0	1	1	0	0
0	1	1	1	1
1	0	0	0	1
1	0	0	1	1
1	0	1	0	1
1	0	1	1	1
1	1	0	0	0
1	1	0	1	0
1	1	1	0	0
1	1	1	1	1

3.4 Bitstream Modification Attack

A bitstream modification attack was attempted using the information gained via reverse engineering – specifically, how the configuration information for a look-up table was stored. The goal was to simulate an attack where an adversary intercepts a bitstream *en route* from the designer to the target system. The adversary then modifies the bitstream, which is loaded on the target system. This demonstrates the feasibility of a more complicated attack than the hardware Trojan described in [2] and further confirms the validity of the look-up table mask.

Experimental Design. The initial logic function chosen was a simple four-input OR gate. The OR gate was implemented using hardware description language operators instead of configuring the look-up table directly. This ensured that the attack scenario would be similar to the actual process involving an intellectual property design. The inputs were connected to four dip switches based on the Lattice preference file constraints and the output was connected to an LED. After confirming that the design was functioning correctly on the target system, the bitstream was modified directly to implement a four-input AND gate and the new design was loaded on the target system, where the functionality was observed.

Modification Results. Table 8 shows the hardware description language code used to generate the four-input OR gate. Table 9 shows the Lattice preference file used to incorporate the design in a look-up table. The design

Table 8. HDL used to generate the OR gate for bitstream modification.

```
module orgate (a,b,c,d,e);
   input a,b,c,d;
   output e;
assign e = a|b|c|d;
endmodule
```

Table 9. LPF constraints used to generate the OR gate for bitstream modification.

```
BLOCK RESETPATHS;
BLOCK ASYNCPATHS;
Locate comp "a" site "j7";
Locate comp "b" site "j6";
Locate comp "c" site "h2";
Locate comp "d" site "h3";
Locate comp "e" site "u19";
Locate comp "orgate" site "r2c40d";
```

was then loaded on the board and the correct functionality was observed. On this board, the LEDs turned off when they were driven high.

Figure 8. Dip switch states showing the correct function of an OR gate.

Figure 8 shows a subset of the states for the OR gate, demonstrating that the LED correctly lit up when the input was 0000 and the LED was turned off for all other inputs. The bitstream was then modified at the indices related to the first look-up table in the R2C40D configurable logic block.

Table 10 shows the indices that were replaced in the bitstream. The underlined indices correspond to the look-up table configuration bits and the indices in bold font correspond to the checksum bits. The checksum bits were identified when attempting to load the modified bitstream on the target system. The programmer tool was able to detect the modifications and returned an invalid file report or an XCF file reading error. (An XCF configuration file contains information about the device, data files targeted and the operations

Table 10. Indices modified to transform an OR gate into an AND gate.

OR Gate			
00000000	00000000	**00100100**	**01101001**
00000000	00010000	**11111010**	**01111100**
AND Gate			
00000111	11110000	**10001111**	**10001011**
00001111	11110000	**00101101**	**10111101**

to be performed [7].) However, when the checksum bits were replaced with the checksum bits obtained by reverse engineering the mask, the programmer tool accepted the file.

Figure 9. Dip switch states showing the correct AND gate function after the attack.

Figure 9 shows the target system after the modified bitstream attack. The LED was turned on for every input except for 1111. This demonstrates that the correct behavior was obtained after loading the modified bitstream on the target system, implying that the attack was successful. Although the presence of the checksum bits increased the difficulty of the modification attack and the use of pre-synthesized checksums was not feasible, the checksum can, in fact, be defeated. In the case of a simple modification, the checksum can be brute-forced because the indices that are verified by the checksum are known.

The other option is to reverse engineer the checksum algorithm. This is accomplished by running the programmer or the Lattice Diamond software through a debugger to observe the operations that compute and verify the checksum. This method has been used in a similar scenario where the encryption schemes used by the Stratix II and Stratix III FPGAs were defeated [16].

4. Experimental Results

The experiments demonstrate that the locations and values of the various configuration options of the Lattice ECP3 LFE3-35EA-8FN484C FPGA could be reverse engineered successfully. In the case of the input/output block, the pullmode locations and values were reverse engineered for every pin. For the slew rate, drive level, input and output configuration options, the locations and values were found for all the pins in Groups 1 through 4, as well as one pin in Group 5. This equates to a total of 139 pins. For the configurable logic blocks, the encoding of the configuration information for a small set of the look-up tables was located and recorded, which was used in the successful bitstream modification attack.

This research also created a bitstream parser. The parser processes a bitstream synthesized for the LFE3-35EA-8FN484C FPGA and outputs configuration information about the reverse-engineered input/output block. The information obtained for each look-up table after it was fully reversed is passed to the parser to obtain additional information such as the percentage of the look-up table utilized and its logic function.

Although the bitstream parser does not provide complete information about a bitstream, it should be of value to the industrial control system community. Consider a scenario where a critical infrastructure asset owner receives an updated bitstream from a vendor. Running the new and old bitstreams through the parser would help detect errors and/or malicious modifications. The addition of new input or output pins would indicate potential covert channels. Large increases in look-up table utilization could indicate the insertion of malicious hardware. Although the bitstream parser was developed specifically for the LFE3-35EA-8FN484C FPGA, the underlying process can be applied to other Lattice FPGAs that use the Lattice Diamond software, and, with some modifications and enhancements, to other FPGAs.

5. Conclusions

FPGAs are commonly used in critical infrastructure assets. Their power-to-cost ratio and their reprogrammability make them particularly attractive for industrial control applications. However, their complexity increases the risk of attacks. This chapter has demonstrated the process of reverse engineering a portion of a previously-unexplored Lattice FPGA, which has been incorporated in a parser that enables the analysis of bitstreams for errors and malicious modifications without revealing any sensitive intellectual property.

Future research will continue the reverse engineering efforts on the switching matrix and also concentrate on other FPGAs. Additionally, research will focus on automating the reverse engineering process for Lattice FPGAs and FPGAs from other vendors.

Note that the views expressed in this chapter are those of the authors and do not reflect the official policy or position of the U.S. Air Force, U.S. Department of Defense or U.S. Government.

References

[1] J. Brenner, Keeping America Safe: Toward More Secure Networks for Critical Sectors, MIT Center for International Studies, Massachusetts Institute of Technology, Cambridge, Massachusetts, 2017.

[2] R. Chakraborty, I. Saha, A Palchaudhuri and G. Naik, Hardware Trojan insertion by direct modification of FPGA configuration bitstream, *IEEE Design and Test*, vol. 30(2), pp. 45–54, 2013.

[3] Z. Ding, Q. Wu, Y. Zhang and L. Zhu, Deriving an NCD file from an FPGA bitstream: Methodology, architecture and evaluation, *Microprocessors and Microsystems*, vol. 37(3), pp. 299–312, 2013.

[4] S. Drimer, Security for Volatile FPGAs, Technical Report UCAM-CL-TR-763, Computer Laboratory, University of Cambridge, Cambridge, United Kingdom, 2009.

[5] U. Farooq, Z. Marrakchi and H. Mehrez, Chapter 2, FPGA architectures: An overview, in *Tree-Based Heterogeneous FPGA Architectures: Application Specific Exploration and Optimization*, Springer, New York, pp. 7–48, 2012.

[6] I. Kuon and J. Rose, Measuring the gap between FPGAs and ASICs, *IEEE Transactions on Computer-Aided Design of Integrated Circuits and Systems*, vol. 26(2), pp. 203–215, 2007.

[7] Lattice Semiconductor, Lattice Diamond 3.4 Help, Hillsboro, Oregon, 2014.

[8] E. Lubbers, Configurable System-on-Chip: Xilinx EDK, University of Paderborn, Paderborn, Germany (`slideplayer.com/slide/5083550`), 2014.

[9] S. Mal-Sarkar, A. Krishna, A. Ghosh and S. Bhunia, Hardware Trojan attacks in FPGA devices: Threat analysis and effective countermeasures, *Proceedings of the Twenty-Fourth Edition of the Great Lakes Symposium on VLSI*, pp. 287–292, 2014.

[10] E. Monmasson, L. Idkhajine, M. Cirstea, I. Bahri, A. Tisan and M. Naouar, FPGAs in industrial control applications, *IEEE Transactions on Industrial Informatics*, vol. 7(2), pp. 224–243, 2011.

[11] A. Moradi, A. Barenghi, T. Kasper and C. Paar, On the vulnerability of FPGA bitstream encryption against power analysis attacks: Extracting keys from Xilinx Virtex-II FPGAs, *Proceedings of the Eighteenth ACM Conference on Computer and Communications Security*, pp. 111–124, 2011.

[12] National Instruments, Introduction to FPGA Hardware Concepts (FPGA Module), Austin, Texas, 2011.

[13] J. Note and E. Rannaud, From the bitstream to the netlist, *Proceedings of the Sixteenth International ACM/SIGDA Symposium on Field Programmable Gate Arrays*, pp. 264–272, 2008.

[14] J. Rodriguez-Andina, M. Moure and M. Valdes, Features, design tools and application domains of FPGAs, *IEEE Transactions on Industrial Electronics*, vol. 54(4), pp. 1810–1823, 2007.

[15] J. Rodriguez-Andina, M. Valdes-Pena and M. Moure, Advanced features and industrial applications of FPGAs – A review, *IEEE Transactions on Industrial Informatics*, vol. 11(4), pp. 853–864, 2015.

[16] P. Swierczynski, A. Moradi, D. Oswald and C. Paar, Physical security evaluation of the bitstream encryption mechanism of Altera Stratix II and Stratix III FPGAs, *ACM Transactions on Reconfigurable Technology and Systems*, vol. 7(4), article no. 34, 2015.

[17] S. Trimberger, Three ages of FPGAs: A retrospective on the first thirty years of FPGA technology, *Proceedings of the IEEE*, vol. 103(3), pp. 318–331, 2015.

[18] C. Wolf, Project IceStorm (`www.clifford.at/icestorm`), 2018.

Chapter 7

PROTECTING INFRASTRUCTURE DATA VIA ENHANCED ACCESS CONTROL, BLOCKCHAIN AND DIFFERENTIAL PRIVACY

Asma Alnemari, Suchith Arodi, Valentina Rodriguez Sosa, Soni Pandey, Carol Romanowski, Rajendra Raj and Sumita Mishra

Abstract Protecting critical infrastructure data is challenging because it typically includes sensitive information that is often needed by analysts to answer crucial questions about the critical infrastructure. For example, in the healthcare sector, epidemiologists need to analyze personally identifiable information to track the spread of diseases or regional emergency services managers may need to view details of all 911 calls made during a hurricane or terrorist incident. In other situations where personally identifying information is not needed to perform analyses, studies have shown that anonymization approaches such as k-anonymity or l-diversity cannot safeguard the information from inadvertent or malicious exposure. Additionally, recent data breaches involving critical infrastructure information demonstrate that current access control mechanisms, including role-based access control, are neither sufficient to secure the information nor adequate to prevent the ensuing loss of privacy. This chapter presents a novel approach that integrates existing access control mechanisms with blockchain and differential privacy to protect infrastructure data.

Keywords: Data protection, data privacy, access control, blockchain

1. Introduction

Sensitive datasets, such as data generated by critical infrastructure assets, often need to be analyzed to recognize trends, optimize resources and determine proper courses of action [12]. However, critical infrastructure data typically includes a great deal of personally identifiable information (PII) in addition to

J. Staggs and S. Shenoi (Eds.): Critical Infrastructure Protection XII, IFIP AICT 542, pp. 113–125, 2018.
https://doi.org/10.1007/978-3-030-04537-1_7

other sensitive data pertaining to locations, building access, perimeter security, etc. Based on their data needs, analysts can be categorized into three groups:

- **Primary Analysts:** These users must have complete access to all the critical infrastructure data and related data products to perform their tasks. For example, an emergency manager in a county in the United States may need to see the details of every call in the county's 911 system.

- **Secondary Analysts:** These users may need access to critical infrastructure data that includes some personally identifiable information and/or sensitive information, but the rest of the data can be restricted using aggregation or anonymization. For example, an employee in a different agency who analyzes resource allocation in a county, only needs to see aggregated information from the dataset with most of the personally identifiable information removed. However, the employee may need access to location information that could become personally identifiable information in sparsely-populated areas of the county.

- **Tertiary Analysts:** These users do not need to see any personally identifiable information, but may need access to aggregated or anonymized information. For example, a member of the local news media should not have access to any sensitive information, but may be allowed to see summary data.

The dilemma is to ensure the maximal protection of critical infrastructure data while providing appropriate access to legitimate uses by the three types of data analysts. In all these cases, system access is permitted, but the access must be controlled.

Current access control methods have proven to be inadequate for sensitive datasets. According to tracking by the Privacy Rights Clearinghouse [13], more than 550 data breaches were publicly reported in 2017. In other words, on average, more than 1.5 data breaches occurred daily in the United States. Because these correspond to the events that were recorded and reported, the actual number of data breaches is likely to be considerably higher. In many cases, the breaches were caused by inadequate access control mechanisms that essentially enabled outsiders or malicious insiders to breach them fairly easily.

Access control mechanisms must be enhanced to provide better data security and protection. This chapter argues that access control should be considered to be only the first layer of data protection. The logical next layer is data anonymization – for example, abstracting individual data items as ranges can obscure sensitive values and concept hierarchies can mask specific attributes. However, most techniques such as k-anonymity and l-diversity cannot prevent the exposure of private information when data is queried [9]. Because anonymization is inadequate, a crucial role can be played by differential privacy [6] in providing overall data protection. Differential privacy makes the presence or absence of an individual or single entity indistinguishable, thereby reducing any benefit of adversarial background knowledge about individuals' data in a dataset. For example, Lin et al. [8] propose an approach that adds random noise to true answers, but even this method is not foolproof. An attacker repeatedly asks the same question and a different answer is provided each time; however, this itself provides a clue that the information is sensitive. Complicating this situation is the fact that real-world data is not independent. This requires the implementation of a comprehensive strategy to hide correlations between attributes [19].

This chapter proposes a layered methodology that enhances access control using blockchain and differential privacy to provide strong data protection for critical infrastructure assets and reduce data privacy losses. The proposed framework develops the appropriate access and differential privacy strategies based on user types and dataset characteristics.

2. Motivating Scenarios

This section provides examples that illustrate how the proposed framework would be applied in different domains. Emergency management and healthcare are chosen as the sample domains, although similar scenarios can be developed for other critical infrastructure domains. While the easiest way to safeguard datasets is to completely restrict them, the proposed framework assumes that analyses of the datasets are beneficial as long as the protection of sensitive data is assured.

2.1 Scenario 1: Emergency Services Sector

Emergency response in the United States is typically handled at the municipal level (village, town or city) until an event overwhelms the local resources [15]. At this point, the emergency response is managed at the county level from an emergency operations center. Data about the emergency event is collected by the countywide 911 system and other repositories (e.g., after-action reports). The collected data is analyzed to identify ways in which municipalities can optimize resource allocation, merge or move fire/police stations, or even suggest changes to roadway intersections to minimize accidents. However, some data – especially 911 call data — contains personally identifiable information such as names, addresses, phone numbers, driver's license numbers, medical status and other sensitive data related to individuals and businesses.

This example considers the three user roles mentioned above. The primary analysts are the county emergency manager and municipal department heads. The secondary analysts are county or municipal personnel who analyze broad event patterns that affect resource usage, such as arsons, accidents and emergency medical calls. The analyses do not require and should not contain personally identifiable information, but would have specific event location information and response unit identification data. Finally, the tertiary internal or external users include lower-level municipal employees and university researchers who perform high-level analyses. These users would not have access to personally identifiable information, specific event locations or response unit identifiers beyond the types of response units (police, fire and emergency medical units).

A more detailed version of this scenario assigns different roles to users depending on their positions. For example, the county emergency manager would have access to all the data regardless of jurisdiction, but a town official may not be granted unrestricted access to data outside the official's municipality. Alternately, an attribute-based control system could accomplish the same purpose.

The benefit to using the proposed framework in this scenario is tighter access control over private data belonging to individuals and sensitive information related to businesses and government entities. Since many government data sources are subject to "freedom of information" type requests, the differential privacy aspect of the framework provides external users with access while protecting critical assets. Safeguarding personally identifiable information is important, but it is just as critical to avoid breaches that might expose the vulnerabilities of business or government installations.

2.2 Scenario 2: Healthcare Sector

In the healthcare sector, information sharing has become crucial to improving healthcare quality and outcomes, as well as lowering costs [17]. The benefits of sharing information must be balanced with security and privacy concerns, especially when healthcare personally identifiable information is involved. The U.S. Health Insurance Portability and Accountability Act (HIPAA) places strict requirements, including access control, for protecting healthcare personally identifiable information [16]. The constant barrage of successful attacks in the healthcare sector and the consequent data breaches reveal that the implemented access control mechanisms are inadequate [13]. Moreover, healthcare organizations incur significant penalties for one-time violations and repeat violations across all HIPAA violation categories [18].

Consider a healthcare scenario similar to the emergency management scenario discussed above. The healthcare scenario has trusted internal users (doctors, nurse practitioners and other medical personnel involved in direct patient care), internal users (medical personnel not involved in direct patient care) and internal/external users such as administrative personnel and researchers. However, the healthcare setting includes aspects that make the scenario more complex than the emergency services scenario.

In the healthcare setting, primary analysts have access to all the information about patients under their care. Unlike the emergency management scenario, the doctor-patient relationship excludes the possibility of a trusted user with unrestricted access to all the data related to patients.

Secondary analysts such as medical technicians would have access to data pertaining to their particular functions for short periods of time. While one would expect the doctor-patient relationship to be ongoing, ancillary medical personnel and even floor nurses would not need to access patient data after the patients are out of their care.

Tertiary analysts in medical administration have no need to see detailed health data such as laboratory reports and nursing notes, although they would need to know patient diagnosis and insurance information, thereby having access to personally identifiable information. External analysts such as medical researchers have no need to access personally identifiable information. Given the complexity of the healthcare scenario, attribute-based access control (ABAC) appears to be a better fit than role-based access control (RBAC) [4]. An attribute-based access control approach would also account for the temporal aspects of the healthcare sector.

In short, privacy requirements along with increased information sharing in the healthcare sector provide additional and compelling motivation for the enhanced access control framework proposed in this chapter.

3. Background

This section provides background information needed to understand the proposed framework. It discusses the key concepts of access control, blockchain and differential privacy that set the stage for the rest of this chapter.

3.1 Access Control

Access control models help ensure that only authorized users are allowed to perform previously-approved operations on objects. Numerous access control models have been developed over the years, each with its advantages and disadvantages. Software systems in the critical infrastructure sectors tend to use some variant of role-based access control [3, 14].

Role-based access control is based on five sets of entities: (i) subjects; (ii) objects; (iii) roles; (iv) operations; (v) and permissions; and two relations: (i) subject-to-role assignment; and (ii) permission-to-role assignment.

Central to role-based access control is the concept of a role, which specifies an organizational job function. Each role can also represent a set of responsibilities (or operations) associated with the job function. Instead of granting permissions individually to each subject, permissions are first associated with roles, following which roles are assigned to subjects based on their job functions.

The strengths of role-based access control arise from its simplicity of authorization administration and support for developing secure systems without requiring actual subjects. Because role-based access control is a static model, its access logic relies on a predefined set of associations of permissions to roles, which makes it unsuitable for use in environments and sectors that change dynamically. Also, role-based access control has inadequate protections against information disclosure and modification [14]. While security researchers have recently proposed models such as attribute-based access control to address problems with role-based access control, the new models have yet to gain widespread acceptance; as a result, role-based access control continues to be the dominant model used in critical infrastructure systems [3].

3.2 Blockchain

The decentralized and cryptographically secure characteristics of a blockchain enable it to serve as an immutable public ledger of records that are linked to each other [11]. Each block in the blockchain is a collection of transactions; for example, a block may be a set of financial transactions used for a cryptocurrency.

In a typical blockchain architecture, the blocks are linked to each other via hashing. All the transactions in a block are digitally signed by the involved parties with their private keys, and anyone can verify the owner using the owner's public key. For a cryptocurrency such as Bitcoin, transaction linkage also helps to keep track of the participants' balances. Each transaction is broadcast across the network and can be validated by each node in the network; nodes outside the network do not have permission to broadcast blocks. After the entire network validates a block with the chosen consensus algorithm that establishes agreement, the block is added to the blockchain by all the local nodes. This action results in all the network nodes having the same consistent data in the form of linked blocks – called the blockchain – without any central authority. An external node that wishes to join the network can build the blocks from the starting block to the most recent one with the help of its peers.

Smart contracts are often used in blockchain technology; these elements are executable code where any logic can be applied on all the nodes in the network [5]. In the context of this research, a smart contract contains the user information (roles and attributes) needed by the access control system.

A blockchain provides a decentralized method for enforcing rules and policies at all the network nodes. It also ensures that all the nodes follow and agree on the decisions, and maintains consistency of the data. Traditionally, access control systems have been centralized as opposed to distributed, with a single point of failure affecting and compromising the entire system. In contrast, a blockchain does not have a single point of failure. Blockchain technology has been used to secure data and preserve its privacy [20]. It can also be used to store access permission information.

3.3 Differential Privacy

Differential privacy as proposed by Dwork et al. [6] seeks to make the presence of an individual indistinguishable regardless of the background knowledge that an adversary may have about the dataset containing the individual's data. Hence, applying any analysis on the dataset gives almost the same results as when a record is added or removed from the dataset [1].

Let q be an arbitrary query with domain M and range P ($q : M \rightarrow P$) and let D and D' be two neighboring datasets that differ in one record. Furthermore, let f_q be a randomized function used to answer the query q. Then, f_q provides ε-differential privacy if for any $s \subseteq \mathit{Range}(f_q)$:

$$\Pr[f_q(D) \in s] = e^{\varepsilon} Pr[f_q(D') \in s]$$

Adding noise to the true answers is a common way to satisfy differential privacy. Consider a query q on a dataset D. If r is the true answer of query q, then the answer to the query that satisfies differential privacy is $r + y$, where y is random noise.

Several approaches have been proposed for generating noise. The most common approach is to draw the noise from a Laplace distribution with mean 0 and scale $\Delta f/\varepsilon$, where Δf is the maximum difference between $f_q(D)$ and $f_q(D')$ and ε is a parameter that controls privacy (as ε becomes smaller, the privacy level increases, but the accuracy decreases) [6].

Counting queries require an aggregating function to retrieve a specific value (count) of records that satisfy certain conditions [2]. Answers to these queries could exacerbate individuals' loss of privacy [7]. Because interactive settings provide better privacy than non-interactive settings, user access to data can be limited dynamically.

An unlimited number of sequential queries could still result in sensitive information being leaked, especially when the queries operate over related attributes. However, this issue can be resolved by setting up a workload of queries ahead of time and submitting them as a batch to adjust the level of added noise based on the given queries. Partitioning mechanisms permit sensitive areas of the vector of counts to have larger amounts of noise than other areas. This helps ensure more accurate answers when the workload has insensitive queries. The mechanism thus considers the sensitivity of the given set of queries, but is otherwise data independent [2].

4. Design and Implementation

This section describes the design and implementation of the proposed framework for enhancing access control using blockchain and differential privacy.

4.1 System Architecture

The system architecture assumes a role-based access control model with three major roles, primary, secondary and tertiary, corresponding to the three types of analysts discussed above. Other access control models are also possible, but role-based access control is sufficient for the purposes of this work. To address the goal of protecting sensitive information, the framework uses layered access as shown in Figure 1. Each layer receives input queries from the previous layer (higher in the figure), and invokes the appropriate access policy depending on the analyst's role.

The system comprises the following layers:

- **Client Layer:** The client layer accepts queries from the different types of analysts and passes the queries along with user credentials to the access control layer.

- **Access Control Blockchain Layer:** The access control blockchain layer is responsible for granting access to the requested data. The layer is

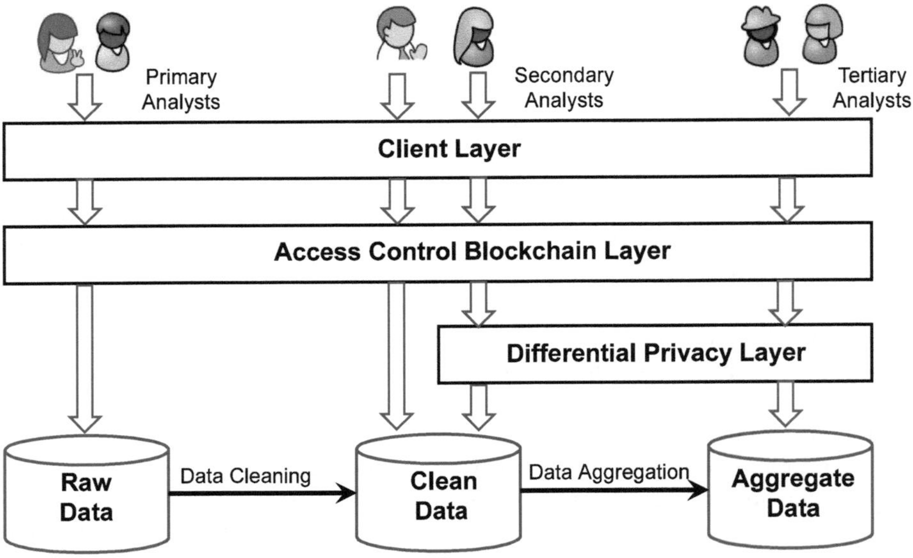

Figure 1. System architecture.

implemented using blockchain technology, where each user/node initiates a transaction on the blockchain network. The transaction is initiated after a smart contract is executed by the client layer. Based on the inputs provided to the smart contract, the user is provided with appropriate access permissions to complete the transaction. The smart contract runs on all the nodes that attempt to gain access to the data tables. The block is then broadcast across the blockchain network. All the network nodes validate the block, come to an agreement based on the chosen consensus algorithm and add the block to the blockchain.

The smart contract code cannot be modified by any of the users and the logic is always executed after a user attempts to access data. The access control system leverages blockchain technology and smart contracts in granting secure access, returning the key used to execute the queries. The main advantage of using smart contracts is that any complex access permission logic can be coded easily.

- **Differential Privacy Layer:** The differential privacy layer implements differential privacy techniques to provide further protection to sensitive information. The access control layer requires secondary and tertiary analysts to provide all their queries as a workload and then invokes the differential privacy layer. Based on the workload of queries, the actual results are modified to ensure individuals' privacy and operational privacy as discussed in Section 4.3.

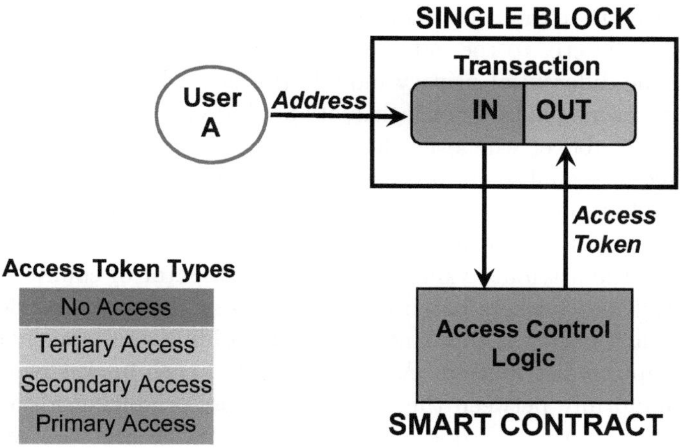

Figure 2. Single block example of smart contract interaction with access control.

4.2 Queries by Different Types of Analysts

In the case of primary analysts, the initial access layer works in a pass-through mode. Specifically, it simply passes the query straight to the raw database without any processing. This path is shown in the left-hand side of Figure 1.

In the case of secondary analysts, differential privacy techniques may be invoked depending on the nature of the query. Specifically, whether the query includes sensitive attributes or combinations of such attributes. If no sensitive attributes are present, then the query is passed through, as in the case of primary analysts. This path is shown in the middle of Figure 1.

In the case of tertiary analysts, the differential privacy layer is always used. The query path is shown in the right-hand side of Figure 1.

4.3 Implementation Details

Based on the architecture shown in Figure 1, role-based access control was implemented on the Ethereum blockchain [5] with the analyst roles stored in a smart contract. Analysts interact with the system via a public address to issue queries. The client layer receives an analyst's public address and then executes a call to the smart contract. The smart contract returns the analyst's role if access is granted; otherwise, access is denied.

Figure 2 illustrates the access control mechanism using a single transaction in a block. The analyst's address is input for the transaction and is used by the access control logic in the smart contract to look-up and return an appropriate access token for the analyst. The access token (primary access, secondary access, tertiary access or no access) returned by the smart contract is the transaction output, which is stored with the issuing analyst's address in a block.

Assuming that the access control layer approves, the client layer request is either sent directly to the data repository (for primary analysts) or is passed through the differential privacy module (for secondary and tertiary analysts). Of course, users who are not legitimate analysts are denied access.

As stated above, whenever the differential privacy module is invoked, the system requires analysts to present all the queries in a single batch or workload. This module employs the workload partitioning mechanism described in earlier work [2]. The mechanism takes the provided set of queries as a workload, along with the attribute values expressed as a vector of counts. The vector is partitioned into buckets based on the ranges of the given queries. The total count of each bucket is then anonymized by adding an amount of noise drawn from a Laplace distribution. After the count of each bucket is anonymized, it is split uniformly between the vector positions, producing a different private vector for answering the queries. The results, which are then returned to the user via the client layer, provide the desired additional privacy.

5. Preliminary Analysis

The proposed generic prototype can be used to implement a variety of access control models provided that the access control logic of the models can be programmed in the smart contract. Blockchain technology and differential privacy provide added protection for sensitive data.

However, the extra protection comes at a cost – in this case, additional overhead from the system components. First, the efficiency of the system is influenced by the complexity of the access control logic for the selected access control model. Depending on the application, the access control logic chosen and implemented can vary from simple to complex, and the execution time overhead varies accordingly.

Second, by requiring each node to process a transaction, blockchains can slow the system and are, therefore, unlikely to be scalable [10]. Additionally, underlying distributed blockchain network parameters such as the network load, consensus mechanism, processing power of the nodes, number of nodes and other distributed network parameters also affect system performance. Figure 3 shows the possible impact of access policies and blockchain overhead on the processing time.

Third, using differential privacy may affect system performance because of the processes that must be performed until the final answers are returned to a user. However, differential privacy may not be universally invoked for all users.

Other concerns regarding the security and privacy of the proposed framework include:

- The access control logic in the smart contract cannot be modified after it is deployed. For this reason, the smart contract code must be foolproof with no bugs and other programming flaws. If there are any issues, an adversary may be able to view the smart contract code and exploit flaws in its code.

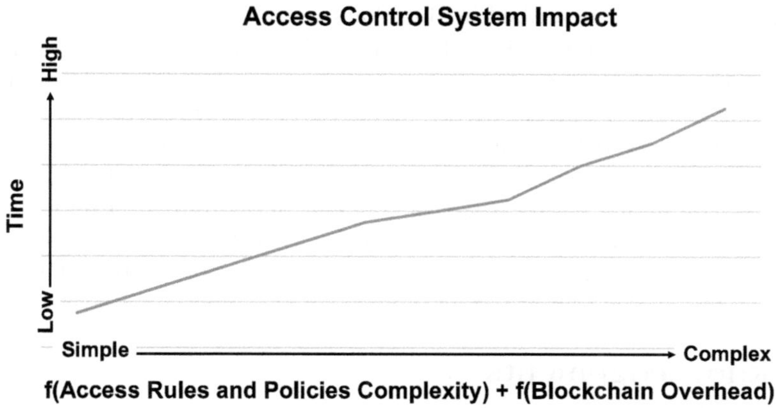

Figure 3. Effects of access control policies and blockchain on processing time.

- The data stored in the blockchain is visible to all the nodes in the network and any node can view the access permission of another node, leading to a potential privacy issue. However, because an analyst's address is stored with his/her access permissions, the system creates new addresses whenever a user query is sent to the system, preventing blockchain users from breaching analyst privacy. Current research is investigating the possible impact of scale on this approach.

- The framework only stores access permission details in the blockchain, not the real data. An application that uses the framework must ensure that the user who wishes to gain access interacts with the access control system to obtain the access permission; also, it should ensure that no adversary can circumvent the access control system. The blockchain ensures that unauthorized users cannot initiate transactions or change data in the ledger.

- The heart of the blockchain is the consensus mechanism. If more than half of the network nodes are not trustworthy, then there is a chance that adversaries may be able to take over the system. However, the possibility of this occurring is remote.

6. Conclusions

Effective information sharing, decision making and allocation are critical precursors to effective response, especially under conditions of widespread stress and overwhelming need. Even in such precarious times, it is important to protect individual, collective and, perhaps, operational privacy, and to secure critical infrastructure assets. Many current information sharing systems depend on outmoded controls that provide little certainty, and exhibit undesirable trade-offs between access control and responsiveness.

The framework described in this chapter addresses these fundamental concerns while supporting optimal decision making in evolving environments. However, a thorough exploration of the layered approach involving systematic testing and parameter optimization remains to be performed. Since questions still remain about system scalability and potential vulnerabilities, future research will focus on prototype testing under a range of parameter settings using a dataset containing twelve years of 911 call data from Monroe County, New York. The raw dataset contains personally identifiable information and sensitive critical asset information, which makes it possible to test the differential privacy module as well as the access control and blockchain layers.

Acknowledgements

Asma Alnemari acknowledges the support of the Ministry of Higher Education of the Kingdom of Saudi Arabia. This research was partially supported by the National Science Foundation under Grant No. DGE-1433736.

References

[1] R. Agrawal and R. Srikant, Privacy-preserving data mining, *Proceedings of the ACM SIGMOD International Conference on Management of Data*, pp 439–450, 2000.

[2] A. Alnemari, C. Romanowski and R. Raj, An adaptive differential privacy algorithm for range queries over healthcare data, *Proceedings of the IEEE International Conference on Healthcare Informatics*, pp. 397–402, 2017.

[3] S. Alshehri, S. Mishra and R. Raj, Using access control to mitigate insider threats to healthcare systems, *Proceedings of the IEEE International Conference on Healthcare Informatics*, pp. 55–60, 2016.

[4] S. Alshehri and R. Raj, Secure access control for health information sharing systems, *Proceedings of the IEEE International Conference on Healthcare Informatics*, pp. 277–286, 2013.

[5] V. Buterin, A Next-Generation Smart Contract and Decentralized Application Platform (`www.github.com/ethereum/wiki/wiki/White-Paper`), 2014.

[6] C. Dwork, F. McSherry, K. Nissim and A. Smith, Calibrating noise to sensitivity in private data, in *Theory of Cryptography*, S. Halevi and T. Rabin (Eds.), Springer, Berlin Heidelberg, Germany, pp. 265–284, 2006.

[7] C. Dwork and A. Roth, The algorithmic foundations of differential privacy, *Foundations and Trends in Theoretical Computer Science*, vol. 9(3-4), pp. 211–407, 2014.

[8] C. Lin, Z. Song, H. Song, Y. Zhou, Y. Wang and G. Wu, Differential privacy preserving in big data analytics for connected health, *Journal of Medical Systems*, vol. 40(4), 2016.

[9] A. Machanavajjhala, J. Gehrke, D. Kifer and M. Venkitasubramaniam, *L*-diversity: Privacy beyond *k*-anonymity, *Proceedings of the Twenty-Second International Conference on Data Engineering*, pp. 24–36, 2006.

[10] L. Mearian, Ethereum explores a fix for blockchain's performance problem, *Computerworld*, January 5, 2018.

[11] S. Nakamoto, Bitcoin: A Peer-to-Peer Electronic Cash System (`bitcoin.org/bitcoin.pdf`), 2008.

[12] President's National Security Telecommunications Advisory Committee, NSTAC Report to the President on Big Data Analytics, Washington, DC, 2016.

[13] Privacy Rights Clearinghouse, Chronology of Data Breaches: Security Breaches 2005 – Present, San Diego, California (`www.privacyrights.org/data-breaches`), 2018.

[14] R. Raj, S. Mishra, C. Romanowski, J. Schneider and S. Alshehri, Modeling threats: Insider attacks on critical infrastructure assets, poster presented at the *IEEE International Symposium on Technologies for Homeland Security*, 2017.

[15] C. Romanowski, R. Raj, J. Schneider, S. Mishra, V. Shivshankar, S. Ayengar and F. Cueva, Regional response to large-scale emergency events: Building on historical data, *International Journal of Critical Infrastructure Protection*, vol. 11, pp. 12–21, 2015.

[16] U.S. Department of Health and Human Services, Standards for Privacy of Individually Identifiable Health Information; Final Rule, *Federal Register*, vol. 67(157), pp. 53182–53273, August 14, 2002.

[17] U.S. Department of Health and Human Services, HITECH Act Enforcement Interim Final Rule, Washington, DC (`www.hhs.gov/ocr/privacy/hipaa/administrative/enforcementrule/hitechenforcementifr.html`), 2017.

[18] M. Winger, HIPAA increases financial penalties for repeat violations to address increasing healthcare data breaches, Zephyr Networks, Laguna Hills, California (`www.zephyrnetworks.com/hipaa-healthcare-data-breaches-financial-penalties`), February 10, 2013.

[19] T. Zhu, P. Xiong, G. Li and W. Zhou, Correlated differential privacy: Hiding information in a non-IID data set, *IEEE Transactions on Information Forensics and Security*, vol. 10(2), pp. 229–242, 2015.

[20] G. Zyskind, O. Nathan and A. Pentland, Decentralizing privacy: Using blockchain to protect personal data, *Proceedings of the IEEE Security and Privacy Workshops*, pp. 180–184, 2015.

Chapter 8

A NEW SCAP INFORMATION MODEL AND DATA MODEL FOR CONTENT AUTHORS

Joshua Lubell

Abstract The Security Content Automation Protocol (SCAP) data model for source data stream collections standardizes the packaging of security content into self-contained bundles for easy deployment. However, no single data model can satisfy all requirements. The source data stream collection data model does not adequately meet the needs of SCAP content authors, and its implementation-specific syntax lacks the ability to express packaging subtleties critical to software developers and content authors. This chapter defines a new implementation-neutral information model that is easier to understand and does a better job at expressing relationships between objects comprising a source data stream collection. A new authoring data model for facilitating the implementation of SCAP content development software applications is derived from the information model. Also described is an application implementing the authoring data model that enables SCAP content developers to create source data stream collections using a friendly and intuitive syntax, which is then transformed into SCAP-standard-conforming content.

Keywords: Security Content Automation Protocol, information model, data model

1. Introduction

The Security Content Automation Protocol (SCAP – pronounced *ess-cap*) is an ecosystem of interoperable Extensible Markup Language (XML) [31] vocabularies, reference data repositories and software tools [24]. System administrators – and increasingly operators of manufacturing facilities – use SCAP to secure servers, workstations, networks and other deployed hardware and software. A central part of the SCAP ecosystem is the source data stream collection format, an XML-expressed data model specified in NIST Special Publication (SP) 800-126 (Technical Specification for the Security Content Automation Proto-

J. Staggs and S. Shenoi (Eds.): Critical Infrastructure Protection XII, IFIP AICT 542, pp. 127–146, 2018.
https://doi.org/10.1007/978-3-030-04537-1_8

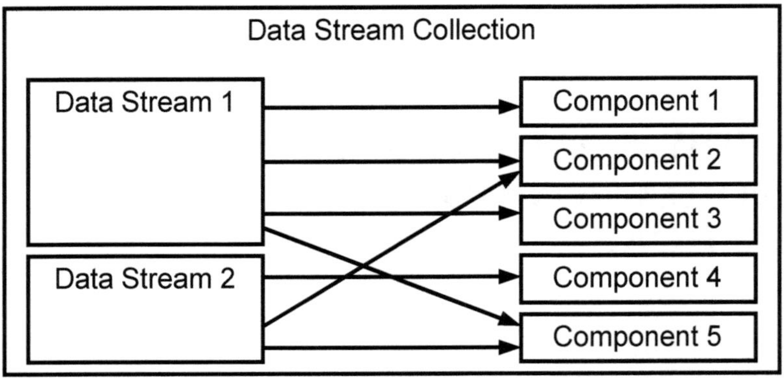

Figure 1. SCAP data stream collection.

col) [28]. This data model is instrumental for enabling the lossless exchange of security content between SCAP-conforming software products. However, no model can single-handedly satisfy the needs of SCAP software developers, content authors and users. This chapter provides an overview of the NIST SP 800-126 source data stream collection data model – highlighting where it succeeds and where it falls short – and then defines supplemental models to address unmet requirements.

The NIST SP 800-126 data model for source data stream collections defines how to package (into a self-contained entity) the collective input required for an SCAP software tool to perform one or more use cases. SCAP use cases include cyber security functions such as configuration checking and vulnerability detection. Self-containment is advantageous because it facilitates SCAP deployment where network connectivity and filesystem access are restricted, as is often the case for industrial control systems and Industrial Internet of Things (IIoT) environments. Self-containment also promotes portability – a single SCAP source data stream collection is easier to distribute reliably to partners, customers and other third parties than an interdependent set of resources. Self-containment also supports digital signing of a source data stream collection as a whole in order to ensure integrity and trustworthiness.

Figure 1 shows the high-level logical relationships within a sample SCAP source data stream collection. The example has two data streams and five components. Each component contains XML data conforming to an XML language that is part of the SCAP ecosystem. Each data stream corresponds to a specific SCAP use case. The arrows pointing from data streams to components are component references. Multiple data streams can reference the same component. For example, both the data streams reference Components 2 and 5.

An SCAP source data stream collection bundles components together such that the components themselves are unmodified from their original states. The packaging operation is thus reversible, allowing for the extraction of SCAP content from a collection and the repackaging of content into a new collection

Table 1. SCAP data model GUID format convention [28].

Object	**Identifier Format Convention**
Data Stream Collection	scap_*reverseDNS*_collection_*name*
Data Stream	scap_*reverseDNS*_datastream_*name*
Component Reference	scap_*reverseDNS*_cref_*name*
Component	scap_*reverseDNS*_comp_*name*

while simultaneously preserving the original content. Reversibility is a desirable property because it promotes interoperable data streams. For example, suppose an SCAP content developer extracts all the components from a source data stream collection, including a security checklist component that conforms to the Extensible Configuration Checklist Description Format (XCCDF) specification [29]. Suppose the user then employs an XCCDF-compliant software tool to select a subset of the checklist rules, assign parameters to the rules and save the resulting XCCDF profile as a separate tailoring component. A tailoring component enables a named profile to be defined separately from the original checklist (without modifying the XCCDF checklist), but it is still explicitly traceable to the original. Next, following best practices for reusing third-party-developed SCAP content [3], the user repackages the extracted components plus the newly-created tailoring component into a new SCAP data stream collection. The reversibility property ensures that none of the components extracted from the old collection and deployed in the new collection are altered.

SCAP encourages content developers to provide globally-unique identifiers (GUIDs) for data stream collections, data streams, components and component references. To this end, the data model requires the identifier format conventions shown in Table 1. An identifier must be an underscore-delimited string beginning with *scap*, followed by a reverse domain name system (DNS) style substring associated with the content author, followed by a substring denoting the object type being identified (*collection*, *datastream*, *cref* or *comp*), and ending with an XML NCName. An NCName [32] is any allowable XML name that does not contain the ":" character. For example, a data stream collection developed by Example Corporation for Ubuntu Linux version 16.04 (also known as Xenial Xerus) could have *scap_com.example_collection_ubuntu-xenial* as its identifier. By promoting GUIDs, the SCAP specification reduces the likelihood of conflicting identifiers in a source data stream collection and that an SCAP content developer would create identifiers that conflict with identifiers created by other developers from the same organization.

The NIST SP 800-126 data model is beneficial for use in applications that consume source data streams, such as configuration scanners and vulnerability detection software. Self-containment of data streams reduces the need for network connectivity. Reversibility preserves the integrity of SCAP components. GUID conventions reduce name collisions. The XML representation of the data model provides additional advantages. It enables software developers to leverage a wide variety of low-cost XML parsers, validators and transformation tools, saving them the trouble of having to implement this functionality in their own software products. Additionally, the XML representation supports the validation of SCAP content and the verification of software purporting to be SCAP-conforming as being in compliance with the NIST SP 800-126 requirements.

However, the characteristics of the NIST SP 800-126 data model that are positives for SCAP software developers can be negatives for developers of SCAP content:

- The GUID formatting conventions result in long and repetitive identifiers. Shorter, context-sensitive identifiers – although dangerous from a deployment standpoint – make a source data stream collection easier for humans to author and understand.

- The XML syntax favors implementation over human readability. For example, the NIST SP 800-126 data model uses the XML Catalogs [19] syntax to define mappings from external uniform resource identifier (URI) references from within a component to the corresponding location within the context of a data stream. The mappings are needed to meet the reversibility and self-containment requirements. Although many XML tools implement XML Catalogs, the syntax is not human-friendly.

- The XML syntax, although naturally hierarchical, is limited in its ability to express the subtleties of part-whole relationships in an SCAP data stream collection. These subtleties are critical to software developers and content authors alike for understanding SCAP data stream collections.

A single data model for source data stream collections is not enough. Although the NIST SP 800-126 data model meets the needs of SCAP configuration scanner and vulnerability detection software developers, it falls short in meeting the needs of developers who create and manage SCAP content. Therefore, SCAP needs the following additional models:

- **Information Model:** The information model for source data stream collections prioritizes human readability over software implementation.

- **Authoring Data Model:** The authoring data model is designed to create new content and transform it to an SCAP-conforming source data stream collection.

As stated by Pras and Schoenwaelder in RFC 3444 [22], an information model and data model are fundamentally different. An information model is expressed at a conceptual level in order to make the design as clear as possible to anyone trying to understand the model, regardless of the implementation context. Therefore, an information model omits implementation details.

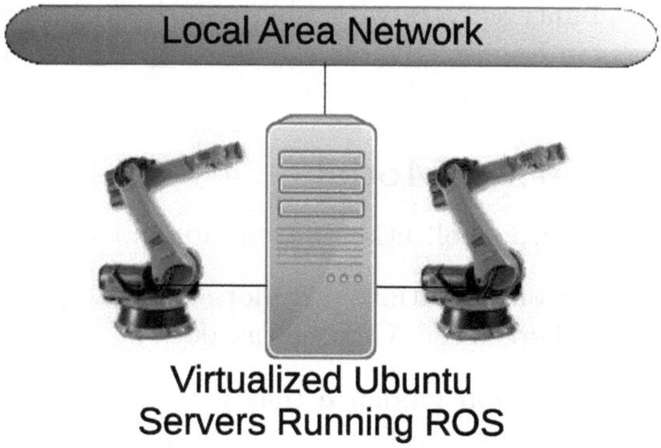

Figure 2. Robots connected to a server running ROS.

In contrast, a data model assumes a specific implementation technology and, therefore, is expressed in an implementation-specific language such as XML. Because a data model is at a lower-level of abstraction than an information model, multiple data models could be derived from a single information model.

This chapter provides an information model for source data stream collections as well an authoring data model. The information model uses the Unified Modeling Language (UML) [16] notation. The authoring data model employs the XML syntax developed using the Darwin Information Typing Architecture (DITA) [20], a standard for authoring, managing, reusing and transforming technical content. Several aspects of the authoring model correspond directly to concepts in the information model described in this chapter, demonstrating the utility of the information model in developing alternative data models. The chapter also describes a software application for creating and transforming an instance of the authoring model into a source data stream collection that conforms to NIST SP 800-126.

The information model, the authoring data model and the authoring and transformation application are all motivated by the growing need for increased SCAP usage in Industrial Internet of Things environments. In this spirit, an example used in the remainder of this chapter is based on a scenario involving a hardware-in-the-loop simulation that is part of a larger industrial control system security testbed [33]. The hardware-in-the-loop simulation involves two robotic arms that interact with a simulated machining process. The simulated manufacturing machines communicate with the robot controllers over a local-area Ethernet network. Each robot is controlled by servers that are deployed as virtual machines within a hypervisor. The controllers run the Robot Operating System (ROS) [7, 26], a software framework widely used in research and increasingly in commercial robotic applications that executes on top of Ubuntu Linux version 16.04. Figure 2 shows the testbed architecture.

The SCAP source data stream collection example used in the context of the testbed scenario incoporates an XCCDF checklist with rules to ensure that AppArmor [2], an Ubuntu Linux kernel enhancement, is installed and configured properly. Ubuntu servers with high security requirements, such as the virtualized servers in Figure 2, commonly use AppArmor. Also the access control method employed by AppArmor works well with ROS [30].

2. Information Model

The source data stream collection information model has the following goals:

- Make compositional relationships more explicit. The UML notation allows for this whereas the XML syntax does not.
- Omit implementation guidance that gets in the way of human understanding, specifically, the GUID conventions. Such guidance is vital for implementations, but it can make models unnecessarily confusing in a pedagogical context.
- Facilitate the development of other models. An information model should pave the way for the development of models that are implementation-focused. The discussion of the authoring model later in this chapter provides examples of how the authoring model elements and attributes correspond to their counterparts in the information model.

Figure 3 shows a UML class diagram representing the source data stream collection information model. A `DataStreamCollection` contains one or more `DataStream` objects and one or more `Component` objects. The `DataStream` and `Component` objects do not exist outside the scope of `DataStreamCollection`, as indicated by the solid diamonds on the links connecting them to the `DataStream Collection`. The `reverseDNS` UML attribute of `DataStreamCollection` has as its value the reverse-DNS string used in SCAP identifiers (see Table 1).

A `Component` contains an object that is a subtype of `XMLDocument`. The `timestamp` UML attribute of a `Component` specifies when the `XMLDocument` was packaged as part of a `DataStreamCollection`. Thus a `Component` is nothing more than a snapshot of `XMLDocument` at a particular point in time. `XMLDocument` is italicized in Figure 3, indicating that it is an abstract class (which cannot be instantiated). `XMLDocument` is a generalization of the five allowable SCAP source data stream component XML document types.

The five subclasses of `XMLDocument` are:

- **CPEDictionary:** This is an XML representation of a platform (hardware, operating system or software application). Each platform has a unique Common Platform Enumeration (CPE) identifier.
- **Benchmark:** This is an XML representation of a security checklist (also called a benchmark), which is valid with respect to the Extensible Configuration Checklist Description Format (XCCDF) specification.
- **OVALDefs:** This is an XML representation of system configuration information, tests and states, which is valid with respect to the Open Vulnerability Assessment Language (OVAL) specification [21]. XCCDF

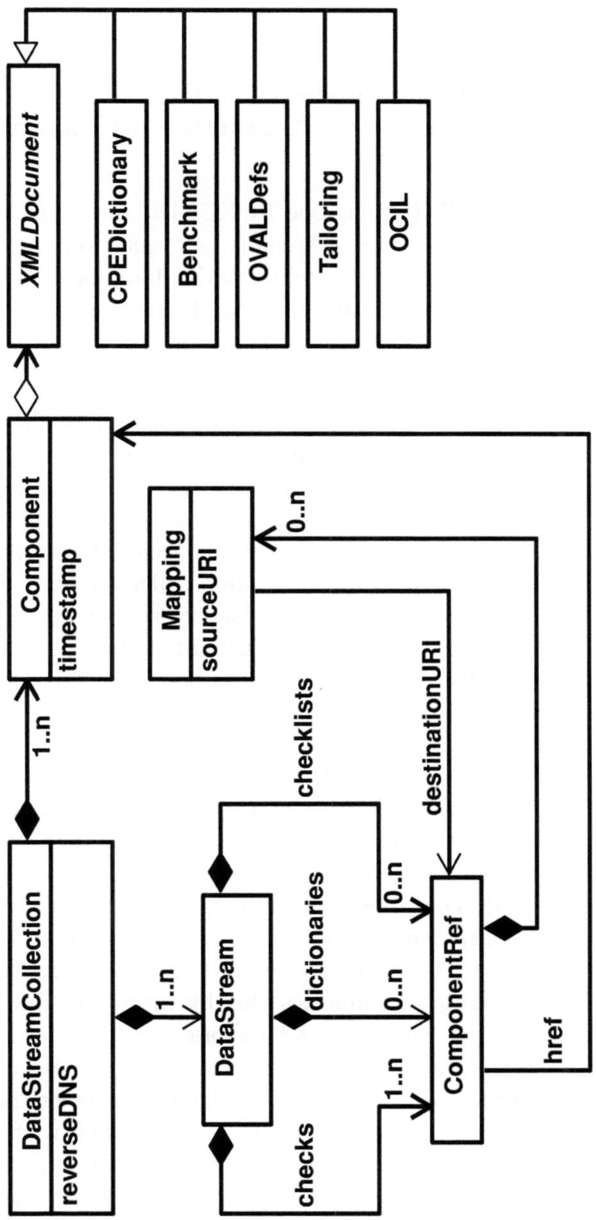

Figure 3. Source data stream collection UML class diagram.

checklist rules use OVAL to determine if the current state of a system satisfies the rule criteria. XCCDF checklist rules and OVAL definitions together typically account for most of the XML data in a source data stream collection.

- **Tailoring:** This is an XML representation of the profiles of a Benchmark, which is valid with respect to the `<Tailoring>` element definition of the XCCDF specification.

- **OCIL:** This is an XML format used by XCCDF rules for checks requiring information collected from a human via a questionnaire. It is valid with respect to the Open Checklist Interactive Language (OCIL) specification. OCIL is used for checking state via human-oriented collection of information that is not feasibly obtained using OVAL-based methods.

The existence of an `XMLDocument` is not limited to its existence in the context of a `Component`, as indicated in Figure 3 by the hollow diamond on the link from `Component` to `XMLDocument`. What this means is that, in addition to being part of the `Component`, the `XMLDocument` can be part of a `Component` in another data stream collection or have a life of its own outside the scope of SCAP. As a consequence, an `XMLDocument` in a source data stream collection may reference another `XMLDocument` in the same collection, but using a URI outside the scope of the source data stream collection. For example, a source data stream collection could incorporate a `Benchmark` and an `OVALDefs` with lives outside the scope of the collection, with the `Benchmark` using an external URI to reference the `OVALDefs`.

The source data stream collection information model handles external URI references in a manner that maintains the SCAP reversibility and self-containment requirements discussed in the introductory section. A `DataStream` contains at least one `ComponentRef` that references a `Component` containing an `OVALDefs` or `OCIL` object, and zero or more `ComponentRef` objects referencing a `Component` containing a `CPEDictionary` object, `Benchmark` object or `Tailoring` object. A `ComponentRef` may contain zero or more `Mapping` objects. A `Mapping` resolves references from within an `XMLDocument` to another `XMLDocument`. The `Mapping` accomplishes this by providing the information needed to translate the URI within the `XMLDocument` referencing the external resource to a URI referencing the `ComponentRef` within the `DataStream` containing the `ComponentRef` to which the `Mapping` belongs. The `sourceURI` UML attribute of the `Mapping` has as its value a URI that matches a referenced URI in the `Component` referenced by the `ComponentRef` that contains the `Mapping`. The `destinationURI` association of the `Mapping` references a `ComponentRef` object.

The UML object diagram in Figure 4 illustrates how the information model in Figure 3 could be used to describe a source data stream collection incorporating the XCCDF checklist introduced above. The XCCDF checklist `xenial-apparmor-xccdf.xml` and its referenced `oval-definitions.xml` are represented as `Benchmark` and `OVALDefs` objects. The `OVALDefs` object is

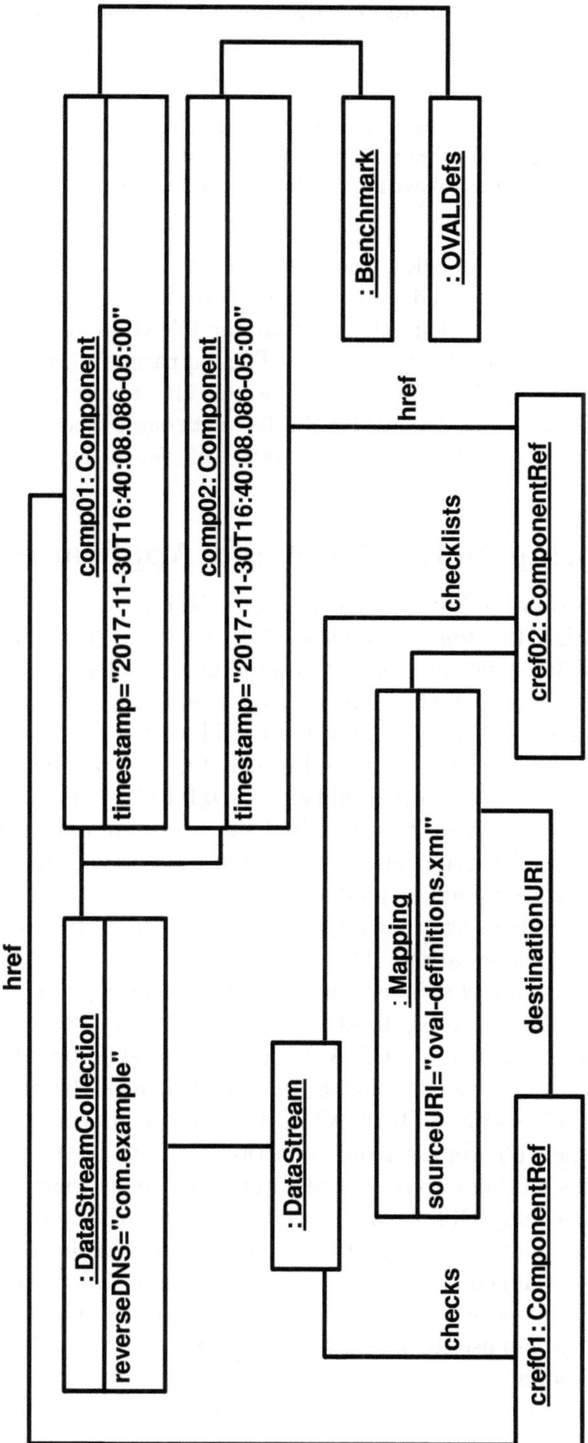

Figure 4. UML object diagram.

contained in `Component` comp01 and the `Benchmark` object is contained in `Component` comp02. `ComponentRef` cref02 contains a `Mapping` object. This mapping object is needed because the XCCDF `<check>` elements of `Benchmark` contain URI references to `oval-definitions.xml`, for example:

```
<check system="http://oval.mitre.org/XMLSchema/oval-definitions-5">
  <check-content-ref href="oval-definitions.xml"
                     name="oval:com.ubuntu.xenial:def:100"/>
</check>
```

This `<check>` element specifies that OVAL definition `oval:com.ubuntu.xenial:def:100` should be used to determine compliance with the XCCDF rule containing the `<check>` element, and that the OVAL definition is located at the relative URI `oval-definitions.xml`. The `Mapping` object says that, instead of looking for the OVAL definition in `oval-definitions.xml`, an SCAP-conforming software product processing the `DataStream` object should locate the OVAL definition within `Component` comp01 (referenced by `ComponentRef` cref01).

3. Authoring Data Model and Application

The Darwin Information Typing Architecture (DITA) [20, 23] has two primary building blocks: the *topic* and *map* XML element types. A topic represents a chunk of information. A map represents a collection of topics or other maps. DITA facilitates the reuse of topics and maps, as well as XML elements and fragments within a topic or map. DITA topic and map types are specializable. Specialization, which is the inverse of generalization, helps avoid inconsistencies and enables interoperability [12]. DITA allows for the definition of new specialized element types based on built-in topic and map types. A specialized DITA information type refines the existing base type and, therefore, must be at least as constrained. By adhering to these constraints, specialized DITA types have the advantage that implementations can easily leverage other DITA-conforming implementations [11].

The authoring data model defines a new DITA element type for source data stream collections, which is specialized from the DITA map base type. This new element type is expressed as a DITA document type shell based on DITA's map document type shell. A document type shell defines the elements and attributes that are allowed in a DITA XML document conforming to the specialized element type. The data stream collection document type shell follows the DITA standard's modular architecture for creating shells, ensuring that the shell can be used with any DITA-conformant XML authoring tool.

The DITA map type was chosen as the basis for specialization because an SCAP source data stream collection is inherently map-like. Like a DITA map, a source data stream collection is essentially a structured collection of references to components. Maps can use the DITA `<topicref>` element to reference external (non-DITA) resources, as well as to aggregate groups of nested `<topicref>` elements. Both these uses of `<topicref>` correspond to concepts from the in-

Table 2. Data stream collection DITA document type shell.

Element	Specializes	Content Model

`<DataStreamCollection>`	`<map>`	`@reverseDNS`
		`@scapName`
		`@schematronVersion`
		`<scapComponent>`*
		`<DataStream>`+

`<scapComponent>`	`<keydef>`	`@keys` `@href`

`<DataStream>`	`<topicref>`	`@scapName`
		`@scapVersion`
		`@useCase`
		`<Dictionaries>`?
		`<Checklists>`?
		`<Checks>`

`<Dictionaries>`	`<topicref>`	`<CpeListRef>`+
`<Checklists>`	`<topicref>`	`<BenchmarkRef>`+ `<TailoringRef>`+
`<Checks>`	`<topicref>`	`<OvalRef>`+ `<OcilRef>`+

`<CpeListRef>`
`<BenchmarkRef>`
`<TailoringRef>`
`<OvalRef>`
`<OcilRef>`
	`<topicref>`	`@keyref`
		`<ExternalLinks>`?

`<ExternalLinks>`	`<topicref>`	`<Uri>`+
`<Uri>`	`<topicref>`	`@keyref`

formation model in Figure 3. A `ComponentRef` object references a subclass of `XMLDocument`, which is an external resource. The `DataStreamCollection` composition link pointing to `DataStream` collects `DataStream` objects. The `dictionaries`, `checklists` and `checks` composition links of a `DataStream` collect `ComponentRef` objects. The `ComponentRef` composition link pointing to `Mapping` collects `Mapping` objects. Therefore, the DITA source data stream element type defines new elements specialized from `<topicref>` to represent data streams, component references, collections of component references and mappings from URI references within external resources to the appropriate component references.

Table 2 shows the XML elements and attributes in the source data stream collection document type shell. The left-hand column contains the element names. The middle column presents the DITA map built-in element specialized to define the element in the left-hand column. All the left-hand columns elements are specializations of `<topicref>`, except for `<DataStreamCollection>` (which specializes `<map>`) and `<scapComponent>` (which specializes `<keydef>`, a built-in map element). The right-hand column shows the content model that constrains each left-hand column element. Names preceded by an @-sign are required XML attributes. An asterisk following an element means zero or more occurrences of the element are allowed. A plus sign means one or more occurrences are allowed. A question mark means zero or one occurrences are allowed. For example, `<CpeListRef>` has a required `@keyref` attribute and may optionally contain a single `<ExternalLinks>` element.

The `<scapComponent>` element of the document type shell contains no subelements. This is because it has no author-provided content. As mentioned above, a `Component` is no more than an `XMLDocument` with a timestamp added to it. Since the timestamp is system-generated, the authoring and transformation application only needs the referenced XML resources external to the data stream collection DITA map to create NIST SP 800-126 data model component elements.

In order to understand how the authoring and transformation application processes an XML instance in a manner that is valid with respect to the authoring data model, consider the following DITA map, which represents the Xenial AppArmor source data stream collection from Figure 4:

```
<DataStreamCollection reverseDNS="com.example" scapName="apparmor"
                      schematronVersion="1.2">
  <scapComponent keys="xccdf_apparmor"
                 href="xenial-apparmor-xccdf.xml"/>
  <scapComponent keys="oval_apparmor"
                 href="oval-definitions.xml"/>
  <DataStream scapName="xenial_apparmor" scapVersion="1.3"
              useCase="CONFIGURATION">
    <Checklists>
      <BenchmarkRef keyref="xccdf_apparmor">
        <ExternalLinks>
          <Uri keyref="oval_apparmor"/>
        </ExternalLinks>
      </BenchmarkRef>
    </Checklists>
    <Checks>
      <OvalRef keyref="oval_apparmor"/>
    </Checks>
  </DataStream>
</DataStreamCollection>
```

The `@reverseDNS` attribute of the `<DataStreamCollection>` element responds directly to its counterpart in Figure 4. `@scapName` provides the *name* portion needed to construct the NIST SP 800-126 data stream collection identifier according to the GUID conventions in Table 1. `@schematronVersion` specifies the version of the Schematron schema to which the source data stream collection conforms. This information is needed because NIST SP 800-126 requires a source data stream collection to be valid with respect to a set of rules defined using Schematron [10], an XML language for expressing and testing natural language assertions about an XML document type.

The source data stream collection type uses `<scapComponent>` to associate a more succinct key name (`@keys`) with an XML document URI (`@href`). This serves multiple purposes. First, it makes source data stream collection DITA maps easier to maintain. Referencing each URI only once in `<scapComponent>` and referencing the associated name elsewhere in `@keyref` XML attributes add a level of indirection, reducing the number of DITA map revisions needed if an XML document URI changes. Second, using the key name in place of the URI improves readability of the XML. Finally – and most importantly – key names serve as the *name* portion of GUIDs generated by the authoring and transformation application when processing `@keyref` XML attributes.

`<DataStream>` has three attributes: (i) `@scapName` provides the *name* portion used by the authoring and transformation application to construct the data stream GUID; (ii) `@scapVersion` specifies the version of SCAP to which the data stream conforms (1.3 is the most recent SCAP version); and (iii) `@useCase` specifies the SCAP use case.

`<Checklists>`, which corresponds to the *checklists* composition link in Figure 4, contains `<BenchmarkRef>` elements. The authoring and transformation application uses `<BenchmarkRef>` to generate a data stream component that holds the contents of `xenial-apparmor-xccdf.xml` and a component reference. The generated component is simply a wrapper element with an application-generated timestamp value that contains the XCCDF XML. As discussed above, the XCCDF `<check>` elements contain URI references to `oval-definitions.xml`. The generated component reference, where *sds:* and *cat:* are XML namespace prefixes mapping to namespaces defined in [28] and [19], respectively, is as follows:

```
<sds:component-ref
    id="scap_com.example_cref_xccdf_apparmor"
    href="#scap_com.example_comp_xccdf_apparmor">
  <cat:catalog>
    <cat:uri name="oval-definitions.xml"
             uri="#scap_com.example_cref_oval_apparmor"/>
  </cat:catalog>
</sds:component-ref>
```

The `@id` value of the `<sds:component-ref>` element is a GUID generated by the authoring and transformation application using the `@keyref` value of the DITA map's `<BenchmarkRef>`. The `@href` value refers to the `@id` of the `<sds:component>` that contains the XCCDF checklist XML. The transformation generates `<cat:catalog>` from the DITA map's `<ExternalLinks>` element and `<cat:uri>` from the DITA map's `<Uri>` element, which corresponds to the `Mapping` object in Figure 4. The authoring and transformation application assigns the `@href` value of the `<scapComponent>` whose `@keys` attribute value matches `<Uri>`'s `@keyref` value to the `<cat:uri>` `@name` attribute. `<cat:uri>`'s `@uri` attribute is assigned a component reference GUID prefaced by # whose *name* substring is the value of the `<Uri>` element's `@keyref` attribute.

`<Checks>` and `<OvalRef>` are transformed similarly to `<Checklists>` and `<BenchmarkRef>`; however, since OVAL definitions do not reference any external URIs, there is no embedded `<ExternalLinks>` element to transform.

The authoring and transformation application was implemented using the DITA Open Toolkit [5], a specialization-aware, output-producing DITA processor. The DITA standard requires output-producing processors to merge topics referenced in a map as well as resolve key references, eliminating the need for custom transformation code to perform the functions. Specialization-aware DITA processors are required to do all of the above for specialized DITA documents by inheriting processing behavior from base types. Therefore, leveraging the DITA Open Toolkit greatly reduced the coding effort required to build the authoring and transformation application.

The DITA Open Toolkit has a modular architecture with an extensible plug-in mechanism for implementing custom document type shells and output formats. Plug-ins can be run in any XML authoring software environment that uses the DITA Open Toolkit. The authoring and transformation application was implemented as a NIST SP 800-126 conformant output plug-in. The source data stream collection document type shell was also implemented as a plug-in. The authoring and transformation application was successfully deployed in a commercial XML editor product, which was then used to create the Xenial AppArmor DITA map example in this chapter as well as other SCAP source data stream collection examples.

4. Related Efforts and Next Steps

Other recent and ongoing research efforts have fostered the development of systems of related models for achieving automation and integration. Kulvatunyou et al. [13] provide examples of standards for smart manufacturing where alternative models were developed to satisfy different implementation contexts. Smart manufacturing requires all engineering information to be represented digitally and to be completely computer interpretable. Two examples provided by Kulvatunyou and colleagues are ISO 10303-242 [9], a standard for computer-aided design (CAD) geometry and product manufacturing data, and the Open Application Group Integration Specification (OAGIS) [17], a suite of information standards for interfacing manufacturing systems with business functions such as sales and finance. ISO 10303-242 includes a low-level data model for CAD geometry and other CAD-related information, as well as a higher-level business object model that represents additional information needed for manufacturing and product support, such as part assemblies and bills of materials. OAGIS defines an abstract implementation-neutral information model for individual transaction standards called business object documents (BODs). OAGIS also defines multiple data models for implementing business object documents, including an XML-based model and a JavaScript Object Notation (JSON) [4] model.

Health Level 7 (HL7) [8] – an organization that promulgates standards for exchange management and integration of healthcare information – has created a standards architecture with an abstract information model from which implementation-specific data models are derived. The HL7 Reference Information Model (RIM) is broad and minimalist, but it provides an integrated view that facilitates the development of interoperable implementation-specific data models [6]. The Clinical Document Architecture, an HL7 standard derived from the HL7 RIM, combines an XML document type with a specialized RIM-based model to precisely specify clinical information requirements [8].

As part of a study on the challenges of automating security configuration checklists in manufacturing environments, Lubell and Zimmerman [15] developed a simple XCCDF checklist modeled in UML. The UML model uses `AND`, `OR` and `NOT` classes to represent Boolean operations in XCCDF `<check>` elements. In a follow-up effort by Lubell [14], a DITA element type developed for XCCDF rules uses specializations of DITA's built-in `<sectiondiv>` topic element to model Boolean operations. This XCCDF rule element type demonstrates the power and versatility of DITA specialization, and was a precursor to the research presented in this chapter.

The DITA XCCDF rule and SCAP data stream collection element types exemplify the recent trend of using DITA to create and manage intelligent content. Traditional content management solutions focus on information that is consumed mainly by humans via print media, the web or (more recently) mobile devices. Intelligent digital content such as SCAP, however, can be delivered to a broader range of targets for multiple purposes – not just to humans for reading – and, therefore, requires higher-precision data models and increased automation [25]. The increasing prevalence of intelligent content is causing content management to evolve from being mainly editorial in nature to a more engineering-focused pursuit [1].

The research discussed in this chapter leads to two questions that merit future study:

- How effective would the proposed information and authoring data models be in reducing the effort needed to develop and deploy SCAP source data stream collections in the hardware-in-the-loop testbed environment discussed in the introduction?

- Would expanding the scope of the information and authoring models to include low-level objects constituting `Benchmark` and `OVALDefs`, in addition to high-level concepts such as `DataStream` and `Component`, enable an authoring solution that is superior to existing approaches?

To answer the first question, the source data stream collection authoring application could be used to support a NIST-industry collaborative effort to establish best practices for securing industrial control systems in the manufacturing sector [27]. Two cyber security capabilities within the project scope – behavioral anomaly detection and industrial control application whitelisting – are addressable using SCAP. For example, a source data stream could check that AppArmor is installed and properly configured to protect an industrial control system from a software application hijacked by malware that causes the application to behave in an aberrant manner. As another example, a source data stream deployed in an industrial control device could enforce application whitelisting by checking if installed software packages are on an approved whitelist. The information model and authoring-model-based application could be used to assemble a source data stream collection from existing XCCDF rules and OVAL definitions for detecting behavioral anomalies and the presence of unauthorized software. The effort expended could then be compared against the effort required for manual source data stream collection, or against third-party software tools that might be available.

Answering the second question would require more effort than answering the first question and would involve the following modeling and implementation steps:

- Develop information models for XCCDF benchmarks and OVAL definitions. Based on these information models:
  - Create DITA specializations corresponding to XCCDF XML schema elements for representing benchmarks, profiles and rules.
  - Create DITA specializations corresponding to OVAL XML schema elements for representing definitions, criteria, tests and endpoint information.
- Implement an authoring and transformation application that assembles the collection of DITA documents representing the XCCDF and OVAL into a source data stream collection conforming to NIST SP 800-126.

The resulting implementation could then be compared against the current approach used to author and manage content for the SCAP Security Guide (SSG) [18], an open-source project whose output is a set of SCAP source data stream collections for Linux distributions and software applications. Contributors of SCAP Security Guide content use an *ad hoc* collection of tools created by the guide developers for authoring content such as XCCDF checklist rules and OVAL definitions. These tools enable contributors to use a shorthand XML syntax that is transformed into standards-conforming XCCDF and OVAL content, which in turn are transformed into a source data stream collection conforming to NIST SP 800-126. As discussed in [14], although the SCAP Security Guide authoring framework has proven successful in producing extensive and widely-used SCAP content, the framework and tools are complex, difficult for contributors to understand and hard for SCAP Security Guide developers to maintain. They also lack the validation capabilities of DITA document shells and authoring convenience of DITA-specialization-aware XML editing software.

Although the DITA-based approach shows promise [14], more thorough implementation and analysis are needed to determine whether or not the preliminary results are scalable to a larger and more representative corpus of security content.

5. Conclusions

This chapter describes two original research contributions: (i) a UML information model representing SCAP source data stream collections; and (ii) an authoring data model specialized from the DITA map element type and derived from the UML information model. The illustrative example involving the secure configuration of servers that control industrial robots demonstrates that the information model is easier to understand than the XML-based data model described in NIST SP 800-126, and is also better at expressing compositional relationships in a data stream collection. A DITA Open Toolkit plug-in implementation of the authoring data model provides a means for creating new SCAP content in an author-friendly manner and producing output that conforms to NIST SP 800-126. The review of related research reveals parallels with information models and data models developed for manufacturing systems and for healthcare enterprises, as well as with emerging trends in the field of content management.

The Industrial Internet of Things is spurring the need to secure an ever-growing variety of devices, operating systems and software. The diversity requires better tools than those currently available for SCAP content authors. The proposed source data stream collection information model and authoring model constitute a first step toward the development of SCAP authoring and content management solutions that meet the challenges.

This chapter is a contribution of the National Institute of Standards and Technology (NIST). Certain commercial and third-party products and services are identified in this chapter to enhance understanding. Such identification does not imply any recommendation or endorsement by NIST, nor does it imply that the materials or equipment identified are necessarily the best available for the purpose.

Acknowledgement

The author wishes to thank the individuals who provided helpful reviews of earlier drafts of this chapter, especially his NIST colleagues, David Waltermire and Timothy Sprock, for their insights regarding SCAP and information modeling.

References

[1] R. Andersen and T. Batova, The current state of component content management: An integrative literature review, *IEEE Transactions on Professional Communication*, vol. 58(3), pp. 247–270, 2015.

[2] M. Bauer, Paranoid Penguin: AppArmor in Ubuntu 9, *Linux Journal*, issue 185, September 1, 2009.

[3] H. Booth, M. Cook, S. Quinn, D. Waltermire and K. Scarfone, Security Content Automation Protocol (SCAP) Version 1.2 Content Style Guide: Best Practices for Creating and Maintaining SCAP 1.2 Content, NISTIR 8058 (Draft), National Institute of Standards and Technology, Gaithersburg, Maryland, 2015.

[4] T. Bray, The JavaScript Object Notation (JSON) Data Interchange Format, RFC 8259, 2017.

[5] DITA Open Toolkit Project, DITA Open Toolkit (`www.dita-ot.org`), 2018.

[6] T. Eggebraaten, J. Tenner and J. Dubbels, A health-care data model based on the HL7 Reference Information Model, *IBM Systems Journal*, vol. 46(1), pp. 5–18, 2007.

[7] C. Fairchild and T. Harman, *ROS Robotics by Example*, Packt Publishing, Birmingham, United Kingdom, 2016.

[8] Health Level Seven International, About HL7 International, Ann Arbor, Michigan (`www.hl7.org`), 2018.

[9] International Organization for Standardization, Industrial Automation Systems and Integration – Product Data Representation and Exchange – Part 242: Application Protocol: Managed Model-Based 3D Engineering, ISO 10303-242:2014, Geneva, Switzerland, 2014.

[10] International Organization for Standardization, Information Technology – Document Schema Definition Languages (DSDL) – Part 3: Rule-Based Validation – Schematron, ISO/IEC 19757-3:2016, Geneva, Switzerland, 2016.

[11] E. Kimber, *DITA for Practitioners, Volume 1, Architecture and Technology*, XML Press, Laguna Hills, California, 2012.

[12] S. Krima and J. Lubell, Flat versus hierarchical information models in PLM standardization frameworks, in *Product Lifecycle Management for Digital Transformation of Industries*, R. Harik, L. Rivest, A. Bernard, B. Eynard and A. Bouras (Eds.), Springer, Cham, Switzerland, pp. 121–133, 2016.

[13] B. Kulvatunyou, N. Ivezic and V. Srinivasan, On architecting and composing engineering information services to enable smart manufacturing, *Journal of Computing and Information Science in Engineering*, vol. 16(3), pp. 031002-1–031002-13, 2016.

[14] J. Lubell, Using DITA to create security configuration checklists: A case study, *Proceedings of Balisage: The Markup Conference*, vol. 19, 2017.

[15] J. Lubell and T. Zimmerman, Challenges to automating security configuration checklists in manufacturing environments, in *Critical Infrastructure Protection XI*, M. Rice and S. Shenoi (Eds.), Springer, Cham, Switzerland, pp. 225–241, 2017.

[16] Object Management Group, OMG Unified Modeling Language Version 2.5.1, Needham, Massachusetts (www.omg.org/spec/UML/2.5.1), 2017.

[17] Open Applications Group, OAGi Integration Specification Release 10.4, Marietta, Georgia (www.oagi.org), 2018.

[18] OpenSCAP Project, SCAP Security Guide: Baseline Compliance Content in SCAP Formats (github.com/OpenSCAP/scap-security-guide), 2018.

[19] Organization for the Advancement of Structured Information Standards, XML Catalogs v1.1, OASIS Standard, Burlington, Massachusetts (www.oasis-open.org/standards#xmlcatalogsv1.1), 2005.

[20] Organization for the Advancement of Structured Information Standards, Darwin Information Typing Architecture (DITA) v1.3, OASIS Standard, Burlington, Massachusetts (www.oasis-open.org/standards#ditav1.3), 2016.

[21] OVAL Project, OVAL Documentation (ovalproject.github.io), 2017.

[22] A. Pras and J. Schoenwaelder, On the Difference Between Information Models and Data Models, RFC 3444, 2003.

[23] M. Priestley and D. Schell, Specialization in DITA: Technology, process and policy, *Proceedings of the Twentieth Annual International Conference on Computer Documentation*, pp. 164–176, 2002.

[24] S. Radack and R. Kuhn, Managing security: The Security Content Automation Protocol, *IT Professional*, vol. 13(1), pp. 9–11, 2011.

[25] A. Rockley and J. Gollner, An intelligent content strategy for the enterprise, *Bulletin of the American Society for Information Science and Technology*, vol. 37(2), pp. 33–39, 2011.

[26] ROS Industrial Consortium, ROS-Industrial, San Antonio, Texas (rosindustrial.org), 2018.

[27] K. Stouffer and J. McCarthy, Capabilities Assessment for Securing Manufacturing Industrial Control Systems, Cybersecurity for Manufacturing, National Cybersecurity Center of Excellence, National Institute of Standards and Technology, Gaithersburg, Maryland, 2017.

[28] D. Waltermire, S. Quinn, H. Booth, K. Scarfone and D. Prisaca, The Technical Specification for the Security Content Automation Protocol (SCAP) Version 1.3, NIST Special Publication 800-126, Revision 3, National Institute of Standards and Technology, Gaithersburg, Maryland, 2018.

[29] D. Waltermire, C. Schmidt, K. Scarfone and N. Ziring, Specification for the Extensible Configuration Checklist Description Format (XCCDF), Version 1.2, NISTIR 7275, Revision 4, National Institute of Standards and Technology, Gaithersburg, Maryland, 2012.

[30] R. White, H. Christensen and M. Quigley, SROS: Securing ROS over the wire, in the graph and through the kernel, presented at the *IEEE-RAS International Conference on Humanoid Robots*, 2016.

[31] World Wide Web Consortium, Extensible Markup Language (XML) 1.0 (Fifth Edition), W3C Recommendation, Massachusetts Institute of Technology, Cambridge, Massachusetts (www.w3.org/TR/REC-xml), November 26, 2008.

[32] World Wide Web Consortium, Namespaces in XML 1.0 (Third Edition), W3C Recommendation, Massachusetts Institute of Technology, Cambridge, Massachusetts (www.w3.org/TR/xml-names), December 8, 2009.

[33] T. Zimmerman, Metrics and Key Performance Indicators for Robotic Cybersecurity Performance Analysis, NISTIR 8177, National Institute of Standards and Technology, Gaithersburg, Maryland, 2017.

III

INFRASTRUCTURE MODELING AND SIMULATION

Chapter 9

MODELING A MIDSTREAM OIL TERMINAL FOR CYBER SECURITY RISK EVALUATION

Rishabh Das and Thomas Morris

Abstract High-fidelity cyber-physical testbeds that mimic the cyber and physical responses of real-world systems are required to investigate the vulnerabilities of industrial control systems. This chapter describes the construction of a large, virtual, high-fidelity testbed that models a midstream oil terminal. The testbed models interconnected tank farms, a tanker truck gantry, a shipping terminal and a 150 km pipeline connection to a refinery. The virtual midstream oil terminal helps experiment with cyber attacks, explore the impacts of cyber attacks in order to prototype and evaluate security controls, and support education and training efforts. The virtual midstream oil terminal is constructed using a novel modular modeling technique that segments the overall system into the physical system, cyber-physical link, distributed controllers, communications network and human-machine interface. Simulation results involving normal operations and cyber attack scenarios are presented. The midstream oil terminal testbed demonstrates that large-scale models of industrial control systems for cyber security research are feasible and valuable.

Keywords: Cyber-physical testbed, oil terminal operations, risk evaluation

1. Introduction

This chapter describes the architecture of a virtual midstream oil terminal testbed. The testbed incorporates five distinct subsystem models: (i) physical system; (ii) cyber-physical link; (iii) programmable logic controller (PLC); (iv) network; and (v) human-machine interface (HMI). The virtual midstream oil terminal is a high-fidelity model of a real midstream oil terminal. The components in the physical system model adhere to American Petroleum Institute (API) standards. The programmable logic controller model is a software ver-

Published by Springer Nature Switzerland AG 2018.
J. Staggs and S. Shenoi (Eds.): Critical Infrastructure Protection XII, IFIP AICT 542, pp. 149–175, 2018.
https://doi.org/10.1007/978-3-030-04537-1_9

sion of OpenPLC [2], which is available in hardware or software. The network model, which is provided by a VMWare workstation, supports the Ethernet, TCP/IP and Modbus/TCP protocols. The human-machine interface is the SCADABr open-source software product, which has been used to monitor and control real and virtual industrial control systems. The human-machine interface is the same software that is used in real midstream oil terminals.

The virtual midstream oil terminal testbed models three tank farms, a tanker truck gantry, a shipping terminal with two ocean-going oil tankers and a 150 km pipeline that is connected to a refinery. The three tank farms hold three liquid petroleum products: (i) gasoline; (ii) diesel; and (iii) aviation turbine fuel (ATF). The gasoline and diesel tank farms have four fixed/floating roof tanks each while the aviation turbine fuel tank farm has three dome roof tanks. Each tank farm includes a network of pipelines that supports recirculation, filling from external sources and transfers to the tanker truck gantry. Each tank farm also includes a set of pumps to move liquid cargo.

The tanker truck gantry incorporates three tanker truck models, each tanker truck with two internal tanks. The trucks must be grounded to initiate a fill operation.

The shipping terminal supports loading and unloading operations. Each ocean-going tanker has six internal tanks. The 150 km pipeline system includes a graduated pipeline that maintains pressure throughout the length of the pipeline.

In total, the physical system model incorporates 217 modeled sensors and actuators. Twelve networked programmable logic controllers are connected to the physical system model to implement distributed control. The programmable logic controllers communicate via Modbus/TCP over a TCP/IP network to the human-machine interface. The human-machine interface remotely polls the programmable logic controllers for system state information and provides supervisory control capability.

The high-fidelity testbed can be used to conduct cyber security research at a larger scale than most industrial control system testbeds available to researchers. Users can simulate cyber attacks and examine the impacts on physical system components. The scale of the virtual midstream oil terminal testbed enables researchers to model cyber attacks that exploit multiple components simultaneously or in sequence. This flexibility supports the reproduction of large-scale and cascading events, as well as analyses of the interdependencies existing between systems. Researchers can also use the pipeline testbed to prototype and evaluate the effectiveness of new cyber security controls.

Cyber security researchers often need data captured from industrial control systems during normal and cyber attack situations. Most industrial control system operators either do not have such data or will not share their data for reasons of sensitivity. The virtual midstream oil terminal can be used to produce the data required for research. Additionally, since the testbed is virtual, the testbed itself and the scripts used to generate interesting cyber attacks in the testbed are readily shared.

The virtual midstream oil terminal can be distributed electronically and can run on virtual machines in a cloud computing environment. This makes the testbed very useful for education and training. Students can use the virtual testbed to explore the functionality of industrial control systems, experiment with cyber attacks and evaluate security controls.

Modeling energy sector systems is highly relevant to cyber security research. Malfunctions of critical components such as oil terminals, pipelines, storage tanks and cargo vessels can cause fires, explosions or harm to the environment, which can impact energy supply and lead to large economic losses. In 2008, hackers successfully suppressed alarms and penetrated the communications network of the Baku-Tbilisi-Ceyhan pipeline [15]. The attack essentially blinded pipeline system operators. The pipeline was intentionally over-pressurized by the hackers, resulting in a rupture and explosion that spilled more than 30,000 barrels of crude oil. It took 24 hours to extinguish the resulting fire and the entire pipeline was not functional for eighteen days. This incident led to a serious political conflict between Georgia and Russia. In 2012, the Shamoon virus, released by the hacktivist group Cutting Sword of Justice, destroyed 30,000 computers at Saudi Aramco, which supplies 10% of the world's oil [11]. Saudi Aramco was forced to work offline for five months.

2. Related Work

Oil terminals and refineries are critical infrastructure assets that demand high operational vigilance. A malfunction, such as a pipeline rupture or vapor leak, can release a cloud that can ignite and cause a large fire or explosion. Zhou et al. [22] have performed an extensive study of 435 fire and explosion accidents in China. Sixty-six major fires and explosions occurred between 2000 and 2013, causing a total of 390 deaths and 950 injuries. The study also reveals that 76.09% of the accidents were caused by vapor clouds from fuel leaks, pipeline ruptures and mechanical failures.

Several power system testbeds have been developed for simulating cyber attacks against power systems [12]. The Testbed for Analyzing Security of SCADA Control Systems (TASSCS) has been developed to evaluate the effects of eight types of cyber attacks [16]. It provides a high-fidelity simulation of a SCADA network that uses the Modbus and DNP3 protocols. TASSCS does not simulate programmable logic controllers; instead, a Modbus server is hosted on a control server. As a result, vulnerabilities associated with programmable logic controllers cannot be examined using TASSCS.

Adhikari et al. [1] have developed a testbed specifically for cyber security research related to bulk electricity transmission systems. The testbed implements wide-area measurement functionality using a real-time digital simulator, hardware-in-the-loop protection relays, phasor measurement units and phasor data concentrators. However, the testbed does not incorporate any programmable logic controllers.

Morris et al. [18] have developed a high-fidelity gas pipeline testbed for collecting data for intrusion detection research. The testbed is modular and portable, but Python programs are used for control instead of employing simulations of actual programmable logic controllers.

DeterLab is a power system testbed used by more than 2,600 researchers [17]. It incorporates 400 general purpose computing nodes and supports simulations of cyber attacks such as SQL injection, TCP SYN flooding and worms. DeterLab enables high-fidelity simulations, but its architecture is not modular. The security of a power system can be analyzed as a whole; however, researchers interested in analyzing specific industrial control system problems such as programmable logic controller functionality, SCADA network communications and physical system vulnerabilities cannot use this testbed. Additionally, the computing power required to operate the testbed significantly reduces its portability.

At this time, no published research exists related to midstream oil terminal testbeds. Therefore, the virtual high-fidelity testbed that models a midstream oil terminal should be of considerable interest to researchers. The testbed simulates real-world programmable logic controllers and is also lightweight and portable.

3. Testbed Architecture

This section describes the architecture of the virtual midstream oil terminal testbed.

3.1 Virtual Testbed Modular Framework

The midstream oil terminal testbed is implemented using a modular framework that is capable of modeling any SCADA system. The framework organizes a SCADA system in terms of five major components: (i) physical system; (ii) cyber-physical link; (iii) digital control system; (iv) communications network; and (v) human-machine interface. Each of the five major components is replaced by a virtual counterpart.

Figure 1 shows how each modularized component of a SCADA system is replaced by its equivalent virtual counterpart. The modular architecture described in this section and used to implement the midstream oil terminal testbed was also employed by Alves et al. [3] to model a laboratory-scale gas pipeline. Alves and colleagues compared a physical gas pipeline against a virtual model of the same pipeline. They demonstrated that the virtual testbed provided high simulation accuracy for normal operations as well as for cyber attack scenarios.

The physical system is an operational system such as an oil terminal, power system, chemical plant or manufacturing plant. In the virtual model, the physics and operational dynamics of the physical system are simulated via Simulink, a graphical programming environment for simulating, analyzing and modeling multi-domain dynamic systems. Simulink provides toolkits that model a variety of physical system components. The physical system model also includes sensors and actuators. Sensors are modeled in Simulink by connecting internal signals to probes. Actuators are modeled by connecting binary inputs

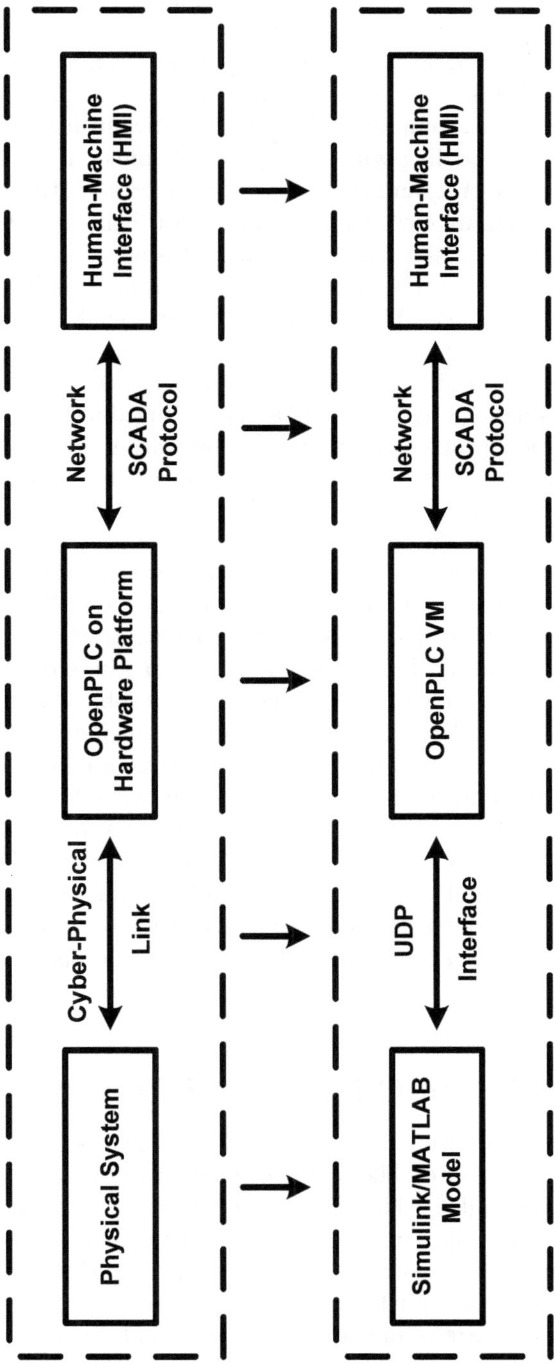

Figure 1. SCADA components and their virtual counterparts.

from the cyber-physical link to control physical components such as valves and switches. The physical system may be modeled using tools other than Simulink when appropriate.

Sensors and actuators are connected to the programmable logic controller via cyber-physical links. A cyber-physical link is as simple as a wire or it may use sensor network communications technologies such as WirelessHart and Zigbee [21]. When modeling wires, the physical connectivity between sensors and actuators and a programmable logic controller is virtualized using UDP sockets. In a real system, each sensor and actuator is independently connected by wires to a programmable logic controller. Likewise, in the virtual model, each sensor and actuator communicates with a programmable logic controller using a unique UDP port. The unique UDP ports enable the programmable logic controller to maintain separate communications with each sensor and actuator, thereby maintaining fidelity with the physical system.

A programmable logic controller is a computing device that monitors and controls the physical process and provides a network link for supervisory monitoring and control at a control center. It connects to sensors and actuators via cyber-physical links. The virtual testbed models programmable logic controllers using OpenPLC [2]. OpenPLC is open-source programmable logic controller software that supports all five IEC 61131-3 standard programming languages and the Modbus/TCP and DNP3 protocols. OpenPLC supports a wide variety of hardware platforms. In the case of a virtual testbed, software versions are executed in virtual machines using Windows or Linux operating systems.

The human-machine interface is a dedicated graphical user interface used by operators to remotely monitor and supervise an industrial process. The human-machine interface communicates with programmable logic controllers using standard communications protocols and provides the operator with the real-time status of the physical system. The human-machine interface may run on a virtual machine or on a separate host computer. Communications between a programmable logic controller and human-machine interface can employ virtual networking provided by a hypervisor or a real network. The human-machine interface software and application-specific user interface for the process control system are typically the same for real-world and virtual versions.

3.2 Midstream Oil Terminal Testbed

The midstream oil terminal testbed was implemented using the modular framework described above. The physical system was modeled using the Simulink SimHydraulics toolkit, which provides constructs for modeling pipes, bends, valves and other hydraulic components. The exact configurations of the various physical system sub-components are described later in this chapter.

The physical system model incorporates 217 sensors and actuators. The sensors and actuators are connected to twelve virtual programmable logic controllers using a virtual wire bridge with a UDP socket for each sensor and actuator. Each virtual programmable logic controller is an OpenPLC instance

Table 1. Components controlled by the programmable logic controllers.

PLC	Controlled Component
1	Marine tanker pipeline loading
2	Marine tanker pipeline unloading
3	Pipeline transfer operation
4	Oil tanker discharging
5	Marine tanker loading
6	Tanker truck gantry
7	Gasoline pump house
8	Diesel pump house
9	Aviation turbine fuel pump house pipeline
10	Gasoline tank farm
11	Diesel tank farm pipeline
12	Aviation turbine tank farm pipeline

that runs on a Debian virtual machine. The programmable logic controller programmation was developed using ladder logic. The actuators and sensors in the Simulink model of the virtual oil terminal communicate with the programmable logic controllers using a software interface hosted by the PLC 1 virtual machine. The software interface distributes the sensor readings to the programmable logic controllers and delivers control commands and information from the programmable logic controllers to the Simulink model.

The midstream oil terminal human-machine interface was created using SCADABr, an open-source, web-based, human-machine interface development environment. The Modbus/TCP protocol is used for communications between the human-machine interface and programmable logic controllers. The attack scenarios simulated in this research assume that the attacker is physically connected to the network that houses the programmable logic controllers. Since programmable logic controllers enable clients to connect to them without authentication, an attacker can connect to any programmable logic controller and query the status of the registers and coils.

Figure 2 presents a high-level layout of the simulated testbed. Table 1 lists each of the twelve programmable logic controllers and the component it controls.

4. Standards and Components

Oil and gas sector operations are divided into three sectors: (i) upstream; (ii) midstream; and (iii) downstream. The upstream sector generally involves exploration and drilling to locate and recover crude oil and natural gas. The midstream sector moves the materials from remote production locations to population centers. The downstream sector refines the materials into petroleum

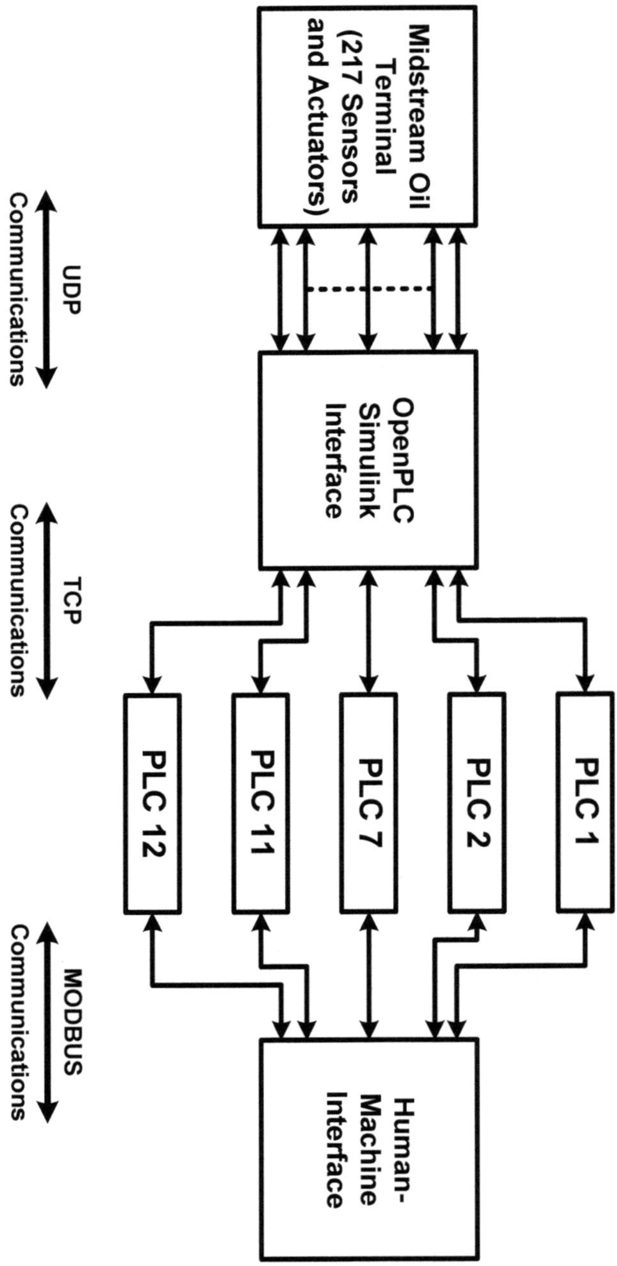

Figure 2. Simulated testbed.

Table 2. Midstream oil terminal component specifications.

Standard	Description
API SPEC 5L [6]	Pipeline specifications (tank farm)
API SPEC 6D [8]	Pipeline valve specifications
API SPEC 6H [5]	Pipeline connector specifications
API SPEC 11L6 [4]	Motor and pump specifications
API SPEC 12B [7]	Liquid cargo tank specifications
API RP 1007 [9]	Tanker truck specifications
API RP 1109 [10]	Pipeline transfer operation specifications

products and distributes the products to the retail market. Tanker trucks, marine tankers, pipelines and storage terminals are employed in all three sectors.

Figure 3 shows an overview of the virtual midstream oil terminal. The midstream oil terminal stores gasoline, diesel and aviation turbine fuel (ATF). Each of the three tank farms has a pump house. The tank farms are connected to a tanker truck gantry, which loads fuel into tanker trucks. The tank farms also load and unload marine tankers (MTs). The tank farms are connected to the marine tankers via a 12 km pipeline. The tank farms are also connected to a shore refinery via a 150 km cross-country pipeline. The network of pipelines and valves is abstracted in Figure 3.

4.1 Midstream Oil Terminal Standards

The American Petroleum Institute (API) promulgates standards for oil terminal equipment and components. The relevant American Petroleum Institute standards were followed to achieve high fidelity between the simulated model and a real midstream oil terminal. Table 2 lists the standards used in the simulation. The specifications and operational guidelines for marine tanker operation documented in the International Safety Guide for Oil Tanker and Terminals (ISGOTT) [14] were also used in the simulation.

4.2 Midstream Oil Terminal Components

This section provides detailed descriptions of the major components and activities of the midstream oil terminal: (i) tank farms; (ii) pump houses; (iii) tanker truck gantry; (iv) pipeline transfer; and (v) vessel operation.

Tank Farms. A tank farm is a network of tanks, valves, pumps and pipes that stores cargo in an oil terminal. The tank farms form the core of a midstream oil terminal because all terminal operations are either from or to tank farms. The presence of a fuel-air mixture makes a tank farm susceptible to fire and explosion due to the storage of volatile cargoes such as diesel, gasoline and aviation turbine fuel. According to a case study performed by Zhou et al. [22],

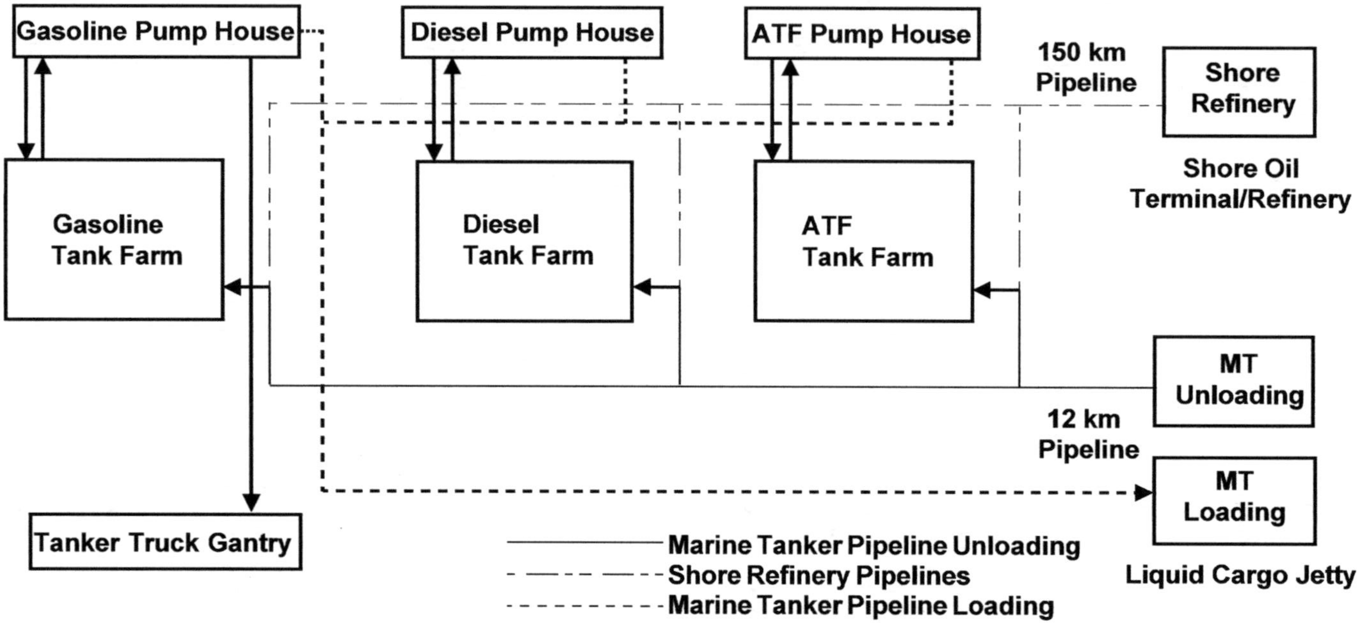

Figure 3. Midstream oil terminal subsystems with pipeline connections.

Table 3. Tank farm specifications.

	Gasoline Tank Farm	**Diesel Tank Farm**	**ATF Tank Farm**
Number of Tanks	4	4	3
Type	Fixed/floating roof	Fixed/floating roof	Dome roof
Height	15 m	15 m	18 m
Diameter	20 m	20 m	18 m
Inlet	16 in	16 in	18 in
Outlet	18 in	18 in	20 in
Inlet/Outlet	16 in	16 in	16 in

76.09% of major accidents in oil terminals were due to the presence of a fuel-air mixture and 25.75% of major accidents originated in tank farms. Due to the critical nature of a tank farm, a number of standards are adopted to ensure safe operation. API SPEC 5L [6] and API SPEC 12B [7] specify tank farm pipeline and valve configurations, respectively.

The modeled midstream oil terminal has three tank farms, one each for gasoline, diesel and aviation turbine fuel. Volatile cargoes such as diesel and gasoline are susceptible to vapor loss [19]. API SPEC 12B [7] requires the use of fixed roof or floating roof tanks for storing these cargoes. Aviation turbine fuel is a type of superior kerosene oil with quality standards that require less than 15 ppm of water to be present in stored or dispatched fuel [20]. To adhere to these requirements, fixed and floating roof tanks cannot be used; instead, dome roof tanks with fixed ceilings are employed for storage.

Table 3 lists the numbers of tanks, tank types, tank heights, tank diameters, inlet diameters, outlet diameters and inlet/outlet diameters for the tank farms modeled in Matlab Simulink for the virtual midstream oil terminal. There are three tank farms in the model, one each for gasoline, diesel and aviation turbine fuel. The gasoline and diesel tank farms have four tanks each while the aviation turbine fuel tank farm has three tanks. The tanks are named according to ISGOTT naming conventions [14]. Each tank is named TK followed by the tank farm number and tank number. For example, the first tank in the diesel tank farm is TK 21 and the second tank in the aviation turbine fuel tank farm is TK 32.

Each tank has three dedicated pipeline connections: (i) receipt; (ii) dispatch; and (iii) recirculation. The receipt pipeline receives cargo from a marine tanker or from the shore terminal via a pipeline transfer. The dispatch pipeline connection is used as an outlet; this pipeline transfers cargo out from the tank to a tanker truck, marine tanker or another tank. The recirculation pipeline connection is used for operations within the tank farm. Operations such as inter-tank transfers using gravity or pumps are performed using the recirculation connection. The recirculation connection can be used as a tank inlet or outlet.

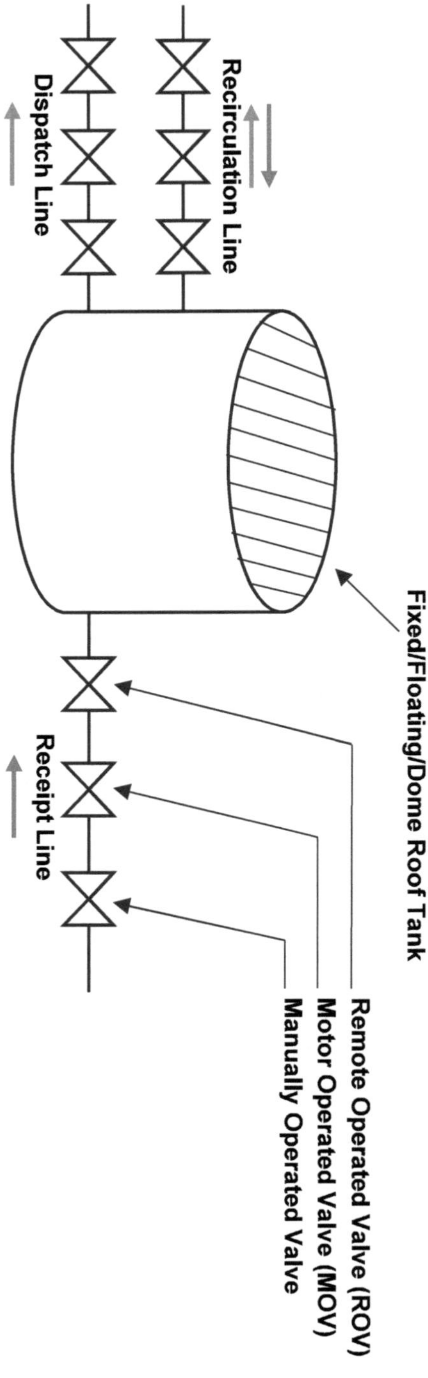

Figure 4. Typical simulated tank.

Table 4. Pump house specifications.

	Gasoline Pump House	Diesel Pump House	ATF Pump House
Pump Specifications	Centrifugal 1 × 100 m³/h 2 × 200 m³/h 2 × 500 m³/h	Centrifugal 1 × 100 m³/h 3 × 250 m³/h 1 × 500 m³/h	Centrifugal 3 × 250 m³/h
Inlet	16 in	16 in	18 in
Outlet	20 in	20 in	24 in
Drive Motor Specifications	Induction 1 × 40 kW (79 A) 2 × 90 kW (180 A) 2 × 200 kW (345 A)	Induction 1 × 40 kW (79 A) 3 × 110 kW (192 A) 1 × 200 kW (345 A)	Induction 3 × 110 kW (192 A)

According to the Oil Industry Safety Directorate (OISD) Standards 169, 118 and 129 and the recommendation by Lal et al. [13], three types of valves, each controlled by a different actuation mechanism, should be used between each tank and its pipeline connection. Hence, in the virtual midstream oil terminal model, each pipeline connection to a tank incorporates three valves. The valve closest to the tank is controlled pneumatically, the second valve is electrically actuated using a motor and the third valve is operated manually. Figure 4 shows a typical modeled tank with three pipeline connections and valves. The pneumatic valve, labeled remote operated valve (ROV), and the motor operated valve (MOV) can be operated remotely from the human-machine interface. The manual valve is operated physically. In the Matlab Simulink model, manual valves are operated by toggling a switch manually.

Pump House. The pump house is the heart of the midstream oil terminal. Each tank farm has a dedicated pump house. The gasoline, diesel and aviation turbine fuel pump houses have five, five and three pumps respectively. The modeled pumps are of various sizes and can be connected in parallel to achieve the desired flow rate. The valves in the pump houses can be remotely configured to dispatch cargo from tanks to marine tankers, tanker trucks or to other tanks in the tank farm. The gasoline and diesel pump houses have dedicated pipelines for transferring cargo to the tanker truck gantry. Per API SPEC 11L6 [4], the pumps use three-phase induction motors that deliver constant torque via a universal coupling connected through a common shaft to the centrifugal pumps. Table 4 shows the detailed specifications for the pumps in the virtual midstream oil terminal.

Tanker Truck Gantry. The tank truck gantry is the most operationally active area of the terminal. The presence of moving trucks and open volatile

Table 5. Tanker truck loading bay specifications.

	Bay 1	**Bay 2**	**Bay 3**
Cargo	Gasoline	Diesel	Gasoline and Diesel
Loading Arm	2×6	2×6	1×6 1×6
Bay	Single cargo express loading bay	Single cargo express loading bay	Mixed cargo loading bay
Valve	Butterfly valve for flow regulation		
Tanker Trucks	2×6 kl tankers Safety features include overfill sensors, tanker truck ground connections, flow regulators for loading arms		

cargoes makes this area susceptible to fires and explosions. More than 51% of major accidents in oil terminals originate in tanker truck gantries [22].

A tanker truck gantry typically has several loading zones with dedicated loading arms for transferring liquid cargoes into tanker trucks. The allowable cargo capacity in a tanker truck is between 2,000 and 16,000 gallons (7,570 and 49,205 liters). At least 3% of a tank must be left empty to provide space for product expansion.

A tanker truck gantry with three loading bays is modeled in the virtual midstream oil terminal. One bay is allocated for gasoline, the second bay is for diesel and the third mixed bay can load gasoline or diesel. Aviation turbine fuel cannot be loaded on a truck.

Each modeled tanker truck has two internal 6 kl tanks. API RP 1007 [9] states that the body of a tanker truck must be electrically grounded during loading operations to prevent static charge accumulation in the tanker truck. Therefore, each modeled tanker truck bay has sensors connected to a programmable logic controller that detects if the tanker truck is not grounded correctly. The programmable logic controller prevents the loading operation if the truck is not electrically grounded. The tanker truck gantry programmable logic controller also regulates product flow using a butterfly valve. An overfill sensor connected to the programmable logic controller stops product flow when the tank truck is full. Table 5 provides the specifications of the three modeled loading bays.

Cross-Country Pipeline. The virtual midstream oil terminal testbed models a 150 km underground cross-country pipeline from a shore-based oil refinery to the tank farm. Pipeline transfer is a cost efficient and safe way to transfer liquid cargo over long distances. Operational hazards are minimized because the volatile cargo is never exposed to the ambient environment. Due to the length of a pipeline, remotely-monitored sensors provide pipeline state in-

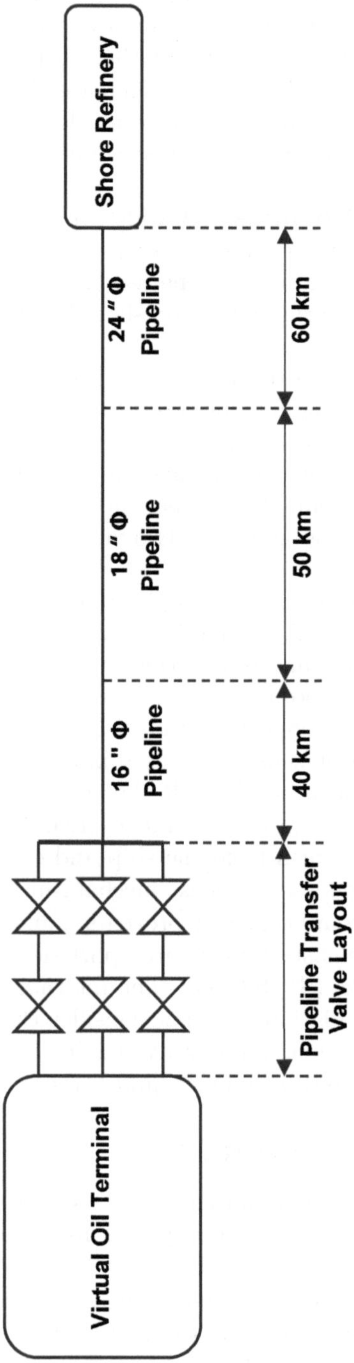

Figure 5. Cross-country pipeline.

formation to operators. A cyber attack that spoofs pipeline sensor readings can disrupt and harm the pipeline transfer operation [15]. The modeled pipeline complies with API RP 1109 [10]. Multiple flow rate and pressure sensors are modeled to enable remote monitoring of the status of the pipeline transfer operation. The diameter of the pipeline decreases farther from the source to compensate for the drop in pressure due to the long-distance pumping operation. Figure 5 shows the layout of the cross-country pipeline.

Terminal-to-Jetty Pipelines. A wide array of liquid and liquefied gas cargoes are transferred across large distances using marine tankers. Marine tanker loading and unloading require the use of many cyber-physical systems, including a marine loading arm (MLA), on shore holding tanks, pumps, on-ship tanks on-ship pipelines.

The testbed simulates two terminal-to-jetty 12 km pipelines. One pipeline is dedicated to vessel loading and the other to vessel unloading. ISGOTT [14] has published safety regulations for oil tanker cargo operations. During cargo operation, double-wall segregation of valves is mandatory, i.e., two valves must separate the operating pipeline from other pipelines. As a result, the modeled terminal-to-jetty pipeline has six valves, two for each cargo type on the terminal side as shown in Figure 6.

The terminal-to-jetty pipelines are coupled to the manifolds of marine tankers using marine loading arms. A marine loading arm is a sophisticated pipeline that connect the shore pipeline to a marine tanker to facilitate cargo transfer. A marine loading arm incorporates safety features that prevent oil spillage and offer a mechanism for the connection and disconnection of the shore pipeline and marine tanker. Position sensors are used in a marine loading arm to sense the orientation of the marine tanker. If the ship drifts away from the jetty, an emergency valve called a power emergency release coupling is actuated to release the marine loading arm from the ship and close the pipeline valves to prevent spillage. This emergency release mechanism is crucial to the dynamic jetty-vessel coupling system because it prevents damage to the loading arm.

Each simulated ship has six tanks; three port-side tanks (P1, P2, P3) and three starboard-side tanks (S1, S2, S3). The internal pipeline connections are not modeled and the simulation does not consider the effects of ballast tanks and ballasting operations that pump sea water into and out of a ship to compensate for the outgoing and incoming liquid cargo.

5. Simulation Results

The midstream oil terminal can simulate a variety of normal cargo operations. The supported normal cargo operations include inter-tank, tank-to-tanker-truck, tank-to-ship, ship-to-tank and refinery-to-tank transfers. In addition to normal cargo operation simulations, cyber attacks can be launched against the cyber systems modeled in the midstream oil terminal. This section describes the simulation results obtained for inter-tank transfers using gravity and using pumps under normal and attack scenarios. Other normal and attack

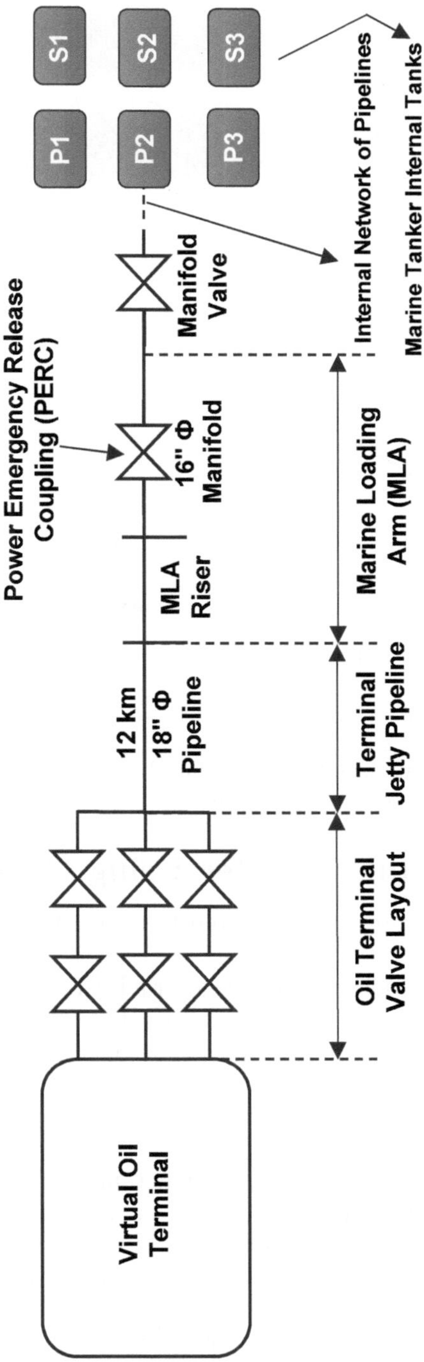

Figure 6. Marine loading operation.

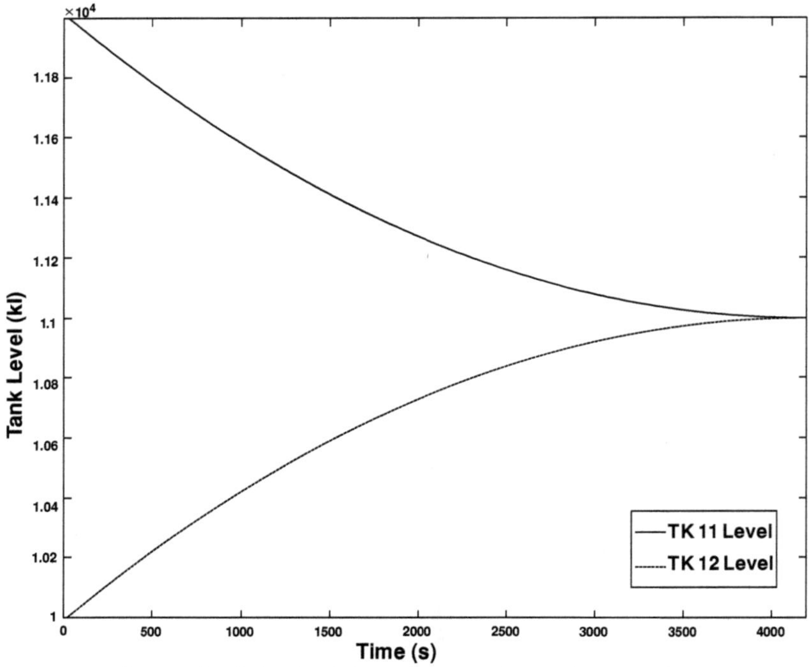

Figure 7. Inter-tank transfer operation using gravity.

scenarios have been simulated and validated using the testbed, but they are not described here for reasons of brevity.

5.1 Inter-Tank Transfer Using Gravity

Inter-tank transfer moves liquid cargo from one tank to another one in a tank farm.

An inter-tank operation may leverage gravity (head) associated with the difference in the liquid levels in the two tanks. For an inter-tank transfer using gravity, the valves between the two tanks are opened to enable cargo to flow from the tank with the higher liquid level to the tank with the lower liquid level. Over time, the tanks reach equilibrium, at which point both the tanks have the same liquid level.

Figure 7 shows the liquid levels in gasoline tanks TK 11 and TK 12 observed from the human-machine interface during an inter-tank transfer operation. Three valves (remote operated, motor operated and manual) in the recirculation pipeline of each tank are involved in the inter-tank transfer. All three valves are opened to initiate transfer and may be closed at any time during the transfer. Figure 7 shows that the flow rate between tanks is not constant. In fact, the flow rate is dependent on the difference between the liquid cargo levels in the tanks – the greater the difference in levels, the greater the flow rate.

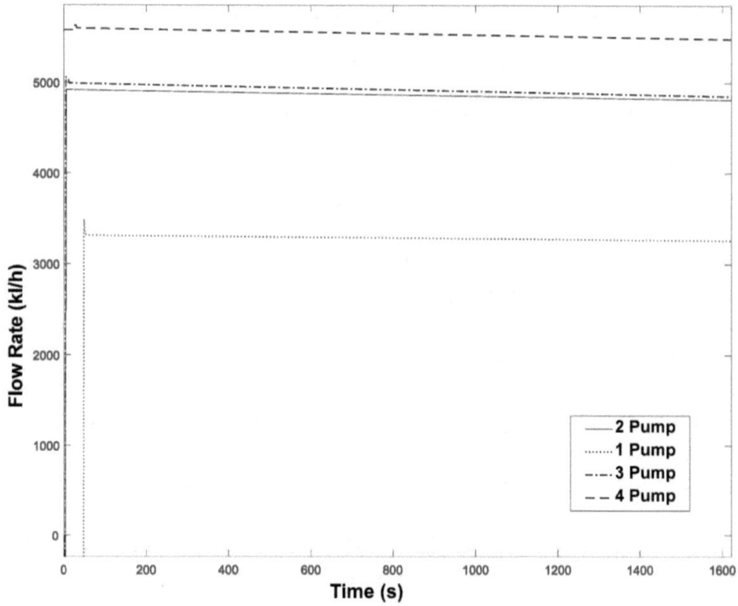

Figure 8. Inter-tank transfer operation using centrifugal pumps in parallel.

5.2 Inter-Tank Transfer Using Pumps

In some cases, the difference in liquid levels in the two tanks (head) may not be adequate to facilitate the transfer of cargo with a sufficient flow rate, or the transfer of cargo may have to go against gravity. In these cases, an inter-tank transfer is accomplished using pumps. To facilitate the operation, the human-machine interface is used to connect the dispatch pipeline of the source tank to the inlets of the relevant pumps and the pump outlets are connected to the recirculation connections of the destination tanks. The human-machine interface is used to start and stop the pumps at the beginning and end of the operation, respectively.

Figure 8 shows sensor readings from the inlet flow rate sensor at the destination tank during inter-tank transfer operations. The inter-tank transfer was repeated four times with one, two, three and four pumps working in parallel to complete the transfers. The graphs are labeled with the numbers of pumps used for the operations. When a single pump is used, a delay of 20 to 30 seconds occurs between the start of the transfer operation and the increase in the flow rate observed at the tank inlet. The delay is primarily because the air inside the pipeline must be pushed out before the cargo can flow. When multiple pumps are used, the air is pushed out much faster, causing the flow rate to increase at a faster rate, which appears to be instantaneous in Figure 8. As the number of pumps used increases, a higher flow rate is seen due to the accumulation of flow from more pumps in parallel. The three-pump case has a slightly higher flow rate than the two-pump case because the third pump has a low rating of $100\,m^3/h$.

5.3 Cyber Attack Scenarios

The midstream oil terminal testbed can be used to simulate network-borne attacks that target programmable logic controllers or the human-machine interface, physical attacks against process components, attacks that alter the programmable logic controller programmation or firmware, attacks that alter the human-machine interface programmation and other attacks on the human-machine interface executables (e.g., buffer overflows and database injection attacks). During a simulated attack, all the testbed components continue to simulate the system, enabling the behavior of the system to be observed and analyzed. This section describes man-in-the-middle (MiTM) and denial-of-service (DoS) attacks during a tanker truck loading operation. Also, it discusses an injection attack against a tank valve during a tanker truck loading operation.

Man-in-the-Middle and Denial-of-Service Attacks. This section presents the simulation results for two cyber attack scenarios. The first is a man-in-the-middle attack during a tanker truck loading operation, which alters sensor data in transit between the programmable logic controller and human-machine interface. This causes the human-machine interface to present incorrect sensor data to the operator. The second attack is a volumetric denial-of-service attack on the human-machine interface. This attack causes the human-machine interface to stop polling the programmable logic controller for system state updates. The actual process state and the state presented by the human-machine interface are plotted for the two attacks. These scenarios highlight the ability of the virtual midstream oil terminal testbed to model network-borne cyber attacks and the ability to observe the actual physical system state and the state as seen by the human-machine interface.

Figure 9 shows the flow rates measured by a sensor in the tanker truck gantry during a tanker truck loading operation. One curve shows the flow rate observed at the human-machine interface and the other shows the actual flow rate. The majority of Figure 9 shows the normal tanker truck loading operation. However, the effects of the two cyber attacks are also observed.

The first attack occurred between 100 and 270 seconds. During this time period, the man-in-the-middle attack compromised the link between the human-machine interface and programmable logic controller, and altered the flow rate measurements transmitted from the programmable logic controller to human-machine interface. Ettercap was used to perform the ARP spoofing attack.

A man-in-the-middle attack is especially dangerous for a pipeline. In the Baku-Tbilisi-Ceyhan pipeline incident [15], attackers suppressed alarms, altered system control in order to affect the process state and blinded operators who were monitoring the pipeline. The man-in-the-middle attack can be used to inject, alter or drop network traffic between the human-machine interface and programmable logic controller in both directions. Injecting control packets

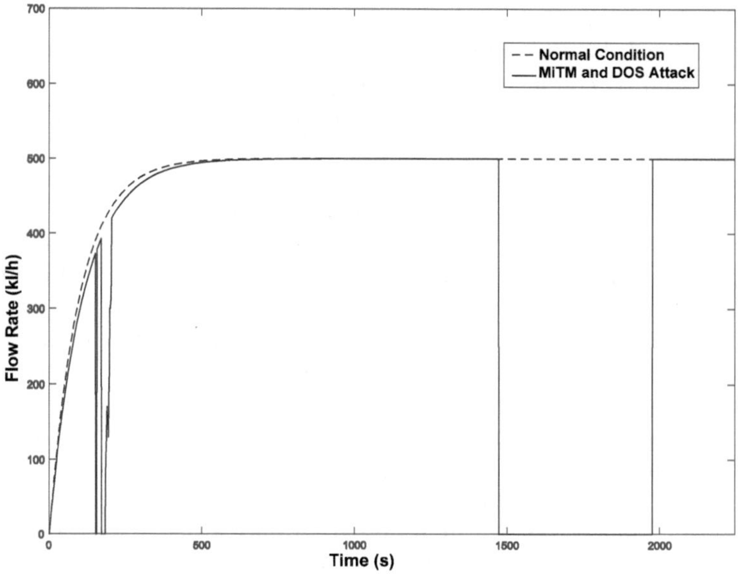

Figure 9. Tanker truck loading flow rates during normal and attack conditions.

can change the process state; altering and/or dropping sensor data can blind operators and upstream controllers. Figure 9 shows that, between 100 and 270 seconds, the flow rate sensor data was altered to show large (spurious) fluctuations in the flow rate. Such an attack could induce the operator to initiate a supervisory control action based on the false sensor data and ultimately move the physical process into an unsafe state.

The second attack involving volumetric denial-of-service occurred between 1,470 and 1,950 seconds. During this time period, the attack targeted the human-machine interface. The open-source Hping3 software was used to perform the attack. Figure 9 shows that from 1,470 to 1,950 seconds, a flow rate of 0 kl/h was presented by the human-machine interface while the actual flow rate remained at 500 kl/h. During the attack, the human-machine interface was overwhelmed and was unable to query the programmable logic controller in order to obtain the current state of the process. This attack prevented the operator from receiving the true state of the system.

Injection Attack. Liquid cargo operations in an oil terminal often involve multiple subsystems. For example, the tanker truck loading operation involves the tank farm, pump house and tanker truck gantry. The liquid cargo stored in a tank farm is transferred into the internal tanks of the tanker trucks using the centrifugal pumps in the pump house. The state reflected by the simulation at any given instant during the cargo operation considers the states of all the interconnected subsystems (tank farm, pump house and tanker truck gantry, vessel operation and pipeline transfer) in the oil terminal. Therefore, during a tanker truck loading operation, if an attacker manages to sabotage any of the oil terminal components, the effects of the attack may be evident across multiple interdependent subsystems. This section discusses the impact of an injection attack on a tanker truck loading operation when the dispatch valve of the gasoline tank in the tank farm is compromised by an attacker.

Three pressure sensors and three flow rate sensors were used to observe the system state. Sensors were positioned at the inlet and outlet of each centrifugal pump, and at the inlet of the loading arm of the tanker truck. Figure 10 shows the normal flow rate (kl/h) at three distinct locations during a cargo transfer operation. The flow rates at the inlet and outlet of the pump rise almost instantaneously and attain a steady state value of 270 kl/h. Since the tanker truck is located some distance away from the pump house, the rise in the flow rate at the tanker truck gantry is delayed. When the cargo reaches the tanker truck, the initial rush produces a spike in the flow rate, which is followed by a drop to the steady state flow rate of 270 kl/h at the loading arm.

Figure 11 shows the measured pressure values at three locations. The pump inlet has the lowest steady state pressure of 1.18 bar while the pump outlet has the highest pressure of 1.8 bar. The difference in pressure is due to the boost provided by the centrifugal pump. After the cargo reaches the pipeline, it starts losing pressure as it travels along the pipeline. When it reaches the tanker trunk gantry, a lower steady state pressure of 1.6 bar is measured at the loading arm. Note that the spikes in pressure measured by the three sensors at the start of the cargo transfer operation are due to the pressure build up in the pipeline.

During the simulated injection attack, the attacker compromised the motor operated valve in the dispatch pipeline of tank TK 12. The valve was toggled three times during the cargo operation, creating spikes in the flow rate and pressure that are unsafe for pipelines and valves. A Python script using the `pymodbus3` library was used to craft the injection packets. A separate attack node, a virtual machine running Kali Linux, was added to the network connecting the human-machine interface and programmable logic controller. The commands to open and close the valves were sent to the programmable logic controller from the attack node. The attack node injected packets every 50 ms. The human-machine interface was configured to send commands that set the states of all the actuators, including the valve, every 500 ms. During each attack session, the attacker closed the valve, waited for two seconds and then reopened the valve. Because the attacker sent commands at a faster rate and the valve has a relatively high latency to open and close, a command to set the valve actuator state sent by the human-machine interface was overridden quickly by the attacker node.

During the first injection, between 37 and 39 seconds, a spike in the flow rate is observed at all three sensors (Figure 12). Similarly, pressure values of 13.2 bar and 11.2 bar are measured at the tanker truck pump outlet and pump inlet, respectively. Since the dispatch valve of the gasoline tank is closed during

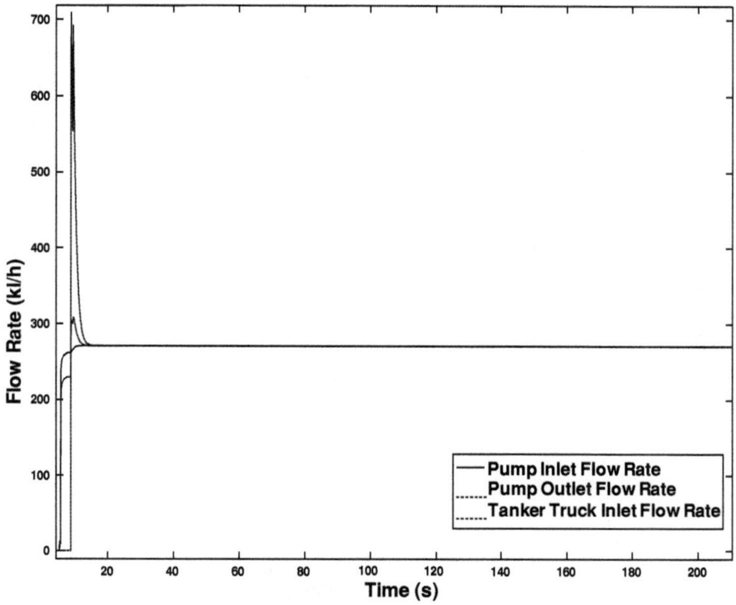

Figure 10. Tanker truck loading flow rate (normal conditions using a single pump).

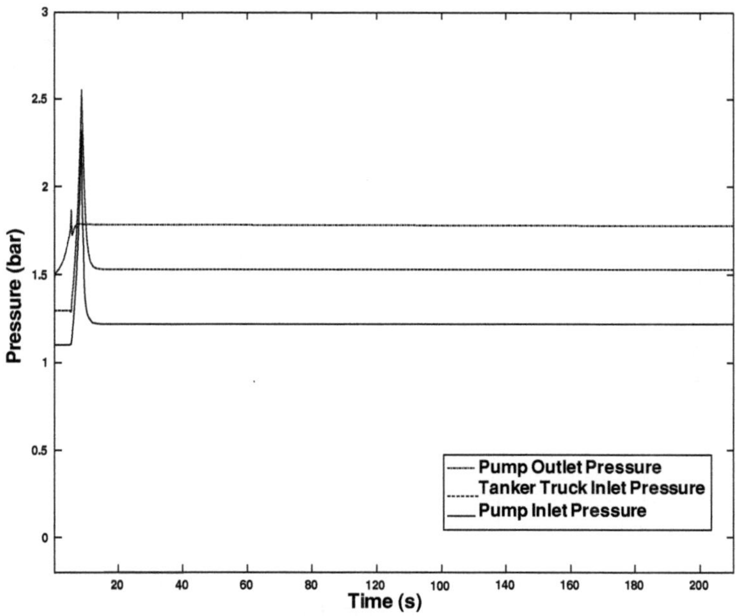

Figure 11. Tanker truck loading pressure (normal conditions using a single pump).

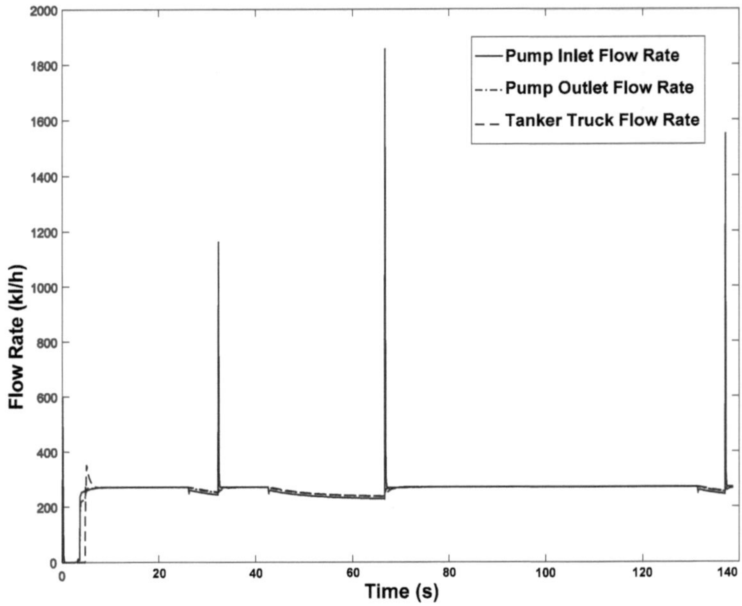

Figure 12. Tanker truck loading flow rate during an injection attack.

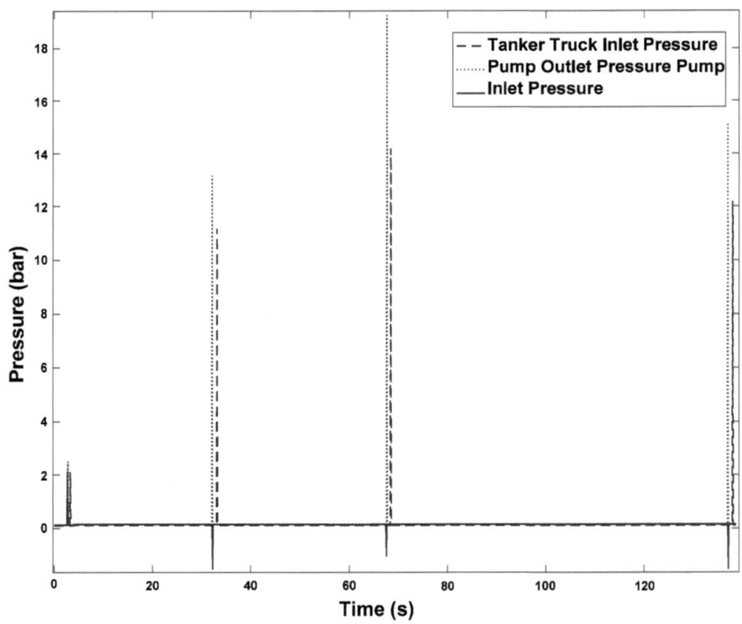

Figure 13. Tanker truck loading pressure during an injection attack.

the attack, the pump creates a negative pressure in the inlet pipeline as shown in Figure 13.

The same injection attack was repeated twice as shown in Figures 12 and 13 between 67 and 69 seconds and between 137 and 139 seconds. During each attack session, the attacker managed to create pressure and flow rate spikes. In the second attack session, the attacker managed to create a very high pressure of 19.8 bar and a flow rate of 1,800 kl/h. Such a high pressure in a closed pipeline system is extremely unsafe and can result in a pipeline rupture.

The injection attack scenario involved an attack on a motor operated valve in the tank farm and the impacts were observed across multiple components of the system. Such a scenario is especially interesting to cyber security researchers because it enables an analysis of the impacts on interdependent components in a midstream oil terminal. In fact, the attack scenario is similar to what occurred in Baku-Tbilisi-Ceyhan pipeline incident [15]. Pressure spikes followed by negative pressure can cause a pipeline to crack or rupture, resulting in the release of hazardous material and, potentially, a fire or explosion.

6. Conclusions

A failure in a midstream oil terminal can result in a catastrophic incident with significant losses of life and property. Cyber threats to critical infrastructure assets such as a midstream oil terminal are dramatically increasing in their number and sophistication. The virtual midstream oil terminal testbed described in this chapter can be used to study cyber security vulnerabilities, examine the impacts of cyber attacks on cyber and physical components, evaluate the effectiveness of security controls and support education and training efforts.

The virtual midstream oil terminal testbed is a large-scale, simulation of multiple interconnected industrial control systems. The entire testbed and all the simulations were executed on a personal computer with an Intel I7 6700K 2,400 MHz processor, 16 GB RAM and a 500 GB solid-state drive running the Windows 10 operating system. Indeed, the virtual midstream oil terminal testbed demonstrates that large-scale models of industrial control systems for cyber security research, education and training are both feasible and valuable.

References

[1] U. Adhikari, T. Morris and S. Pan, WAMS cyber-physical testbed for power system cybersecurity study and data mining, *IEEE Transactions on Smart Grid*, vol. 8(6), pp. 2744–2753, 2017.

[2] T. Alves, OpenPLC Project (`www.openplcproject.com`), 2018.

[3] T. Alves, R. Das and T. Morris, Virtualization of industrial control system testbeds for cybersecurity, *Proceedings of the Second Annual Industrial Control System Security Workshop*, pp. 10–14, 2016.

[4] American Petroleum Institute, Specification for Electric Motor Prime Mover for Beam Pumping Unit Service, API SPEC 11L6, First Edition, Washington, DC, 1993.

[5] American Petroleum Institute, Specification for End Closures, Connectors and Swivels, API SPEC 6H, Second Edition, Washington, DC, 1998.

[6] American Petroleum Institute, Specification for Line Pipe, API SPEC 5L, Forty-Third Edition, Washington, DC, 2004.

[7] American Petroleum Institute, Specification for Bolted Tanks for Storage of Production Liquids, API SPEC 12B, Fifteenth Edition, Washington, DC, 2008.

[8] American Petroleum Institute, Specification for Pipeline Valves, API SPEC 6D, Twenty-Third Edition, Washington, DC, 2008.

[9] American Petroleum Institute, Loading and Unloading of MC 306/DOT 406 Cargo Tank Motor Vehicles, API RP 1007, Washington, DC, 2011.

[10] American Petroleum Institute, Line Markers and Signage for Hazardous Liquid Pipelines and Facilities, API RP 1109, Fifth Edition, Washington, DC, 2017.

[11] C. Bronk and E. Tikk-Ringas, The cyber attack on Saudi Aramco, *Survival: Global Politics and Strategy*, vol. 55(2), pp. 81–96, 2013.

[12] M. Cintuglu, O. Mohammed, K. Akkaya and A. Uluagac, A survey of smart grid cyber-physical system testbeds, *IEEE Communications Surveys and Tutorials*, vol. 19(1), pp. 446–464, 2017.

[13] Independent Inquiry Committee, Independent Inquiry Committee Report on the Indian Oil Terminal Fire in Jaipur on 29th October 2009, Ministry of Petroleum and Natural Gas, Government of India, New Delhi, India, 2010.

[14] International Chamber of Shipping, Oil Companies International Marine Forum and International Association of Ports and Harbors, *International Safety Guide for Oil Tankers and Terminals*, Witherby and Company, London, United Kingdom, 2006.

[15] R. Lee, M. Assante and T. Conway, Media Report of the Baku-Tbilisi-Ceyhan (BTC) Pipeline Cyber Attack, ICS Defense Use Case (DUC), SANS Industrial Control Systems, SANS Institute, Bethesda, Maryland, 2014.

[16] M. Mallouhi, Y. Al-Nashif, D. Cox, T. Chadaga and S. Hariri, A testbed for analyzing security of SCADA control systems (TASSCS), *Proceedings of the Conference on Innovative Smart Grid Technologies*, 2011.

[17] J. Mirkovic and T. Benzel, Teaching cybersecurity with DeterLab, *IEEE Security and Privacy*, vol. 10(1), pp. 73–76, 2012.

[18] T. Morris, A. Srivastava, B. Reaves, W. Gao, K. Pavurapu and R. Reddi, A control system testbed to validate critical infrastructure protection concepts, *International Journal of Critical Infrastructure Protection*, vol. 4(2), pp. 88–103, 2011.

[19] M. Nasir, S. Sultan, S. Nefti-Meziani and U. Manzoor, Potential cyber-attacks against global oil supply chain, *Proceedings of the International Conference on Cyber Situational Awareness*, 2015.

[20] Office of Aircraft Services, *Aviation Fuel Handling Handbook*, CreateSpace, Seattle, Washington, 2015.

[21] B. Reaves and T. Morris, Analysis and mitigation of vulnerabilities in short-range wireless communications for industrial control systems, *International Journal of Critical Infrastructure Protection*, vol. 5(3-4), pp. 154–174, 2012.

[22] Y. Zhou, X. Zhao, J. Zhao and D. Chen, Research on fire and explosion accidents of oil depots, *Chemical Engineering Transactions*, vol. 51, pp. 163–168, 2016.

Chapter 10

A CYBER-PHYSICAL TESTBED FOR MEASURING THE IMPACTS OF CYBER ATTACKS ON URBAN ROAD NETWORKS

Marielba Urdaneta, Antoine Lemay, Nicolas Saunier and Jose Fernandez

Abstract Efficient and safe transportation of people and goods are key requirements in a modern economy. Traffic control systems are installed at complex intersections to ensure the safe and efficient flow of traffic. However, there are concerns that an adversary could launch cyber attacks that exploit flaws in traffic control systems to cause mayhem and accidents.

This chapter presents a co-simulation framework for cyber-physical systems that enables researchers to execute cyber attacks on traffic control systems and measure their impacts on road traffic. The approach integrates an emulated supervisory control and data acquisition master station with a microscopic traffic simulation tool that provides all the functions of a traffic signal control system. The impacts of cyber attacks on road traffic are measured from the outputs provided by the traffic simulation. Experimental results for a corridor of six coordinated signalized intersections are presented, where the impacts are measured in terms of vehicle travel time and queue length. The results reveal that the physical impacts of compromising a single intersection could be felt at other intersections in the road network. This type of emergent result could only have been observed using a co-simulation framework.

Keywords: Road networks, traffic control systems, cyber attacks, testbed

1. Introduction

Traffic congestion is a growing problem and road safety is a major issue in cities around the world [4]. Traffic congestion impacts the economy and the urban environment as well as the quality of life and health of inhabitants. To mitigate congestion, cities are constantly seeking measures that improve and expand their traffic infrastructures and public transportation systems.

Published by Springer Nature Switzerland AG 2018.
J. Staggs and S. Shenoi (Eds.): Critical Infrastructure Protection XII, IFIP AICT 542, pp. 177–196, 2018.
https://doi.org/10.1007/978-3-030-04537-1_10

A road traffic infrastructure comprises road networks and traffic control devices such as signs, markings and traffic signals, which regulate and control traffic at intersections. Traffic signals and sensors are often connected to centralized systems that collect real-time traffic data, which is analyzed in order to design and implement control strategies. The control strategies seek to optimize traffic conditions and increase network capacity and user safety. Also, they attempt to reduce delays, stops, fuel consumption and pollutant emissions.

Modern traffic signal control systems typically incorporate traffic light controllers, sensors, communications networks and a computer-based central system that controls traffic signals and monitors traffic conditions and equipment status [17]. However, as newer technologies are introduced, traffic signal control systems are exposed to increased cyber risks. For example, wireless technologies are used in modern communications networks and by traffic detection systems due to their low maintenance costs and high scalability [6, 19]. However, the cyber risks are also increased.

Despite its benefits, wireless technology renders traffic signal control systems vulnerable to cyber attacks. In particular, wireless communications networks can be accessed remotely. Once a communications network is accessed, the control network is exposed and vulnerable to exploitation as demonstrated by Cerrudo [3] and Ghena et al. [7]. In particular, the researchers exploited vulnerabilities related to weak or no authentication, absence of encryption and wireless access to network components and traffic light controllers. The researchers were able to control traffic signals by capturing and modifying wireless communications, sending fake data and commands to traffic light controllers and connecting to controllers in order to alter their programming.

The feasibility of cyber attacks on traffic control systems demands the investigation of their impacts on road congestion as well as the economic, environmental and social consequences. An experimental environment that faithfully reproduces cyber attacks on traffic control systems and their effects on road traffic would be most useful to municipal authorities, urban designers and homeland security personnel. The environment would support the evaluation of defensive strategies for communications and control networks, and help establish measures for mitigating the physical impacts of attacks. Furthermore, it would facilitate the determination of the best mitigation strategies based on attack impact, enhancing decision making during actual attacks.

This chapter describes a co-simulation-based testbed that enables these capabilities. The testbed incorporates a microscopic traffic simulation package and an emulated supervisory control and data acquisition (SCADA) master station that provides traffic control system functionality. The principal innovation is the creation of a low-cost, reusable and reconfigurable testbed that integrates road traffic control and traffic behavior simulation components to enable the evaluation of cyber attacks and their impacts on road traffic. Unlike other approaches that only include one of the two components, the co-simulation approach significantly enhances the evaluation of cyber security issues because attacks can be conducted against the central control station and the traffic light controllers. Indeed, it is believed that this is the first cyber-physical testbed based on a co-simulation framework that has been created to advance security research activities in the road traffic control domain.

2. Traffic Control Systems

This section describes the key notions related to traffic control systems drawn from various sources [8, 11, 17, 18].

Road traffic comprises pedestrians, cyclists, vehicles, trucks and on-road public transportation systems that concurrently share public roads. The components form traffic movements (or traffic flows) when they move together on the same roadway and in the same direction. At an intersection, two or more traffic movements are considered to be in conflict when their trajectories cross each other at the same level. In such a situation, it is necessary to establish which traffic flow has priority over the other (e.g., yield- or stop-controlled intersections) and when each traffic flow is allowed in the intersection. This assignment is called priority or right-of-way.

Traffic signals are equipped with controllers that switch the lights that inform road users when they have the right to move. Controllers may also be connected to vehicle-presence and pedestrian-presence detectors for real-time adaptation to traffic demand, and to a traffic management center that monitors and controls road traffic conditions and equipment status at intersections.

Traffic signal controllers follow a set of rules that establishes the order in which right-of-way is assigned to the different traffic movements. In addition, the rules establish the duration of the green light for each movement. The element that contains all the rules is called the timing plan and is used by traffic engineers to regulate traffic. The timing plan incorporates control parameters such as cycle length, phases, splits and intervals. A cycle is a complete sequence of phases in which right-of-way is given to all the traffic movements. The time required to complete this sequence is called the cycle length. A phase is the part of a cycle that is assigned to a traffic movement or to multiple traffic movements simultaneously. The part of the cycle assigned to each phase is called a split. The portion of a cycle during which lights do not change is called an interval. Clearly, an attacker with the ability to alter controller configuration (i.e., the timing plan) could disrupt traffic flow.

Traffic signals operate as part of a coordinated system or as isolated nodes. When working in coordination with other signalized intersections, the time (or offset) between the beginning of the cycle of each successive signalized intersection is computed so that vehicles do not have to stop at intermediate intersections.

In contrast, isolated traffic signals are not coordinated and are oblivious to how neighboring intersections are configured. Traffic regulation at isolated intersections employs pre-timed control, actuated control or a combination of the two. Pre-timed traffic lights use pre-elaborated timing plans in which the numbers, sequences and durations of the phases are fixed. Pre-elaborated plans are computed based on historic traffic conditions at intersections. Actuated

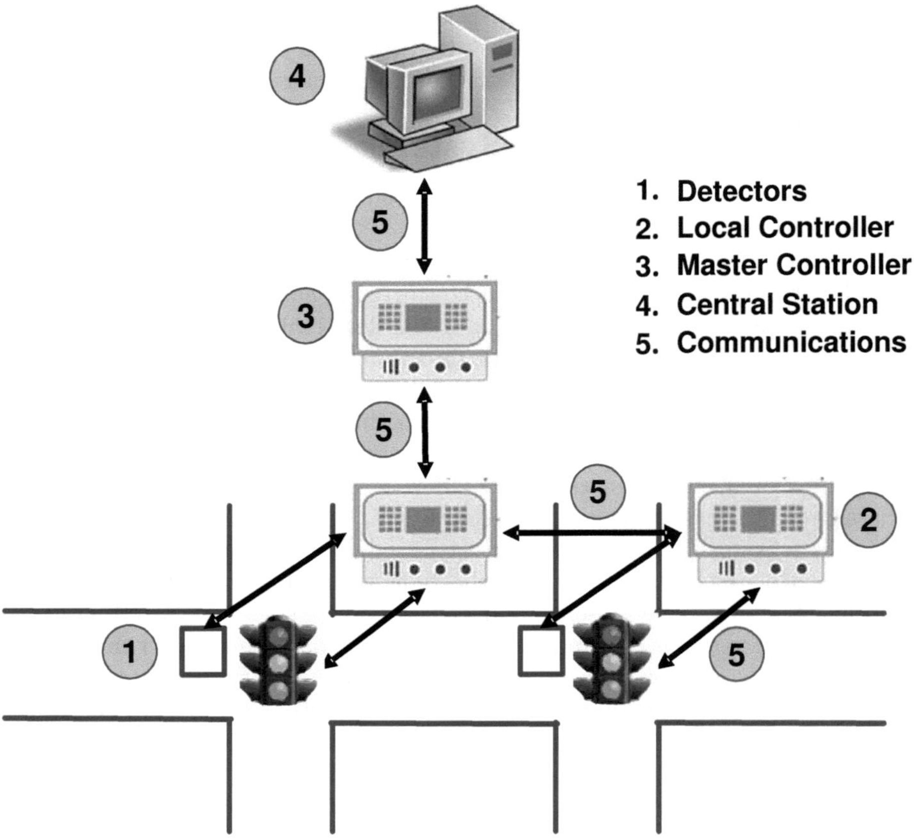

Figure 1. Traffic signal control system [11].

traffic lights use traffic condition information collected by sensors to activate phases when vehicles or pedestrians are detected.

Figure 1 shows the hardware components and architecture of a typical traffic signal control system. It comprises detectors, local controllers, on-street master controllers, a traffic management center and communications networks. Detectors are used to determine vehicle presence and pulse duration, which are needed to compute vehicle volume, occupancy, speed, etc. Local controllers are responsible for switching head lights at intersections using stored timing plans and schedules provided by operators. The controllers receive traffic data from detectors, process the data to obtain volume and occupancy parameters, and send the parameters to on-street master controllers.

Master controllers located at intersections are connected to all the local controllers belonging to the same control area to facilitate communications with the traffic management center. The master controllers are responsible for selecting traffic-responsive timing plans, processing and storing the data collected by detectors, and monitoring the equipment status at intersections.

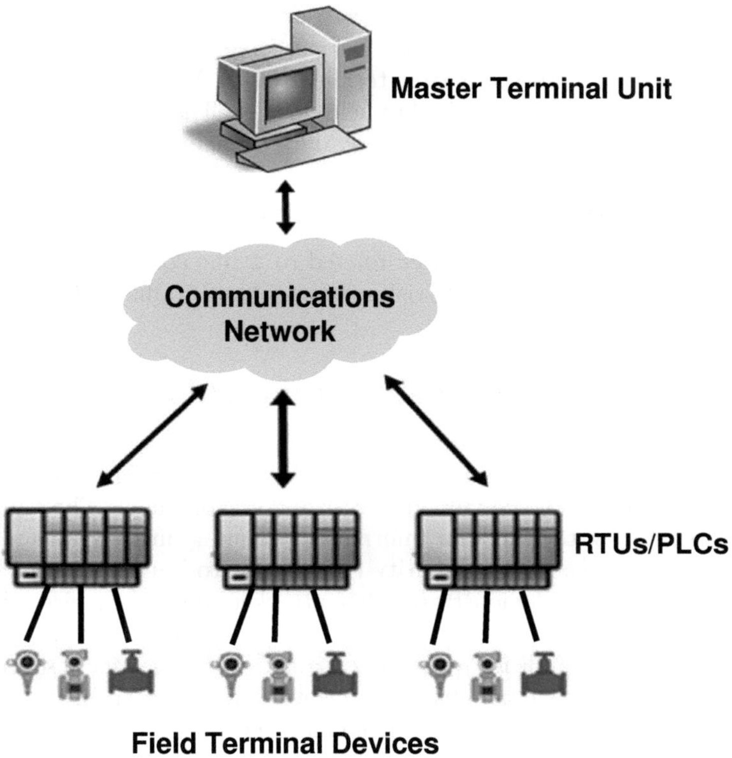

Figure 2. SCADA system.

They communicate with the traffic management center in the case of critical alarms, on a regular predetermined basis and when requested by operators.

The main functions of the traffic management center are to gather and display information about traffic conditions and intersection equipment status. In addition, it calculates the timing plans and schedules. After the timing plans and selection schedules are generated, they can be downloaded to on-street master controllers. Operators at the traffic management center can issue commands to master controllers, for example, to set the time or upload information saved in the master controllers.

The traffic signal control system has the same distributed architecture, control and monitoring elements as a SCADA network. Figure 2 shows a typical SCADA network for an industrial process. The SCADA network has a central station or master terminal unit (MTU) at the highest control level. The master unit processes the data collected from field devices, saves the data and displays it on a human-machine-interface (HMI) to enable operators to monitor and control the industrial process. The master terminal unit is connected to remote terminal units (RTUs) and/or programmable logic controllers (PLCs). The remote terminal units and programmable logic controllers are data acquisition and control devices that are connected to measurement and control points in the field. They collect the measurement data, convert it to a suitable format and send it to the master terminal unit. Additionally, they pass commands from the master terminal to field devices. The communications network provides the required connectivity and data exchange functionality.

3. Related Work

This section discusses research related to traffic control system vulnerabilities, experimental scenarios for risk assessment and traffic control system threat assessment.

3.1 Traffic Control System Vulnerabilities

To demonstrate the exposure of control systems to cyber threats, Luallen [16] asked a group of cyber security students to study an industrial control system in order to find its known vulnerabilities and exploit them. The students leveraged the Internet to search for information about security flaws and proceeded to use a commercial cyber security training kit to launch attacks against the system. This work demonstrates that attackers do not require advanced expertise to attack cyber-physical systems. Valuable information about targets – including vulnerabilities – can be obtained from Internet resources such as technical reports, vendor websites and control system user forums. Having obtained information about a target, commercial products or open-source tools can be used to exploit the vulnerabilities.

Cerrudo [3] and Ghena et al. [7] have described several security flaws in traffic control systems currently deployed in the United States. Although they studied different systems, they findings were very similar: (i) lack of authentication or poor authentication mechanisms to prevent unauthorized access to traffic light controllers; (ii) lack of encryption of data and commands; (iii) use of default credentials supplied by vendors to access traffic light controllers and communications network devices such as switches, access points and repeaters; and (iv) authentication credentials published on vendor websites that are hardcoded in the systems and are not modifiable. Cerrudo and Ghena and colleagues demonstrated that they could gain access to system components and change traffic light states on command.

Krotofil and A.D. [14] state that launching a successful attack on a cyber-physical system involves five fundamental steps: (i) gain access to the system; (ii) discover the system; (iii) take control of the system; (iv) cause damage or disruption to the physical process; and (v) clean up all the evidence pointing to the cyber attack.

To illustrate their approach, Krotofil and A.D. created an experimental cyber-physical testbed that reproduced a traffic light control system for a four-way intersection. The testbed integrated a commercial control system and a cyber security training kit. Credentials provided by the vendor were used to gain access to the system. Having gained access, they acquired knowledge about the system configuration and behavior using tools available on the system for diagnosis, development and visualization. Additionally, they reverse engineered binary files and communications messages to deduce information about the monitoring system and the corresponding elements in the physical system. This enabled them to manipulate the traffic lights at will. To ensure stealth, they manipulated system data so that operators would not notice the unauthorized changes to the traffic lights during the attacks.

The three research efforts demonstrate that flaws in deployed traffic signal systems could be exploited by adversaries. However, the research efforts did not measure the impacts of the attacks on traffic congestion and traffic safety.

3.2 Experimental Scenarios for Risk Assessment

Experimental setups based on co-simulation frameworks have been used to assess the security of various cyber-physical systems. Huang et al. [10] have employed such a framework to evaluate the impact of cyber attacks on a chemical reactor system. Their objective was to measure the impacts of the attacks on the physical process being controlled. Therefore, when conducting attacks, they modeled and monitored the chemical reactor so that they could determine the attacks with the greatest impact. Huang and colleagues discovered that, under steady-state conditions, attacks such as denial-of-service had minor impacts whereas the combination of denial-of-service and integrity attacks could damage the chemical reactor system. They also determined that the costs resulting from the attacks varied depending on the controllers and sensors targeted during the attacks.

Krotofil [13] has developed an open-source framework for controlling a chemical plant based on the well-known Tennessee Eastmann and Vinyl Acetate Monomer models. The previous Matlab models were redeveloped as Simulink models. Krotofil used the framework to develop cyber attacks that targeted sensors and actuators in the plant. Following this, she coupled it to the industrial control network and launched cyber attacks that captured and modified data and commands exchanged between the control system and physical plant.

Bernieri et al. [1] have used a co-simulation framework to evaluate the impacts of cyber attacks on the monitoring elements of a water supply control system. They conducted integrity and availability attacks on the water supply system and employed FACIES [9], an online fault detection and intrusion detection system, to evaluate attack detection performance. The experiments demonstrated that the fault diagnosis system was able to detect replay attacks and attacks that targeted the states of actuators. However, the system failed to identify flooding attacks and attacks that targeted sensor data. More significant was the fact that poor detection performance could induce operators to make unnecessary or erroneous decisions that negatively impacted the physical process.

Lemay et al. [15] have used co-simulation in a testbed that evaluates the effects of cyber attacks on the cyber and physical components of an electric power grid. They employed a virtualized cluster approach that emulates an information technology network with high fidelity [2] and interfaced it with an electrical power flow simulator to model the industrial control network of an electrical grid. The testbed reproduced network attacks such as denial-of-service and data falsification (or injection) attacks, as well as malware infections. Moreover, it efficiently evaluated their impacts on the control network and the power grid.

Testbeds employing co-simulation frameworks are useful for modeling cyber-physical systems and evaluating the effects of cyber attacks. However, no such testbed has, as yet, been developed to assess the security of road traffic control systems.

3.3 Traffic Control System Threat Assessment

Ernst and Michaels [5] have presented a threat assessment framework that evaluates the impacts of vulnerabilities that provide access to field devices in a traffic control system. Their framework considers four access levels whose security flaws may be exploited: (i) vehicle detector; (ii) corridor synchronization; (iii) traditional Internet; and (iv) physical access. Ernst and Michaels employed the Simulation of Urban Mobility (SUMO) package [12] to simulate a road network comprising a corridor with six signalized intersections. They simulated attacks on the first three access levels and measured the attack impacts in various traffic demand scenarios.

Ernst and Michaels used the traffic simulation to investigate how attacks on traffic control system elements would impact road congestion. However, this simulation-only approach does not incorporate the important cyber component of the traffic signal control system. Since the resulting simulation has to rely on broad assumptions of the impacts of cyber attacks, it cannot be used to evaluate network defenses.

4. Testbed Functional Requirements

The goal of this research was to develop an experimental testbed that would enable security researchers to execute cyber attacks on traffic control systems and evaluate the impacts of the attacks on road traffic in real time. To accomplish this goal, it was decided to develop a co-simulation framework that incorporates a two-level distributed control system for an urban road network.

The co-simulation framework would couple a monitoring and control system (e.g., SCADA system) with a microscopic road traffic simulation. The SCADA system would provide the real-time monitoring and control functions required for a large road network. The microscopic traffic simulation would model a road network and traffic conditions to support the development of road traffic control strategies. Additionally, the microscopic traffic simulation would provide data about various road network entities such as pedestrians, vehicles, public transport systems and traffic lights at a suitable level of granularity.

The traffic simulation must provide adequate outputs that enable measurements of the economic, environmental and social effects of road congestion resulting from cyber attacks on the modeled road network. Example outputs include fuel consumption, greenhouse gas emissions, pollutant emissions, noise emissions, vehicle densities, vehicle travel times and vehicle waiting times. All this information could be provided by the microscopic traffic simulation.

Finally, a mechanism must be incorporated that properly couples the cyber and physical components of the traffic control system. This mechanism would handle the time difference between the supervisory and control system sampling time and the traffic simulation step time (if any). Additionally, it would support seamless data exchange between the control system and road traffic simulation.

5. Testbed Architecture

The testbed is designed to support research activities by the cyber security community. To ensure availability, reusability and adaptability, a number of open-source software applications were employed to construct the testbed. Figure 3 presents the testbed architecture and components.

5.1 Monitoring and Control

The high-level control component of the system was reproduced using the ScadaBR 1.0 CE open-source SCADA software, a web-browser-based application that supports access to monitoring, control and automation equipment using various protocols (`www.scadabr.com`). In particular, ScadaBR: (i) provides the monitoring and control functions of a master terminal unit; (ii) displays and saves information about traffic conditions and traffic light states received from the low-level control system; (iii) enables operators to send commands to change traffic light operation modes (e.g., NORMAL/DISABLE); and (iv) runs a Modbus client to communicate with each control and data acquisition device in the low-level control system. ScadaBR can be configured to implement all the functions of a traffic management center that monitors and controls several traffic lights.

The low-level control system was implemented using Python scripts that emulated programmable logic controller functions. The scripts read master terminal unit commands and road network data, converted the data to the proper format and transmitted it to the required control system level. Moreover, the scripts implemented the logic that controlled traffic signals in the network, thereby acting as traffic light controllers. Programmable logic controllers were designed to control all the traffic signals at each signalized intersection. Each programmable logic controller script ran a Modbus/TCP server to communicate upstream with ScadaBR using the Modbus/TCP server functionality provided by the Modbus TK Python library. In addition, each programmable logic controller ran a TCP client to communicate downstream with the road traffic simulation via a communications server.

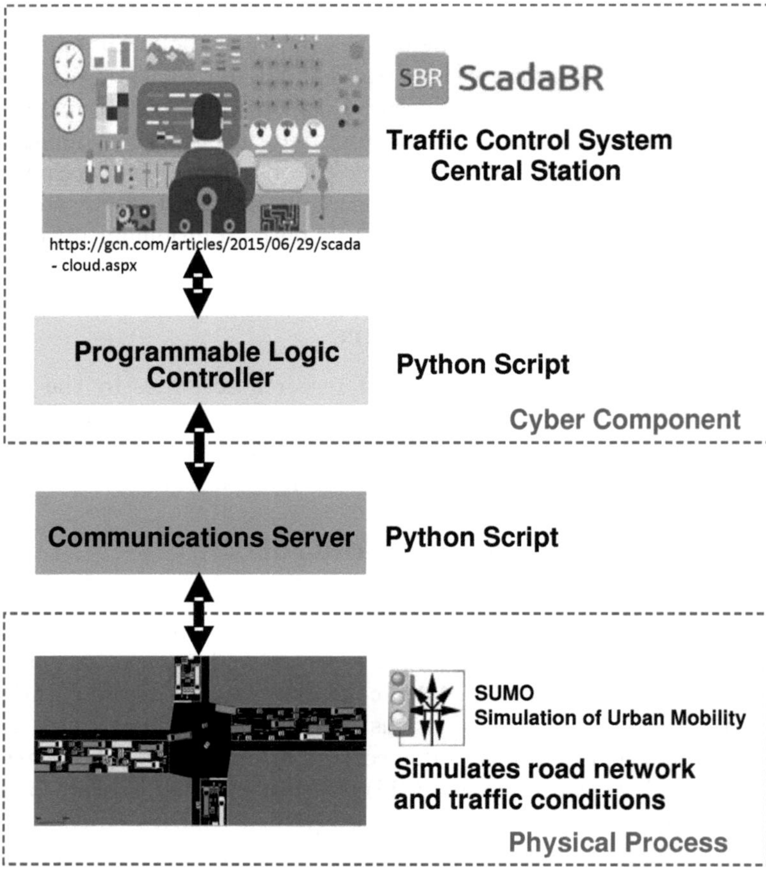

Figure 3. Testbed architecture and components.

5.2 Road Traffic Simulation

The physical process controlled in the testbed is road traffic. The open-source SUMO package developed by the German Aerospace Center [12] was employed to simulate road traffic. SUMO offers the flexibility of creating large-scale road networks from common formats such as shapefiles and Open Street Map files. A SUMO road network incorporates signalized intersections and traffic light plans. Additionally, origin/destination matrices can be converted to single vehicle trips and loaded in the SUMO simulation.

At each time step, SUMO generates outputs that provide information about all the simulated elements in the road network, including vehicles, intersections, roads, lanes, traffic lights and inductive loops. Data produced at this level of granularity is adequate for the monitoring component. Also, SUMO generates noise emission, pollutant emission and fuel consumption outputs required to quantify the economic, environmental and societal effects of road congestion.

SUMO incorporates a Python traffic control interface (TraCI) for interacting with external applications via TCP socket connections. This enables SUMO to connect to other monitoring and control systems. The interface also enables users to set and modify the simulation conditions at any time. For example, the user could change vehicle speeds, driver behavior, road priority and traffic light state as well as force vehicles to change lanes. These features were used to enforce state changes dictated by the control component.

SUMO performs a discrete-time simulation with adjustable step durations from 1 ms and upwards. It also offers two simulation alternatives, one without visualization and the other with visualization via a graphical interface.

5.3 Communications Server

A Python TCP multi-threaded communication server was developed to couple the monitoring and control system with the physical process. Multi-threading enabled the server to handle and serve multiple concurrent incoming client requests at the same time. Moreover, it dealt with communications synchronization issues arising from differences between the programmable logic controller sampling interval and the simulation time step.

At every simulation step, the server received data and requests from SUMO and the programmable logic controllers. The data received from SUMO pertained to each signalized intersection and its traffic light states provided by the simulation. This data was stored in separate tables according to the signalized intersection and its programmable logic controller; the data was transmitted upon request to the corresponding programmable logic controller.

Data received from a programmable logic controller identifies the signalized intersection and the traffic light states set during the simulation. This data was stored in a table that matched each programmable logic controller with its signalized intersection. The data was transmitted to SUMO upon request.

The SUMO traffic control interface was employed to execute a script running a TCP client. At each simulation step, the client transmitted the simulation results to the server and requested new commands from the programmable logic controller. SUMO adjusted the state of the traffic lights according to the information received from the server.

6. Validation and Experimental Setup

This section discusses the initial validation of the co-simulation framework and the experimental setup.

6.1 Initial Validation

For configuration and testing purposes, a preliminary setup was created that connected all the components of the proposed co-simulation framework. This preliminary setup was used to validate: (i) proper integration of all the components; (ii) proper system operation; and (iii) correct conversion/transmission

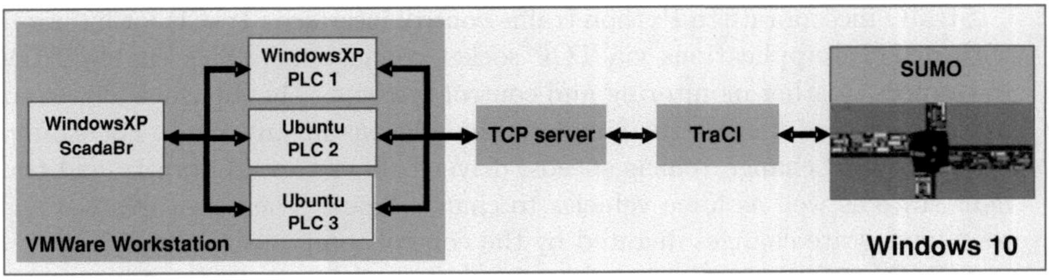

Figure 4. System used for initial validation.

of information from the master terminal unit to the traffic simulation, and vice versa.

The first simulation scenario involved a road network with three signalized intersections spaced 100 m apart and running in the pre-timed or semi-actuated mode. The traffic light control logic replicated the logic specified by Krotofil [14]. The control logic implemented a finite-state machine with eight states and nine transition conditions to model the traffic lights. It employed four control signals: (i) AUTO; (ii) DISABLE; (iii) MAIN ROAD; and (iv) SIDE ROAD. These signals enabled the traffic light operation modes to be set by the master terminal unit. When the operation mode was set to AUTO, the traffic lights commuted automatically based on the finite state machine program. In this case, the traffic lights operated in the pre-timed control mode with fixed control parameters; the timing plans could be changed by modifying the timing conditions and the state sequence in the finite state machine program. When the operation mode was set to DISABLE, the lights changed to yellow in all directions at the intersection; they remained in this state until the DISABLE signal was no longer set. When either the MAIN ROAD or SIDE ROAD signal was set, the traffic lights operated in the semi-actuated control mode. This assigned the green light to the corresponding road (MAIN or SIDE) until vehicles were detected on the opposite road.

All the system components were installed and configured on a desktop computer running the Windows 10 operating system (Figure 4). The SUMO software, simulation update script and communications server ran directly on the computer. ScadaBR and the programmable logic controllers executed in virtual machines. Specifically, ScadaBR and PLC 1 ran on Windows XP virtual machines whereas PLC 2 and PLC 3 ran on Ubuntu Linux virtual machines. All the virtual machines were created using VMWare Workstation software.

After validating the integration and operation of the preliminary setup, cyber attacks were launched to evaluate the fidelity of the testbed. For this purpose, a Kali Linux virtual machine was connected to the same network as the programmable logic controllers and ScadaBR. The Kali Linux machine was then used to conduct man-in-the-middle (MiTM) packet captures and packet injection attacks.

```
▷ Frame 202: 66 bytes on wire (528 bits), 66 bytes captured (528 bits) on interface 0
▷ Ethernet II, Src: Vmware_c0:00:08 (00:50:56:c0:00:08), Dst: Vmware_b6:59:6c (00:0c:29:b6:59:6c)
▷ Internet Protocol Version 4, Src: 192.168.88.1, Dst: 192.168.88.21
▷ Transmission Control Protocol, Src Port: 5430, Dst Port: 502, Seq: 1, Ack: 1, Len: 12
◢ Modbus/TCP
    Transaction Identifier: 988
    Protocol Identifier: 0
    Length: 6
    Unit Identifier: 1
◢ Modbus
    .000 0100 = Function Code: Read Input Registers (4)
    Reference Number: 12
    Word Count: 7
```

a

```
▷ Frame 204: 77 bytes on wire (616 bits), 77 bytes captured (616 bits) on interface 0
▷ Ethernet II, Src: Vmware_b6:59:6c (00:0c:29:b6:59:6c), Dst: Vmware_c0:00:08 (00:50:56:c0:00:08)
▷ Internet Protocol Version 4, Src: 192.168.88.21, Dst: 192.168.88.1
▷ Transmission Control Protocol, Src Port: 502, Dst Port: 5430, Seq: 1, Ack: 13, Len: 23
◢ Modbus/TCP
    Transaction Identifier: 988
    Protocol Identifier: 0
    Length: 17
    Unit Identifier: 1
◢ Modbus
    .000 0100 = Function Code: Read Input Registers (4)
    [Request Frame: 202]
    Byte Count: 14
  ▷ Register 12 (UINT16): 1
  ▷ Register 13 (UINT16): 1
  ▷ Register 14 (UINT16): 1
  ▷ Register 15 (UINT16): 1
  ▷ Register 16 (UINT16): 8
  ▷ Register 17 (UINT16): 6
  ▷ Register 18 (UINT16): 4
```

b

Figure 5. ScadaBR request and PLC 1 response during normal operations.

The scenario assumed that an attacker had gained access to the communications network and intercepted the data exchanged between the master terminal unit and the controller. Since the Modbus protocol does not incorporate authentication and encryption mechanisms, the attacker could inject control packets that would be accepted by the traffic controller. Furthermore, with the help of Internet resources, it would be easy to reproduce the content of Modbus messages and create arbitrary control messages for transmission to the controller.

The man-in-the-middle packet capture attack was executed using a Python script, which performed an address resolution protocol (ARP) cache poisoning that targeted ScadaBR and PLC 1. The attack enabled the adversary to impersonate the ScadaBR and PLC 1, and intercept the messages exchanged by them. Figures 5(a) and 5(b) show a ScadaBR request and the corresponding PLC 1 response during normal operations, before ARP cache poisoning.

```
▷ Frame 2814: 66 bytes on wire (528 bits), 66 bytes captured (528 bits) on interface 0
▷ Ethernet II, Src: Vmware_c0:00:08 (00:50:56:c0:00:08), Dst: Vmware_b8:3c:ab (00:0c:29:b8:3c:ab)
▷ Internet Protocol Version 4, Src: 192.168.88.1, Dst: 192.168.88.21
▷ Transmission Control Protocol, Src Port: 5670, Dst Port: 502, Seq: 1, Ack: 1, Len: 12
◢ Modbus/TCP
    Transaction Identifier: 1016
    Protocol Identifier: 0
    Length: 6
    Unit Identifier: 1
◢ Modbus
    .000 0100 = Function Code: Read Input Registers (4)
    Reference Number: 12
    Word Count: 7
```

a

```
▷ Frame 2862: 77 bytes on wire (616 bits), 77 bytes captured (616 bits) on interface 0
▷ Ethernet II, Src: Vmware_b6:59:6c (00:0c:29:b6:59:6c), Dst: Vmware_b8:3c:ab (00:0c:29:b8:3c:ab)
▷ Internet Protocol Version 4, Src: 192.168.88.21, Dst: 192.168.88.1
▷ Transmission Control Protocol, Src Port: 502, Dst Port: 5670, Seq: 1, Ack: 13, Len: 23
◢ Modbus/TCP
    Transaction Identifier: 1016
    Protocol Identifier: 0
    Length: 17
    Unit Identifier: 1
◢ Modbus
    .000 0100 = Function Code: Read Input Registers (4)
    [Request Frame: 2814]
    Byte Count: 14
  ▷ Register 12 (UINT16): 2
  ▷ Register 13 (UINT16): 1
  ▷ Register 14 (UINT16): 2
  ▷ Register 15 (UINT16): 1
  ▷ Register 16 (UINT16): 6
  ▷ Register 17 (UINT16): 5
  ▷ Register 18 (UINT16): 0
```

b

Figure 6. Intercepted ScadaBR request and PLC 1 response during the attack.

Figure 6(a) shows a request generated by ScadaBR and intercepted by the attacker (MAC address `00:0c:29:b8:3c:ab`) who impersonated PLC 1. Figure 6(b) shows the corresponding response generated by PLC 1 and intercepted by the attacker who impersonated ScadaBR.

The packet injection attacks were executed by a separate Python script that sent Modbus commands from the attacker's machine to PLC 1. Figure 7(a) shows a request generated by the attacker to set the mode of the traffic light to DISABLE (function code Write Single Coil and register reference number 3). Figure 7(b) shows the response generated by PLC 1 confirming the setting of the register value.

6.2 Experimental Setup

Following the initial validation, it was decided to execute a cyber attack on a coordinated traffic light system. This was accomplished by recreating the

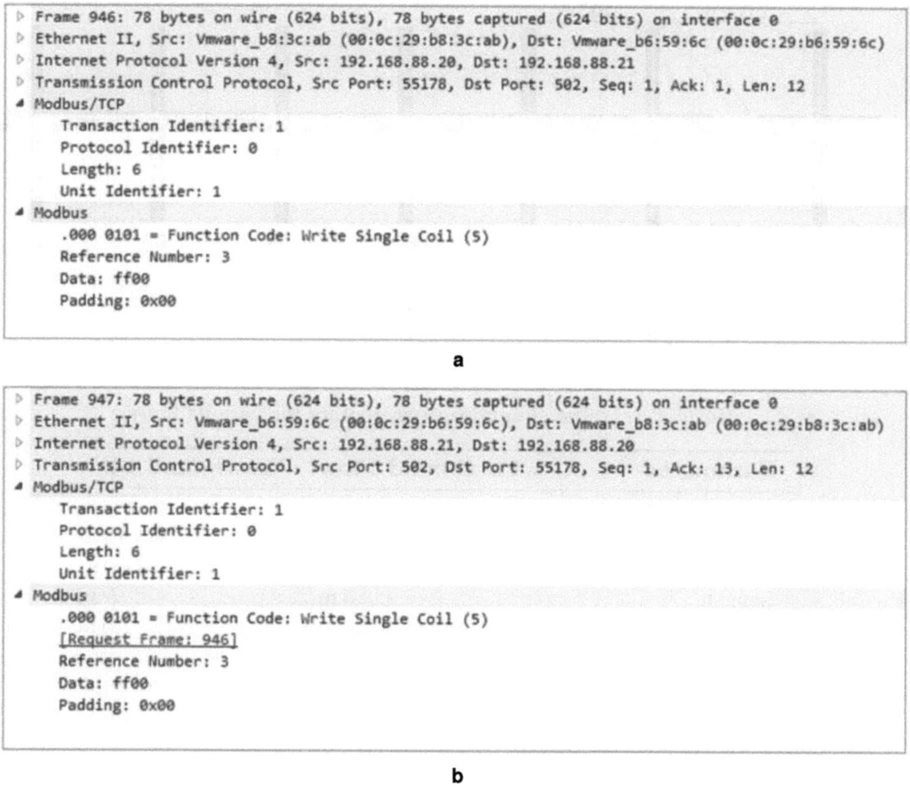

Figure 7. Messages exchanged during the packet injection attack.

road corridor used by Ernst and Michaels [5]. The experimental setup shown in Figure 8 comprised six coordinated signalized intersections, each spaced 100 m apart. An additional intersection was placed 2,000 m from the east entry of the corridor to generate vehicle platoons. As in the case of the Ernst and Michaels model, no turns were allowed and, to keep the model simple, each road had only one lane in each direction. Nonetheless, the model was adequate to demonstrate the impacts of the attacks on a corridor of signalized intersections.

The corridor in the experimental setup was coordinated to favor eastbound flows. Table 1 shows the simulation parameters. Table 2 shows the timing plan parameters used in the experimental setup.

In order to achieve coordination in the corridor, intersection C1 was chosen as the master intersection. Intersections C2 through C6 were coordinated with offsets of 5.8 s, 11.6 s, 17.4 s, 23.2 s and 29 s, respectively. One programmable logic controller was set up to manage intersection C1 while another was used to manage intersection C5, which was the target of the attacks. The control logic for the four remaining intersections (C2, C3, C4 and C6) was implemented by

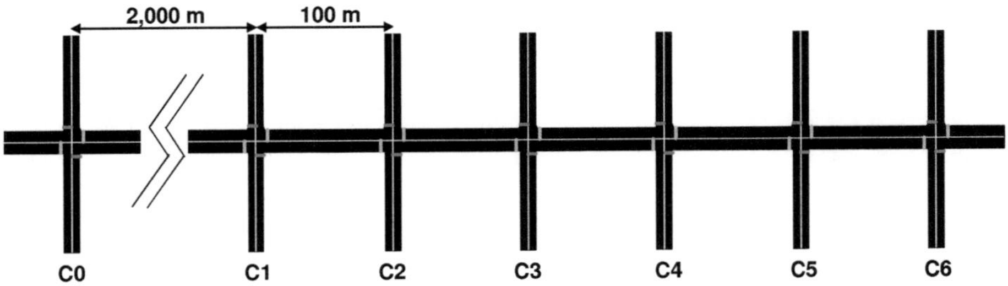

Figure 8. Road network used in the experimental setup.

Table 1. Traffic simulation parameters for various flows.

Parameter	**Eastbound Flow**	**Westbound Flow**
Maximum Speed	16.67 m/s	16.67 m/s
Acceleration	4.5 m/s^2	4.5 m/s^2
Deceleration	0.8 m/s^2	0.8 m/s^2
Length	5 m	5 m
Minimum Gap	2.5 m	2.5 m
Sigma	0.5	0.5
Demand	1,000 vehicles/h	500 vehicles/h
Car Following Model	Krauss	Krauss

Table 2. Timing plan parameters for the coordinated corridor.

Cycle Length	98 s
Main Road Green Duration	60 s
Side Road Green Duration	20 s
Yellow Duration	6 s
All Red Duration	3 s

SUMO instead of simulated programmable logic controllers. The decision not to use fine-grained emulation for these four intersections was made to conserve computing resources. There is no loss of generality because nothing, apart from computational power, would prevent the virtualization of all the programmable logic controllers if they were required.

After configuring the corridor, packet injection attacks were launched on signalized intersection C5. The same Kali Linux machine and script used in the initial validation were used to send Modbus/TCP messages to change the programming of the traffic lights at the intersection. Specifically, the main green light time was changed to 22 s and the side green light time was changed to 10 s.

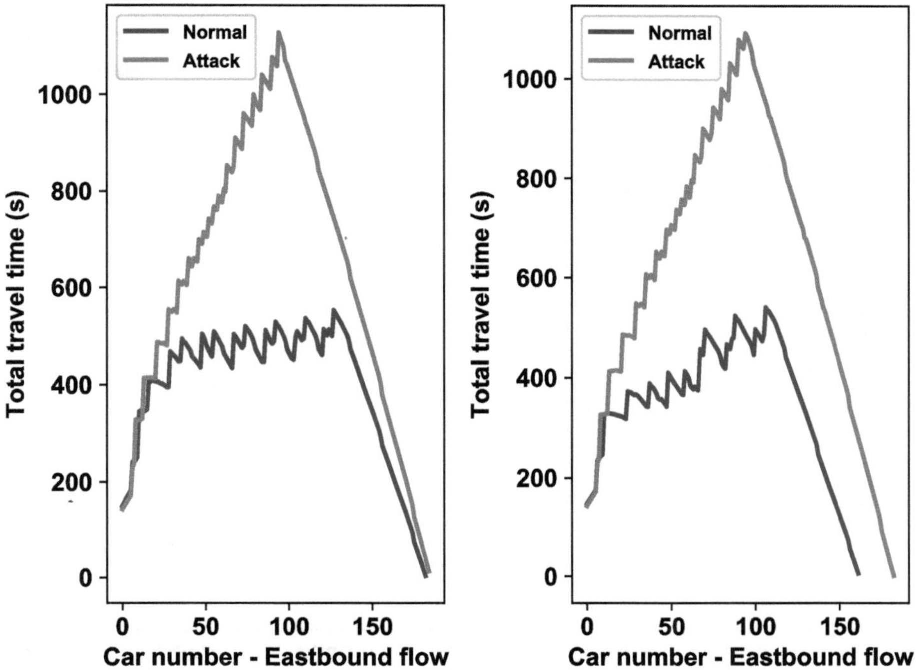

Figure 9. Eastbound vehicle travel times for two simulation runs.

7. Experimental Results

The attack impacts were measured in terms of travel time and queue length. The travel time for each vehicle in the main corridor in the eastbound direction and going through all the intersections was recorded. The travel time was plotted as a function of vehicle number in the order of vehicles entering the intersection. The queue lengths were measured at each simulation step and reported for each corridor section. Following this, the mean queue length for each section was computed based on the results of five simulation runs.

Figure 9 shows the travel time results for two simulation runs under normal conditions and during the attacks.

Figure 10 shows the mean values of the queue length for each corridor section between intersections C0 and C6 under normal conditions and during the attacks.

The results reveal that the travel times increased two to three times during the attacks. The queue lengths increased even more – they were practically non-existent under normal conditions (about two vehicles at most intersections) and increased four to five times (up to eleven vehicles). The effect on queue length was greater for intersections in the middle of the corridor, with queue spillback from downstream intersections.

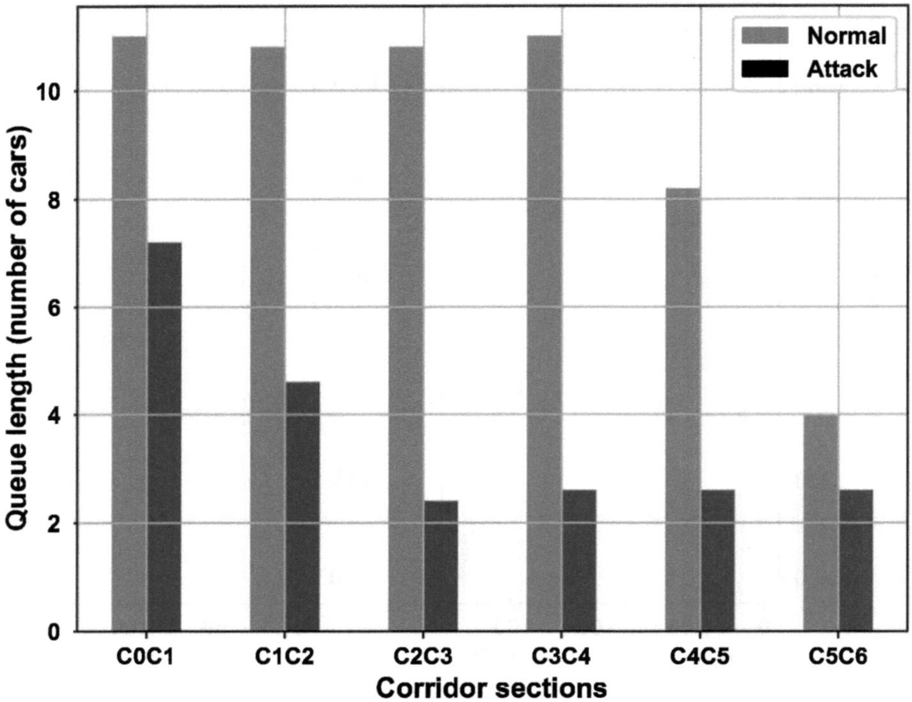

Figure 10. Mean queue length values for each corridor section.

The results also demonstrate that the co-simulation approach is very useful for evaluating the physical impacts of real cyber attacks. Moreover, unlike the work by Ernst and Michaels [5], no assumptions had to be made about the effects of cyber attacks on the traffic light control components.

8. Conclusions

The testbed described in this chapter successfully integrates a microscopic traffic simulation with an emulated SCADA master station to reproduce a traffic control system for a coordinated corridor of signalized intersections. This testbed is well-suited for evaluating the impacts of cyber attacks on traffic control systems. The impacts were measured in terms of traffic performance measures such as travel time and queue length rather than information technology performance metrics. In the man-in-the-middle attack scenario considered in the experiments, the travel time was increased two to three times and the vehicle queue length was increased four to five times over normal operations. Moreover, attacking one intersection produced impacts at other intersections in the road network. The results highlight the importance of understanding the local and global impacts of cyber attacks that target road networks. These emergent results could have only been observed in a co-simulation framework of the type implemented in the testbed.

The testbed incorporates generic simulation and control software. While this has supported the execution of certain attacks and the evaluation of their impacts, it limits investigations of complex attacks and their impacts. This limitation can be overcome by enhancing the fidelity and the capabilities of the testbed by modeling real-world road networks and traffic demand scenarios, and by replacing ScadaBR with real traffic control software. Nevertheless, the testbed can help identify the critical signalized intersections in road networks and the attacks that produce the greatest impacts on traffic conditions. This information can be used to implement security and mitigation strategies, as well as to develop plans for reducing the negative impacts of attacks on traffic performance.

Future research will employ the testbed to evaluate the resilience of road networks to cyber attacks. This will involve the modeling of large road networks and replicating advanced cyber attacks on centrally-controlled traffic control systems whose impacts cannot be evaluated using a traffic simulator alone.

References

[1] G. Bernieri, E. Etcheves Miciolino, F. Pascucci and R. Setola, Monitoring system reaction in a cyber-physical testbed under cyber attacks, *Computers and Electrical Engineering*, vol. 59, pp. 86–98, 2017.

[2] J. Calvet, C. Davis, J. Fernandez, W. Guizani, M. Kaczmarek, J. Marion and P. St-Onge, Isolated virtualized clusters: Testbeds for high-risk security experimentation and training, *Proceedings of the Third International Conference on Cyber Security Experimentation and Test*, 2010.

[3] C. Cerrudo, Hacking US (and UK, Australia, France, etc.) traffic control systems, *IOActive*, Seattle, Washington, April 30, 2014.

[4] G. Cookson and B. Pishue, INRIX Global Traffic Scorecard, INRIX Research, Kirkland, Washington, 2017.

[5] J. Ernst and A. Michaels, Framework for evaluating the severity of a cyber vulnerability of a traffic cabinet, *Transportation Research Record: Journal of the Transportation Research Board*, vol. 2619, pp. 55–63, 2017.

[6] S. Faye, C. Chaudet and I. Demeure, Control of Urban Road Traffic by a Fixed Network of Wireless Sensors, Technical Report 2012D002, Telecom ParisTech, Paris, France, 2012.

[7] B. Ghena, W. Beyer, A. Hillaker, J. Pevarnek and J. Halderman, Green lights forever: Analyzing the security of traffic infrastructure, *Proceedings of the Eighth USENIX Workshop on Offensive Technologies*, 2014.

[8] R. Gordon and W. Tighe, Traffic Control Systems Handbook, Publication No. FHWA-HOP-06-006, Federal Highway Administration, Washington, DC, 2005.

[9] C. Heracleous, E. Etcheves Miciolino, R. Setola, F. Pascucci, D. Eliades, G. Ellinas, C. Panayiotou and M. Polycarpou, Critical infrastructure online fault detection: Application in water supply systems, *Proceedings of the Ninth International Conference on Critical Information Infrastructures Security*, pp. 94–106, 2014.

[10] Y. Huang, A. Cardenas, S. Amin, Z. Lin, H. Tsai and S. Sastry, Understanding the physical and economic consequences of attacks on control systems, *International Journal of Critical Infrastructure Protection*, vol. 2(3), pp. 73–83, 2009.

[11] P. Koonce, Traffic Signal Timing Manual, Publication No. FHWA-HOP-08-024, Federal Highway Administration, Washington, DC, 2008.

[12] D. Krajzewicz, G. Hertkorn, P. Wagner and C. Rossel, SUMO (Simulation of Urban Mobility) – An open-source traffic simulation, *Proceedings of the Fourth Middle Eastern Symposium on Simulation and Modeling*, pp. 183–187, 2002.

[13] M. Krotofil, Rocking the pocket book: Hacking chemical plants for competition and extortion, presented at *Black Hat USA*, 2015.

[14] M. Krotofil and A.D., Hack like a movie star: Step-by-step guide to crafting SCADA payloads for physical attacks with catastrophic consequences, presented at *ZeroNights*, 2015.

[15] A. Lemay, J. Fernandez and S. Knight, An isolated virtual cluster for SCADA network security research, *Proceedings of the First International Symposium on ICS and SCADA Cyber Security Research*, pp. 88–96, 2013.

[16] M. Luallen, Critical Control System Vulnerabilities Demonstrated – And What to Do About Them, InfoSec Reading Room, SANS Institute, Bethesda, Maryland, 2011.

[17] Ministry of Transportation Ontario, *Book 19 – Advanced Traffic Management Systems*, St. Catherines, Canada, 2007.

[18] Minnesota Department of Transportation, *Traffic Signals 101*, St. Paul, Minnesota, 2018.

[19] M. Tubaishat, Y. Shang and H. Shi, Adaptive traffic light control with wireless sensor networks, *Proceedings of the Fourth IEEE Consumer Communications and Networking Conference*, pp. 187–191, 2007.

Chapter 11

PERSISTENT HUMAN CONTROL IN A RESERVATION-BASED AUTONOMOUS INTERSECTION PROTOCOL

Karl Bentjen, Scott Graham and Scott Nykl

Abstract Widespread use of fully autonomous vehicles is near. However, the desire of human beings to maintain control of their vehicles – even limited control – is unlikely to ever go away. Several protocols (e.g., AIM, Semi-AIM and H-AIM) have been developed to safely and efficiently manage reservation-based intersections with a mixture of fully autonomous, semi-autonomous and non-autonomous vehicles. However, these protocols do not incorporate the dynamic of a human maintaining control of a semi-autonomous vehicle when approaching and crossing an intersection. This chapter lays the foundation for the extensions required for human-control of semi-autonomous vehicles, the ultimate goal being a protocol that maintains the efficiency of a fully autonomous environment while allowing human control of vehicles when navigating an intersection. This chapter also proposes information feedback mechanisms for human response, such as displays that provide the intersection arrival time, goal velocity, lane maintaining assistance and other warnings. Additionally, it describes a synthetic environment that enables the testing of intersection protocols that support human interaction.

Keywords: Semi-autonomous vehicles, intersections, reservations, human control

1. Introduction

Self-driving vehicles are already on the road, in some cases without backup drivers [4]. Before long, the traffic infrastructure, specifically intersections, will be required to manage autonomous vehicular traffic in an efficient manner. To address this need, Dresner and Stone [8] introduced a reservation-based intersection protocol called Autonomous Intersection Management (AIM), designed for an environment with strictly autonomous vehicles. The AIM protocol was subsequently modified to incorporate semi-autonomous vehicles that allow lim-

J. Staggs and S. Shenoi (Eds.): Critical Infrastructure Protection XII, IFIP AICT 542, pp. 197–212, 2018.
https://doi.org/10.1007/978-3-030-04537-1_11

ited human control [2]. Another modified version of AIM, known as Hybrid-AIM (H-AIM), further accommodates human-operated vehicles without direct communications between vehicles and the intersection [10]. An additional category to be considered is vehicles that can communicate with the intersection manager, but that are driven by humans.

This chapter lays the foundation for the extensions required for human-control of semi-autonomous vehicles, the ultimate goal being a protocol that maintains the efficiency of a fully autonomous environment while allowing human control of vehicles when navigating an intersection. In particular, it attempts to identify how persistent human control can be introduced in an autonomous intersection. It also describes the AFTR Burner synthetic environment [9] and baseline experiments that establish the viability of the reservation-based intersection protocol. Proposed feedback and control mechanisms to enable human control are also detailed, along with a proof-of-concept system that enables humans to maintain vehicular control when navigating autonomous intersections.

2. Background and Motivation

This section provides an overview of autonomous, semi-autonomous and non-autonomous vehicles. Also, it discusses the requirements for the safe and efficient management of a traffic intersection with autonomous and semi-autonomous vehicles, as well as for an environment where all the vehicles are fully autonomous with no human control.

2.1 Autonomous Vehicle Taxonomy

The U.S. National Highway Traffic Safety Administration (NHTSA) and the Society of Automotive Engineers (SAE) International have developed a taxonomy of vehicles and their levels of autonomy [11]. Table 1 lists the five levels of autonomy and provides brief descriptions. Levels 4 and 5 cover autonomous vehicles that are capable of driving themselves; however, these levels provide options for human drivers to assume control of their vehicles.

This research specifically focuses on the ability – or desire – of a human to maintain control of a vehicle, especially when approaching and traversing an intersection. The intersection is designed such that traditional or legacy non-autonomous vehicles (Level 0) are not normally allowed due to the lack of traffic signals at the intersection and/or the vehicles lack vehicle-to-anything (V2X) communications capabilities. If desired, the intersection may be designed to degrade to a standard intersection when a legacy vehicle approaches, but this problem is outside of scope of this research.

2.2 Reservation Concept

Human safety is paramount when designing a protocol for managing autonomous traffic. Dresner and Stone [7] introduced the concept of a reserva-

Table 1. Automation levels set by SAE International [11].

Automation Level	Description
Human Driver Required	
Level 0: No Automation	The vehicle is completely non-autonomous; the driver performs all the driving tasks.
Level 1: Driver Assistance	The vehicle has some driving assist features such as traditional cruise control, but the driver controls the vehicle.
Level 2: Partial Automation	The vehicle has combined automated functions such as acceleration and steering, but the driver must remain engaged with the driving task and monitor the environment at all times.
Level 3: Conditional Automation	The driver is a necessity, but is not required to monitor the environment; the driver must be ready to take control of the vehicle at all times upon request.
Human Driver Not Required	
Level 4: High Automation	The vehicle can perform all the driving functions under certain conditions, including limitations on locations and environments; the driver may have the option to control the vehicle.
Level 5: Full Automation	The vehicle can perform all the driving functions under all conditions; the driver may or may not have the option to control the vehicle.

tion to address the issue of safely scheduling the passage of autonomous vehicles through an intersection. They used the reservation concept to develop the AIM protocol that can manage an autonomous intersection in a safe and efficient manner. This is accomplished by ensuring that vehicles do not collide and by reducing the delays experienced by vehicles at the intersection compared with a traditional intersection with traffic signals [8].

The AIM protocol uses the reservation concept to safely eliminate traffic signals as long as all the vehicles are fully autonomous with V2X capabilities. The AIM protocol works well in an environment comprising only fully autonomous vehicles. However, while such an environment will surely be realized in the future, there will be a long transition period during which vehicles with all levels of autonomy will have to be integrated safely and efficiently.

The AIM protocol is designed for an environment where at least 90% of the vehicles are fully autonomous and operate without human control. It is speculated that autonomous vehicles with V2X capabilities will not exceed 90% of the vehicular population until at least the year 2045 [3]. Until this time, protocols will be implemented to handle the integration of all levels of autonomous vehicles. All the autonomous vehicles will incorporate human control to some extent.

2.3 Other Intersection Protocols

In 2015, Au et al. [2] published the SemiAIM protocol, an extension of the AIM protocol that incorporates semi-autonomous vehicles. In the SemiAIM protocol, human drivers relinquish control of their vehicles before entering an intersection. However, the protocol requires the use of traffic signals. Semi-autonomous vehicles that fail to receive confirmed reservations must come to a stop and treat the intersection as a traditional traffic signal intersection. The traffic signals are also used by non-autonomous vehicles, which are allowed in the SemiAIM protocol.

Sharon and Stone [10] developed the H-AIM protocol, which is more efficient than the AIM protocol when there is a low concentration of autonomous and semi-autonomous vehicles. The enhanced protocol assumes that the intersection can detect incoming non-autonomous vehicles and enables autonomous vehicles to receive reservations that do not conflict with the possible paths of non-autonomous vehicles. The protocol also depends on traffic signals for human-driven vehicles that do not have V2X communications capabilities.

The SemiAIM and H-AIM protocols do not provide the option for a human-driven vehicle with V2X capabilities to request and receive a reservation, but they do allow a human to maintain persistent control over his/her vehicle. However, a vehicle at Level 2 or higher automation level permits a human to control the steering and/or velocity while navigating through an autonomous intersection without traffic signals.

2.4 Persistent Human Control

It is safe to assume that there will always be humans who want to drive their vehicles and be in control. Indeed, a recent study by Abraham et al. [1] revealed that 48% of the people surveyed would never purchase a car that completely drives itself. Therefore, it is necessary to consider a future environment where all the vehicles are at least semi-autonomous (whether they require a human driver or not), all the vehicles have vehicle-to-vehicle (V2V) and V2X communications capabilities, and autonomous intersections do not have traditional traffic signals. Some sort of backup signal capability may exist, but not for managing traffic on a regular basis.

A protocol such as AIM could prove to be the protocol of choice in such an environment, especially if, like the AIM protocol, it is already shown to be safe and efficient. However, the protocol would have to be modified to enable the human behind the wheel to maintain control over the steering and/or velocity of the vehicle. The majority of the protocol changes would occur at the vehicle side of transactions instead of at the intersection side. Au et al. [2] discuss some of the feedback and control features that would be necessary to implement the modifications. In fact, they recommend the use of a "button" to make a reservation request and an "OK" indicator that would tell the driver to relinquish control of the semi-autonomous vehicle before it enters an intersection.

This control dynamic shared by the human and the intersection gives rise to a form of blended control. The intersection dictates when a specific reservation is possible, but the human has ultimate control over the movement of the vehicle. Of course, the vehicle would have to provide feedback such as lane-keeping and velocity warnings to maintain the tight trajectory constraints.

2.5 Synthetic Environment

Developing protocols that blend human control with automated systems requires an environment in which testing can be conducted safely. This research selected the three-dimensional (3D) virtual world called AFTR Burner, the successor to the STEAMiE engine, which utilizes the Open Dynamics Engine (ODE) for physics simulation and collision detection [9]. Incorporating the physics engine in the testing environment enables the human performance introduced by a protocol to be demonstrated and evaluated without endangering human participants and without incurring significant costs.

3. Proposed Design

This section establishes the viability of the proposed synthetic environment for handling an intersection that manages autonomous vehicular traffic. It also details the key protocol components that support human control in the environment. In particular, the section discusses the assumptions and the reservation concept, establishes a baseline and identifies the features required for persistent human control. It includes details about the reservation concept from the AIM protocol for safety, which is the primary goal. Also, it discusses details about V2X communications such as message timing and content, and feedback features needed to guide human drivers safely through an intersection.

3.1 Assumptions

The following assumptions were made in this research:

- **Latency:** The messaging protocol is abstracted to function calls between the vehicle and intersection world object classes. Latency (delay) between a sender and receiver is not modeled and is, therefore, assumed to be zero.

- **Signal Loss:** Signal loss in V2X communications is not modeled in the synthetic environment. While the potential for lost communications is real, this topic is left for future research.

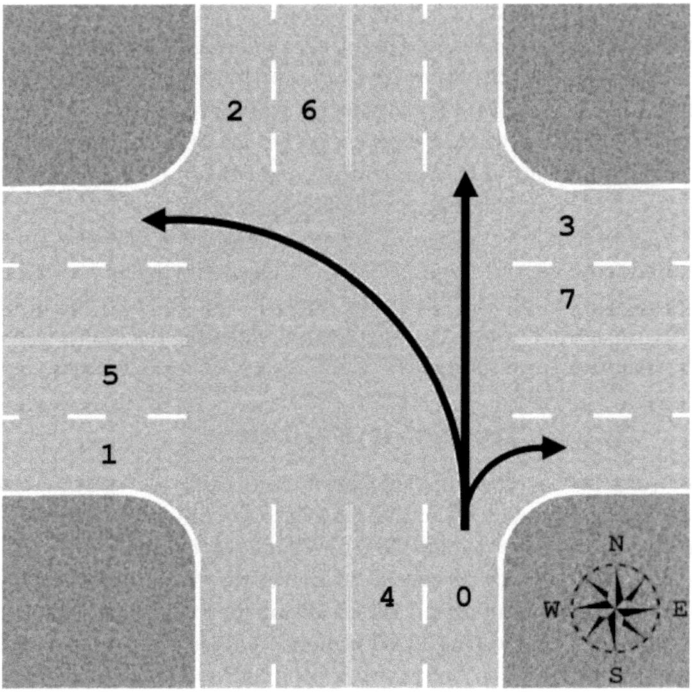

Figure 1. Legal turn direction options from a single lane with lane numbers.

- **Static Lanes:** No lane changes are permitted in an intersection. Turning vehicles move into their respective destination lanes (e.g., left turns from inside lanes terminate at inside lanes). Figure 1 illustrates the possible turning directions from each northbound lane. Note that Lane 0 northbound may turn into Lane 1 eastbound, but not to Lane 5 eastbound.

- **Turning Paths:** Turning paths are smooth or uniform, not abrupt or sharp (Figure 1).

- **Safety Buffer:** In an intersection, the occupied region includes a buffer of approximately 25% of the vehicle length and width for human-operated and autonomous vehicles. This parameter could be the subject of future research that balances safety and efficiency.

- **Stopping Distance:** The stopping distance is set to 25 m from the beginning of the intersection in every direction. This distance represents the beginning of a region where a vehicle must stop if no reservation is confirmed. A vehicle in this region is expected to follow its confirmed reservation.

- **Bounding Box:** The vehicular collision detection system uses a rectangular prism bounding box. This type of bounding box simplifies the

Figure 2. Bounding box of a sedan in the synthetic environment.

computations of the physics engine during collision detection while representing the shapes of vehicles. Figure 2 shows the bounding box of a sedan in the synthetic environment.

- **Velocity:** The maximum velocity before and after an intersection is 8 m/s for fully autonomous vehicles in the synthetic environment. Although this constraint could be relaxed in a future implementation, fully autonomous vehicles are assumed to have a constant velocity of 8 m/s in all directions at an intersection.
- **Single Intersection:** The synthetic environment has a single intersection with two inbound lanes from each cardinal direction.
- **Ambient Environment:** The synthetic environment has no obstructions – visual or otherwise (i.e., the environment is clear with high visibility).
- **Vehicles Only:** The synthetic environment has no obstacles, except for the intersection and other vehicles (i.e., no cyclists, pedestrians, animals or other moving entities).
- **Reservation Order:** A vehicle in a lane may request a reservation if and only if the vehicle directly in front of it already has a reservation. This is determined and enforced via V2V communications.

3.2 Reservations

The primary goal of an autonomous intersection without a traditional signaling system (traffic lights) is safety. Dresner and Stone [6] introduced the

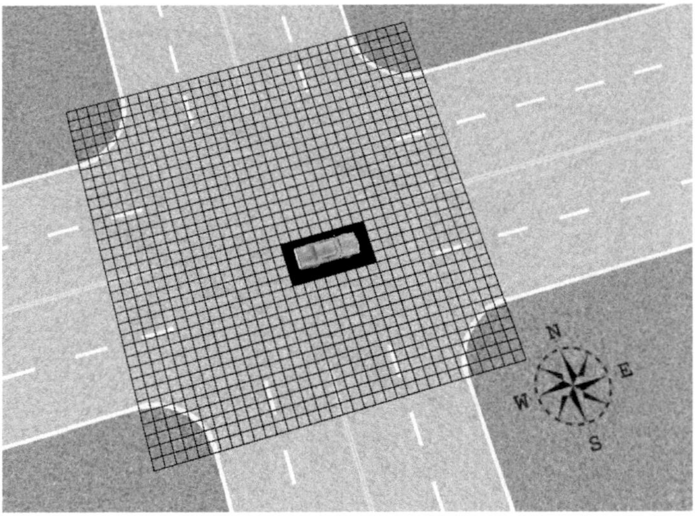

Figure 3. Reservation grid at an instant in time.

reservation concept used here. It divides the intersection into an arbitrary number of squares. The number of squares in each dimension is called the granularity of the reservation grid. A granularity of 34 was employed in this research – an intersection was divided into $34 \times 34 = 1,156$ individual squares.

Figure 3 shows a visual representation of the reservation grid in the synthetic environment at an instant in time when a vehicle was passing through the intersection. The darkened squares represent the space occupied by the vehicle.

The intersection maintains all the reservations for all the vehicles. When a vehicle makes a request, the arrival time, arrival lane, departure lane and velocity are used to determine if, at any instant in time, the proposed reservation overlaps with one or more previously-confirmed requests. If an overlap exists, then the request is denied. Interested readers are referred to [8] for a detailed description of the reservation system.

The reservation system is key to maintaining safety because no reservations are granted if overlaps are detected. The system also eliminates the need for traditional traffic signals because reservations are granted instead of shining green lights. In fact, this approach is highly efficient compared with traditional traffic signals and signs [8].

3.3 Synthetic Environment

The AFTR Burner virtual world was chosen to create the synthetic environment used in this research. Naturally, an algorithm for managing traffic that approaches an intersection is required as well. Although the AIM protocol has not been used in its entirety, many of its concepts are incorporated in a simplified manner in the synthetic environment. For example, the notion of the time-space reservation grid for the intersection manager and the pseudocode for the driver agent managing the autonomous vehicle and messaging protocol are both adapted to the synthetic environment.

A baseline comparison was performed to demonstrate that the intersection management algorithm is viable for handling fully autonomous vehicles. The comparison was conducted between the synthetic environment and the AIM simulator developed by Dresner and Stone [8]. The total average delay experienced and the number of safety violations (i.e., collisions) were selected as response variables in order to determine viability.

The baseline incorporated five trials for each of three traffic levels – 100, 200 and 300 vehicles per lane per hour. The maximum vehicular speed was set to 8 m/s and vehicles in the AIM simulator were limited to sedans. Using the data collection feature of the AIM simulator, the same traffic patterns were used in the synthetic environment to match the response variables in the trials. The relevant data collection items contained in the output included vehicle identification numbers, vehicle generation times, starting lane identifiers, destination identifiers, as well as the simulation exit times for autonomous vehicles.

After the fifteen trials in the AIM simulator and the synthetic environment were completed, the data collection files were compared. A two-tailed z-test with a significance level of $\alpha = 0.05$ was employed.

Table 2 summarizes the results. In every trial, the p-value is at least 0.05. Therefore, the null hypothesis H_o that the total average delays experienced in the AIM simulator and the synthetic environment are the same fails to be rejected. These results suggest that under the assumptions made, the implementation of the intersection manager, which is modeled after the AIM protocol, is roughly equivalent in its operation.

3.4 Messaging

The types of messages exchanged by a vehicle and intersection are modeled closely after those developed by Dresner and Stone [8]. Request messages are sent from a vehicle to the centralized intersection road-side unit (RSU). These messages provide information about the proposed arrival time, starting lane, destination direction and vehicle type. The vehicle type also includes the vehicle size and whether the vehicle is human-controlled. This modification enables the intersection to increase the safety buffer size around a vehicle to provide more flexibility with regard to arrival times and velocities. The road-side unit responds to a vehicle with confirmation messages, rejection messages and acknowledgement messages. A vehicle also can send cancellation messages.

The ovals represent the starting and ending states, rectangles represent processes or actions taken, and diamonds represent decisions. Dashed lines with arrows indicate (wireless) communications between the vehicle and intersection manager.

Table 2. Synthetic environment intersection management baseline results.

H_o: The total average delays experienced in the AIM simulator and the synthetic environment are the same.
H_a: The total average delays experienced in the AIM simulator and the synthetic environment are not the same.

100 Vehicles/Lane/Hour	**Trial 1**	**Trial 2**	**Trial 3**	**Trial 4**	**Trial 5**
AIM Simulator Delay (s)	0.1600	0.1648	0.1562	0.1568	0.1418
Synthetic Environment Delay (s)	0.1846	0.1503	0.1579	0.1933	0.1449
Two-Tailed z-Test p-Value	0.55	0.73	0.97	0.47	0.93
200 Vehicles/Lane/Hour	**Trial 1**	**Trial 2**	**Trial 3**	**Trial 4**	**Trial 5**
AIM Simulator Delay (s)	0.3846	0.3767	0.3690	0.2601	0.2764
Synthetic Environment Delay (s)	0.3327	0.3818	0.2813	0.3152	0.3048
Two-Tailed z-Test p-Value	0.40	0.93	0.05	0.28	0.55
300 Vehicles/Lane/Hour	**Trial 1**	**Trial 2**	**Trial 3**	**Trial 4**	**Trial 5**
AIM Simulator Delay (s)	0.6619	0.5724	0.5479	0.6420	0.6019
Synthetic Environment Delay (s)	0.5698	0.6905	0.5334	0.6782	0.5298
Two-Tailed z-Test p-Value	0.23	0.17	0.83	0.71	0.34

3.5 Human Controls and Feedback Displays

Enabling a human to maintain control of a semi-autonomous vehicle while navigating an intersection without traditional signals is not a trivial problem. In addition to traditional controls such as an accelerator, brake pedal, steering wheel, turn signals, mirrors and speedometer (to list a few), controls and/or displays must be provided to enable persistent human control at an autonomous intersection. This section describes the additional controls and displays that are required.

Figure 4 shows a high level view of the message decision making flow between an autonomous vehicle and intersection.

Currently, traditional road signs communicate information to drivers about upcoming hazards and roadway features such as sharp bends, intersections and speed limits. An in-dash indicator that notifies a human driver of an intersection that has come into range would be needed. In addition to this indicator, a button option [2] could be implemented to initiate vehicle communications with the autonomous intersection. After the button is pressed by the driver, the vehicle communicates with the intersection to arrange a reservation for passage, initiating the messaging flow shown in Figure 4.

When there is traffic congestion, the reservation request sent by a vehicle may be denied. In this scenario, there is a requirement to inform the driver

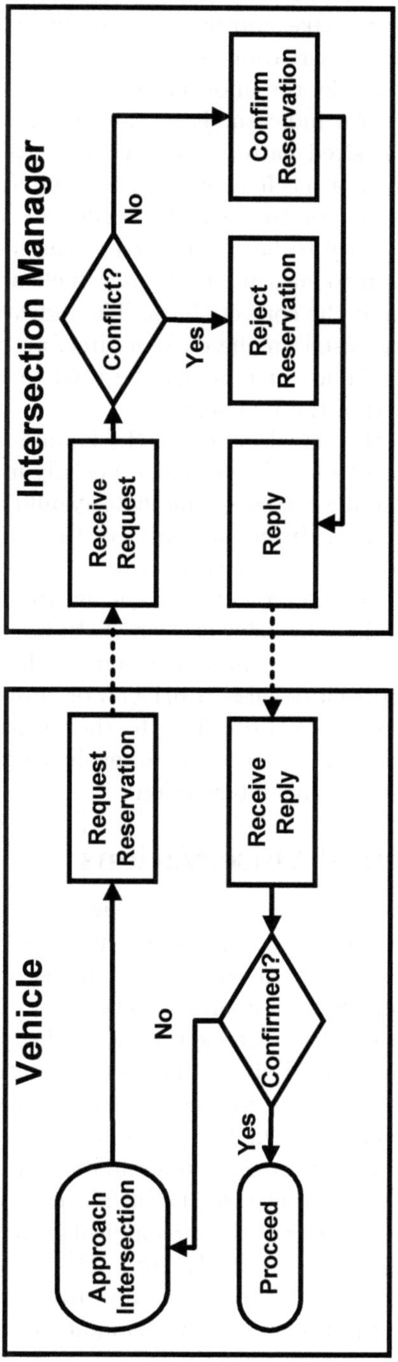

Figure 4. High-level view of a reservation messaging process.

about the reservation denial. For simplicity, a reservation denial would require the driver to slow the vehicle. At this point, the driver may press the button again or the vehicle may make another request automatically. This would continue until a successful reservation is made before entering the intersection.

After a reservation is made, the information supplied to the vehicle must be displayed to the driver. An indicator is needed to show that the reservation has been made for the desired path and the velocity to be maintained in the intersection. This requires a mechanism that communicates to the driver the goal velocity at arrival and/or the velocity needed to arrive at the required time, along with the velocity to be maintained in the intersection. This feedback mechanism is pivotal to ensuring that the vehicle arrives at and traverses through the intersection at the correct times. The indicator must continuously update the goal velocity based on the current time, arrival time and distance to the intersection. The indicator may also be used to communicate the goal velocity to be maintained in the intersection.

Finally, regardless of the vehicle velocity, the human must be able to maintain the correct lateral control of the vehicle, especially in the intersection. The corresponding path maintainer feedback indicator would notify the driver if the vehicle is too far left or right from the center of the current lane, along with the designated path through the intersection.

Table 3 summarizes the human controls and feedback devices required for persistent human control. The next section discusses the manner in which feedback information should be displayed to human drivers. Armed with the human controls and feedback devices, a driver would able to safely enter and navigate an autonomous intersection. Due to the security concerns, the synthetic environment provides the best venue for evaluating the ability of humans to safely traverse an autonomous intersection.

4. Experimental Observations

This section discusses the observations made when testing the proposed protocol that leverages additional human control and feedback devices. The experiments described in this section are notional and serve as proofs-of-concept instead of actual tests involving human subjects.

Figure 5 shows a screenshot of the synthetic environment with the human feedback mechanisms mentioned above. The screens outlined with thick black borders mimic the side-view and rear-view mirrors. The remaining displays present feedback information. On the left-hand side and moving from top to bottom are: the current time in the simulation, the arrival time of the confirmed reservation, the current velocity (m/s) and the goal velocity (m/s). The compass in the upper center of the screen displays one of the eight cardinal or intercardinal headings (i.e., N, NE, E, etc.). On the right-hand side of the screen and moving from top to bottom are: the current simulation name (used for reference purposes), the reservation status indicator and the digital lateral offset.

Table 3. Human controls and feedback devices required for persistent human control.

Item	**Description**
In-Range Indicator	This device informs the driver that an autonomous intersection is within range.
Request Reservation Button	This device initiates V2X communications to request a reservation from the intersection.
Denied Reservation Indicator	This device informs the driver that the requested reservation was denied.
Granted Reservation Indicator	This device informs the driver that the requested reservation was successful and provides the assigned velocity in the intersection.
Goal Velocity Indicator	This active device informs the driver of the velocity to be maintained to keep the reservation; the device may also be used to maintain the correct velocity in the intersection.
Maintain Path Indicator	This active feedback device informs the driver about the left/right position correctness based on the lane or planned path in the intersection.

Figure 5. Screenshot of the synthetic environment with human control.

Table 4. Reservation feedback mechanism states.

Vehicle Location	Reservation Status	Indicator Color
Out of Range	N/A	Clear
In Range	Unconfirmed	Yellow
In Range	Confirmed	Green
Within Stopping Distance	Unconfirmed	Red

Table 4 presents the reservation feedback mechanism states. The driver in the experiment with a current speed of 8.01 m/s was required to increase the velocity slightly to arrive at the intersection on time, as indicated by the goal speed (8.42 m/s) in the feedback display.

Analog versions of the goal speed and lateral feedback indicators are presented on a heads-up-display (HUD) in the direct line of sight of the human driver. The goal velocity is indicated by a green box that hovers around a horizontal black line. If the goal velocity is higher than the current velocity, then the green box hovers above the line; the green box hovers below the line if the goal velocity is lower than the current velocity.

The analog lateral feedback operates similarly. If the human veers to the right or left of the planned path, then the green box hovers horizontally to the left or right of the vertical black line, respectively.

Extreme deviations from the goal velocity and vehicle path turn the green box to a red box. The mechanisms in the heads-up-display are translucent to minimize obstructions to the driver's view.

The design and placement of feedback devices are important. The digital speedometer and goal speed indicator should be close to each other. In fact, an analog display may be better than a digital display. An analog speedometer could have the goal velocity indicated in a separate colored dial located directly above the current velocity dial. Drivers may prefer to have the option of choosing digital versus analog as well. Extensive testing is required to determine the optimal design and placement of the feedback devices.

Maintaining the center of the correct lane appears to be a straightforward task. However, maintaining the correct position in an intersection is more difficult. The path maintainer feedback device helps keep the proper placement of the vehicle, but it is largely reactive in nature. As a proactive measure, it would be prudent to mark the paths of turning lanes, as is done in many traditional intersections.

5. Conclusions

This chapter has laid the foundation for the extensions required for human-control of semi-autonomous vehicles, the ultimate goal being a protocol that maintains the efficiency of a fully autonomous environment while allowing human control of vehicles when navigating an intersection. The reservation-based autonomous intersection protocol derived from the AIM protocol [8] and implemented in the synthetic environment proved to be roughly equivalent to the AIM protocol given the assumptions made; this result was established by the baseline experiments. The limited feedback mechanisms enable a manually-controlled, semi-autonomous vehicle to safely approach, enter, traverse and exit an autonomous intersection, despite the fact that the intersection does not have traditional traffic control signals. In such a scenario, all the control signals must be transmitted to the vehicle via V2X communications at a rate of up to ten signals per second.

Introducing persistent human control has been shown to be feasible given the feedback mechanisms and controls. The AFTR Burner virtual world provides an appropriate synthetic environment. This highly-configurable synthetic environment supports extensive testing of the reservation-based autonomous intersection protocol as well as the integration of semi-autonomous vehicles.

Future research will attempt to determine the minimum amount of information required for a human driver to safely maintain vehicular control, and the optimal types and placement of the driver interaction and feedback mechanisms. Other research topics include maintaining vehicular velocities and paths, establishing safety buffer zones and integrating autonomous, semi-autonomous and legacy vehicles in a busy intersection while ensuring safe and efficient traffic flow.

Note that the views expressed in this chapter are those of the authors and do not reflect the official policy or position of the U.S. Air Force, U.S. Army, U.S. Department of Defense or U.S. Government.

References

[1] H. Abraham, B. Reimer, B. Seppelt, C. Fitzgerald, B. Mehler and J. Coughlin, Consumer Interest in Automation: Preliminary Observations Exploring a Year's Change, White Paper 2017-2, AgeLab, Massachusetts Institute of Technology, Cambridge, Massachusetts, 2017.

[2] T. Au, S. Zhang and P. Stone, Autonomous intersection management for semi-autonomous vehicles, in *Routledge Handbook of Transportation*, D. Teodorovic (Ed.), Routledge, New York, pp. 88–104, 2016.

[3] P. Bansal and K. Kockelman, Forecasting Americans' long-term adoption of connected and autonomous vehicle technologies, *Transportation Research Part A: Policy and Practice*, vol. 95, pp. 49–63, 2017.

[4] J. Doubek, Study backs getting driverless cars on the road, as Waymo ditches backup drivers, *National Public Radio*, November 10, 2017.

[5] J. Douceur, The Sybil attack, *Proceedings of the First International Workshop on Peer-to-Peer Systems*, pp. 251–260, 2002.

[6] K. Dresner and P. Stone, Multiagent traffic management: A reservation-based intersection control mechanism, *Proceedings of the Third International Joint Conference on Autonomous Agents and Multiagent Systems*, pp. 530–537, 2004.

[7] K. Dresner and P. Stone, Multiagent traffic management: An improved intersection control mechanism, *Proceedings of the Fourth International Joint Conference on Autonomous Agents and Multiagent Systems*, pp. 471–477, 2005.

[8] K. Dresner and P. Stone, A multiagent approach to autonomous intersection management, *Journal of Artificial Intelligence Research*, vol. 31, pp. 591–656, 2008.

[9] S. Nykl, C. Mourning, M. Leitch, D. Chelberg, T. Franklin and C. Liu, An overview of the STEAMiE educational game engine, *Proceedings of the Thirty-Eighth Annual Frontiers in Education Conference*, pp. F3B-21–F3B-25, 2008.

[10] G. Sharon and P. Stone, A protocol for mixed autonomous and human-operated vehicles at intersections, in *Autonomous Agents and Multiagent Systems (AAMAS 2017)*, G. Sukthankar and J. Rodriguez-Aguilar (Eds.), Springer, Cham, Switzerland, pp. 151–167, 2017.

[11] Society of Automotive Engineers International, Surface Vehicle Recommended Practice, J3016 SEP2016, Warrendale, Pennsylvania, 2016.

IV

INDUSTRIAL CONTROL SYSTEMS SECURITY

Chapter 12

A HISTORY OF CYBER INCIDENTS AND THREATS INVOLVING INDUSTRIAL CONTROL SYSTEMS

Kevin Hemsley and Ronald Fisher

Abstract For many years, malicious cyber actors have been targeting the industrial control systems that manage critical infrastructure assets. Most of these events are not reported to the public and their details along with their associated threats are not as well-known as those involving enterprise (information technology) systems. This chapter presents an analysis of publicly-reported cyber incidents involving critical infrastructure assets. The list of incidents is by no means comprehensive. Nevertheless, the analysis provides valuable insights into industrial control system threats and vulnerabilities, and demonstrates the increasing trends in the number and complexity of cyber attacks.

Keywords: Industrial control systems, cyber security, incidents, threats, trends

1. Introduction

Industrial control systems are embedded devices that operate critical infrastructure assets. These devices are typically unique to operational technology as opposed traditional (enterprise) information technology. This chapter describes the significant incidents involving industrial control systems along with their threats and vulnerabilities, and demonstrates the increasing trends in the number and complexity of attacks.

Cyber threats on industrial control systems manifest themselves in several ways. This chapter discusses the principal types of threats, which include directed attacks, malware attacks, cyber intrusion campaigns and cyber threat group activities.

Tables 1 and 2 detail the significant cyber incidents involving industrial control systems that are referenced in this study. The threat types, which include directed attacks, malware attacks, cyber intrusion campaigns and cyber threat group activities, are presented in chronological order. The open-source analy-

Published by Springer Nature Switzerland AG 2018.
J. Staggs and S. Shenoi (Eds.): Critical Infrastructure Protection XII, IFIP AICT 542, pp. 215–242, 2018.
https://doi.org/10.1007/978-3-030-04537-1_12

Table 1. Industrial control system incidents.

Year	Type	Name	Description
1903	Attack	Marconi wireless hack	Marconi's wireless telegraph presentation was hacked using Morse code.
2000	Attack	Maroochy Water Services breach	Wireless attack released more than 265,000 gallons of untreated sewage.
2008	Attack	Turkish pipeline explosion	Attackers may have exploited vulnerable security camera software to access the pipeline control network.
2010	Malware	Stuxnet malware	World's first publicly-known digital weapon.
2010	Malware	Night Dragon malware	Attackers used sophisticated malware to target global oil, energy and petrochemical companies.
2011	Malware	Duqu/Flame/ Gauss malware	Advanced malware that targeted specific organizations, including industrial control system vendors.
2012	Campaign	Gas pipeline cyber intrusion campaign	Active series of cyber intrusions that targeted the natural gas pipeline sector.
2012	Malware	Shamoon malware	Malware targeted major energy companies in the Middle East, including Saudi Aramco and RasGas.
2013	Attack	Target Stores attack	Hackers gained access to Target's sensitive financial systems via a contractor that maintained its HVAC industrial control systems.
2013	Attack	New York dam attack	U.S. Justice Department claimed that Iran conducted a cyber attack on the Bowman Dam in Rye Brook, NY.
2013	Malware	Havex malware	Malware attacks targeted industrial control systems.

sis is based on information provided by cyber security companies, independent security researchers, news media, published reports and government sources.

Attribution of attacks, as discussed in the open-source literature, is included for reader awareness. The list of incidents is by no means comprehensive. However, it covers the most significant incidents that have impacted industrial control systems and critical infrastructure assets. In some cases, the attacks focused directly on industrial control systems. In other cases, industrial control systems were indirectly targeted or impacted.

2. Cyber Incidents

This section discusses the cyber incidents listed in Tables 1 and 2. The incidents are discussed in chronological order.

Table 2. Industrial control system incidents (continued).

Year	Type	Name	Description
2014	Attack	German steel mill attack	Cyber attack on a steel mill caused massive damage.
2014	Malware	BlackEnergy malware	Malware targeted human-machine interfaces of control systems.
2014	Campaign	Dragonfly/Energetic Bear campaign no. 1	Ongoing cyber espionage campaign mainly targeting the energy sector.
2015	Attack	Ukraine power grid attack no. 1	First successful cyber attack on a country's power grid.
2016	Attack	Kemuri Water Company attack	Attackers accessed programmable logic controllers and altered water treatment chemicals.
2016	Malware	Return of Shamoon malware	Thousands of computers at Saudi Arabia's civil aviation agency and at Gulf State organizations were wiped in another Shamoon attack.
2016	Attack	Ukraine power grid attack no. 2	Attackers tripped breakers in 30 substations, turning off electricity to approximately 225,000 customers.
2017	Malware	CRASHOVERRIDE malware	Malware that caused the Ukraine power outage was finally identified.
2017	Group	APT33 Group campaign	Cyber espionage group targeted the aviation and energy sectors.
2017	Attack	NotPetya malware	Malware targeted Ukraine by posing as ransomware, but there was no way to pay ransom to decrypt files.
2017	Campaign	Dragonfly/Energetic Bear campaign no. 2	Symantec claimed that the energy sector was being targeted.
2017	Malware	TRITON/Trisis/ HatMan malware	Malware targeted industrial safety systems in the Middle East.

2.1 Marconi Wireless Hack

The world's first cyber incident likely involved the hacking of secure wireless communications. In 1903, the Italian radio pioneer, Guglielmo Marconi, prepared to present the first public demonstration of long-distance wireless communications using Morse code. The live demonstration intended to show that a wireless message could be sent securely from a cliff-top radio station in Poldhu, Cornwall (United Kingdom) to London, some 300 miles away.

However, before Marconi could begin his demonstration, the theater's brass projection lantern that displayed his slides began to click. To an untrained ear, it probably sounded as if the projection system was having technical difficulties. However, Marconi's assistant, Arthur Blok, recognized that the clickity-click coming from the lantern was Morse code [17]. The Morse code spelled out the following unexpected message:

Rats, rats, rats, rats.
There was a young fellow of Italy,
Who diddled the public quite prettily.

The message went on to mock Marconi. The demonstration had been hacked! But it was not apparent who the mysterious hacker was and why he hacked Marconi's demonstration.

A few days later, a letter in *The Times* confessed to the hack [46]. The hacker was British music hall magician, Nevil Maskelyne. It turned out that Maskelyne wanted to disprove Marconi's claim that his wireless telegraph device could send messages securely. The magician, much like today's security researchers, wanted to reveal a security hole for the public good.

Vulnerabilities in industrial control systems are often identified and reported by independent cyber security researchers. Nevil Maskelyne may well have been the first to publicly report a vulnerability in modern technology.

2.2 Maroochy Water Services Breach

In March 2000, Maroochy Water Services, a utility operated by the Maroochy Shire Council in Queensland, Australia, experienced problems with its new wastewater system. Communications sent by radio frequency (RF) signals to wastewater pumping stations failed. Pumps did not work correctly and alarms that were supposed to notify system engineers of faults did not activate as expected [18].

An engineer who was monitoring signals in the system discovered that someone was interfering with them and deliberately causing the problems. The water utility hired a team of private investigators who located the attacker and alerted police.

On April 23, 2001, police chased the automobile of 49-year-old Vitek Boden and ran him off the road. In his car, the police found a laptop and supervisory control and data acquisition (SCADA) equipment he had used to attack systems at Maroochy Water Services [6]. Investigations revealed that Boden's laptop was used when the attacks had occurred. Software for controlling the sewage management control system was discovered on his hard drive [60].

Boden had used a radio transmitter and his laptop to control some 150 sewage pumping stations. Over a three-month period, Boden released millions of gallons of untreated sewage into waterways and local parks [59]. The judge in the case ruled that the act was Boden's revenge for failing to obtain a security position with the Maroochy Shire Council [18].

In his post-incident analysis report, Robert Stringfellow, the civil engineer responsible for the water supply and sewage systems at Maroochy Water Services during the time of the breach, noted that:

- It is very difficult to protect against insider attacks.

- Radio communications commonly used in SCADA systems are generally insecure or improperly configured.

- SCADA devices and software should be secured to the extent possible using physical and logical controls.

- SCADA systems must record all device accesses and commands, especially those involving connections to or from remote sites [59].

The Maroochy Water Services breach is an example of a cyber attack that can be launched on an industrial control system to cause physical damage. In this (rare) case, the attacker was identified and prosecuted.

2.3 Turkish Pipeline Explosion

The 2008 Turkish pipeline explosion has been attributed to a cyber intrusion, but it was actually caused by a physical attack. In August 2008, a segment of the Baku-Tbilisi-Ceyhan (BTC) oil pipeline in Refahiye, eastern Turkey exploded during the Georgian War. Media reports attributed the explosion to a cyber nexus [14–16, 57].

Bloomberg [57] published the original report of the attack on December 10, 2014. However, a subsequent story in a major German newspaper casts significant doubt on a cyber attack causing the explosion [65]. An analysis by Lee [41] concludes that the pipeline explosion was not caused by cyber means. In fact, Lee notes "there are numerous reported and unreported cases of failures at [industrial control system] facilities where a cyber incident is to blame. Without the appropriate data, there will simply not be any lessons learned or resolution [as] to the root cause."

This event is included to make readers aware that this incident is often inaccurately cited as one of the first cyber incidents involving industrial control systems. It is also included to highlight the fact that cyber attribution for physical events can be difficult to ascertain.

2.4 Stuxnet Malware

When it was identified in 2010, Stuxnet was arguably the most sophisticated malware ever encountered [38]. It infected control system networks and may have damaged one-fifth of Iran's uranium hexafluoride centrifuges [79].

Turner [68], a Symantec executive, testified before the U.S. Senate Homeland Security Committee that Stuxnet was a wake-up call to critical infrastructure asset owners and operators around the world. Stuxnet reportedly targeted specific equipment operating in Iran's Natanz uranium enrichment facility [40, 76]. The U.S. Department of Homeland Security's Industrial Control Systems Cyber Emergency Team (ICS-CERT) issued multiple advisories about the Stuxnet malware, which also infected systems in the United States [22].

Stuxnet was dangerous because it self-replicated and spread throughout multiple systems via multiple means, which included:

- Removable drives by exploiting a vulnerability that allowed auto execution.
- Local-area networks (LANs) by exploiting a vulnerability in the Windows Print Spooler.
- Server Message Block (SMB), which provides shared access to files, printers and other devices, by exploiting a vulnerability in the Microsoft Windows Server Service.
- Network file sharing by copying and executing itself.
- Siemens WinCC human-machine interface (HMI) database server by copying and executing itself.
- Siemens Step 7 by copying itself into Step 7 projects so that it automatically executed when a Step 7 project was loaded.

Stuxnet exploited four unpatched Microsoft vulnerabilities, two vulnerabilities for self-replication and two for privilege escalation. These vulnerabilities were previously unknown and are referred to as zero-day vulnerabilities.

One of Stuxnet's significant features was its ability to install itself without being detected. This was accomplished using digitally-signed code produced by legitimate software developers, which had been stolen from two Taiwanese companies. Stuxnet leveraged these digital certificates to contact a command and control (C2) server that enabled the attackers to download and execute updated code.

Stuxnet was also stealthy in that it hid its binaries using a Windows rootkit. It attempted to evade detection by altering several security products if they were found on the targeted system. It also hid modified code in Siemens programmable logic controllers via a rootkit of sorts. Additionally, it modified the data sent from programmable logic controllers so that the human-machine interface displayed incorrect information to plant operators, making them believe that the system was operating normally.

Stuxnet was a precision weapon that looked for specific software to compromise and specific equipment to target. It terminated itself if it did not find the software and equipment as it propagated. When Stuxnet found what it sought, it modified and sabotaged Siemens programmable logic controller code by injecting ladder logic code.

The important lesson learned from Stuxnet is that a well-financed, sophisticated threat actor can likely attack any system. The ability to detect and recover from a cyber attack is also an important takeaway. This is because it is not possible to protect a system from all attacks.

2.5 Night Dragon Malware

Night Dragon is the name given by McAfee to the tactics, techniques and procedures (TTPs) used in coordinated, covert and targeted cyber attacks that were initiated in November 2009 and made public in 2010 [47]. The attackers in China utilized Night Dragon command and control servers in the United States and The Netherlands to target global oil, energy and petrochemical companies. The attacks involved social engineering, spear-phishing, exploitation of Microsoft Windows operating system vulnerabilities, Microsoft Active Directory compromises and the use of remote access Trojans (RATs) in targeting and harvesting sensitive operations-related data and project-financing information about oil and gas field bids and operations [47].

McAfee [47] reported that after the attackers had control of a targeted system, they exfiltrated password hashes and used a common cracking tool to obtain the passwords and access sensitive information. The exfiltrated files related to operational oil and gas production systems as well as financial documents pertaining to oil and gas field exploration and bidding. In some cases, the files were copied to and downloaded from company web servers by the attackers. In other cases, the attackers collected data from SCADA systems.

ICS-CERT issued an initial alert in February 2011 to warn U.S. critical infrastructure asset owners and operators of the Night Dragon threat [21]. The Night Dragon attacks were not sophisticated, but they demonstrated that simple techniques, applied by a skillful and persistent adversary, are enough to break into energy sector companies. More importantly, the attacks demonstrated that they could compromise industrial control systems. Equally concerning is that the tools used by the attackers enabled them to take complete control of systems using remote desktop capabilities. The attackers leveraged the tools to steal valuable information, but they could just as easily have seized control of human-machine interfaces, which would have enabled them to remotely control critical energy systems.

2.6 Duqu/Flame/Gauss Malware

In 2011, Hungarian cyber security researchers with the Laboratory of Cryptography and Systems Security at the Budapest University of Technology and Economics discovered the Duqu malware during an incident response investigation [1]. The Duqu malware was designed to gather information. According to the Hungarian researchers, Duqu bears a striking similarity to Stuxnet in terms of its design philosophy, internal structure and mechanisms, implementation details and the estimated amount of effort needed to create it.

Duqu leveraged a stolen digital certificate from a Taiwanese company, just as Stuxnet did. In both cases, the stolen certificates enabled the attackers to install malware on target systems. In fact, the digital certificates used by Duku and Stuxnet were stolen from businesses located in the same industrial park in Taiwan [80].

According to reports published by Symantec [61] and Kaspersky Lab [36], the Duqu executables share some code with Stuxnet and were compiled after the last Stuxnet sample was recovered. Duqu attempted to disguise its transmissions as normal HTTP traffic by appending the encrypted data to be exfiltrated in a JPG file [31].

Working with other international researchers, the same Hungarian researchers who identified Duqu also identified the Flame or sKyWIper malware. According to the researchers [1], Flame is extremely complex malware that steals information using:

- Microphones installed on systems.
- Web cameras.
- Keystroke logging.
- Extraction of geolocation data from images.

Flame could send and receive commands and data via Bluetooth, and it stored the collected data in SQL databases. It used network connections and USB flash drives for communications. Flame-infected computers masqueraded as proxies for Windows Update using a fake Microsoft certificate and employed an advanced collision attack on the MD5 hash function [1]. Kaspersky Lab researchers also found chunks of code from a 2009 Stuxnet variant inside Flame [36].

Kaspersky Lab subsequently identified malware they named Gauss, which is believed to be related to Duqu and Flame because they all used the same framework [1, 37]. The Gauss malware was also designed to steal information. In particular, it collected the following information from compromised systems:

- Passwords, cookies and browser history obtained by injecting its modules into browsers to intercept user sessions.
- Computer network connections.
- Processes and folders.
- BIOS and CMOS RAM information.
- Local, network and removable drive information.

Gauss also infected USB drives with a spy module to propagate to and steal information from other computers. It interacted with command and control servers to download additional modules and to send the collected information back to the attackers. ICS-CERT issued the initial reports on Duqu [31], Flame [28] and Gauss [29] in 2012.

The important takeaway from the Duqu, Flame and Gauss malware infections is that sophisticated threat actors perform reconnaissance to collect as much information as they can to further their objectives. The attackers used a number of methods to spread their information-stealing code, and they leveraged all the available information to learn about their targets. It is important to emphasize that the first step in the "cyber kill chain" is reconnaissance [44]. Information-stealing malware such as Duqu, Flame and Gauss are used by sophisticated attackers to initiate their cyber kill chain.

2.7 Gas Pipeline Cyber Intrusion Campaign

Beginning in late December 2011, ICS-CERT [19] identified an active series of cyber intrusions by a sophisticated threat actor that targeted natural gas pipeline companies. Analysis of the malware and artifacts associated with the intrusions revealed that the activities were part of a single campaign that leveraged spear-phishing. The spear-phishing attempts tightly focused on key personnel in pipeline companies. The emails were carefully crafted to appear as if they were sent by trusted company employees [19].

ICS-CERT issued an alert (ICSA-12-136-01BP) to the U.S. Computer Emergency Readiness Team (US-CERT) Control Systems Center secure portal library about the threat; information about the attacks was also disseminated to sector organizations and agencies to ensure broad distribution to asset owners and operators [20]. ICS-CERT recommended the implementation of defense-in-depth mechanisms and practices, and educating users about social engineering and spear-phishing attacks [25]. Organizations were also encouraged to review an ICS-CERT incident handling brochure for tips on preparing for and responding to incidents.

ICS-CERT, in coordination with the Federal Bureau of Investigation (FBI), U.S. Department of Energy, Electricity Sector Information Sharing and Analysis Center (ES-ISAC), Transportation Security Administration (TSA) and the Oil and Natural Gas and Pipelines Sector Coordinating Council's Cybersecurity Working Group, conducted a series of action campaign briefings during the 2013 fiscal year to respond to the growing number of cyber incidents involving U.S. critical infrastructure assets. Fourteen classified or unclassified briefings were given to more than 750 total attendees in cities across the country to assist critical infrastructure asset owners and operators in detecting intrusions and developing mitigation strategies [53]. These information sharing efforts made the energy sector more aware of the efforts undertaken by federal agencies to identify threats and help protect critical infrastructure assets.

2.8 Shamoon Malware

On August 15, 2012, the Shamoon malware attacked the computer systems of Saudi Aramco, the largest energy company in the world. The attackers carefully selected the one day of the year that they knew they could inflict the most damage – the day that more than 55,000 Saudi Aramco employees stayed home to prepare for one of Islam's holiest nights – Lailat al Qadr or the Night of Power, which celebrates the revelation of the Quran to Muhammad [54].

The Shamoon malware overwrote data and displayed an image of a burning American flag on more than 30,000 computers. Shamoon was designed to steal information, but it incorporated a destructive module that rendered infected systems unusable by overwriting the master boot record, partition tables and most of the files with random data. The overwritten information was not recoverable. Symantec discussed the malware in one of its official blogs on August 16, 2012 [62]. ICS-CERT also issued a report on the malware [30].

Eleven days later, on August 27, 2012, Shamoon hit its second target, the Qatari natural gas company, RasGas, which is one of the largest liquefied natural gas companies in the world [77]. Despite its destruction of tens of thousands of computers, there is no evidence that Shamoon directly impacted industrial control systems at Saudi Aramco or RasGas.

After infecting a computer, the Shamoon malware attempted to spread to other devices in the local network. Shamoon was programmed to download and run executables from a command and control server, enabling the attackers to manage operations, spread the infection and download additional tools on the victim computers for network traversal.

ICS-CERT [24] has provided guidance on best practices for continuity of operations when dealing with destructive malware like Shamoon. Saudi Aramco and RasGas learned the hard way that malicious actors can and do launch destructive attacks. A key takeaway from the Shamoon experience is that, in addition to protection, organizations must focus on recovering from destructive cyber attacks.

2.9 Target Stores Attack

Cyber intrusions into industrial control systems typically occur by attackers gaining access to corporate networks and then pivoting to control networks. However, the opposite occurred on November 15, 2013, when hackers broke into the computing network of a contractor that maintained Target's heating, ventilation and air conditioning (HVAC) control systems [75].

The cyber attackers, who sought to steal credit card data from Target Stores, first stole the login credentials of an HVAC contractor employee. This was accomplished by sending phishing emails. The victim was fooled by the email and clicked on the bait, enabling the installation of a variant of the Zeus banking Trojan, which provided the attackers with the login credentials needed to access the HVAC systems in Target Stores. Next, the attackers gained access to Target's business network from its building control systems, following which they uploaded malicious credit-card-stealing software to cash registers across Target's chain of stores [39].

According to a U.S. Department of Homeland Security report [9], the attack was part of a widespread operation that used the Trojan.POSRAM tool. The code is based on an earlier malicious tool called BlackPOS, which is believed to have been developed in Russia in 2013. However, the new variant was highly customized to evade detection by anti-virus programs [74, 78].

The breach exposed approximately 40 million debit and credit card accounts. Customer names, credit/debit card numbers, expiration dates and CVV data were stolen. Seventy million customers were affected. The attack itself, along with security upgrades and lawsuits, cost Target about $309 million [45]. Financial institutions whose debit/credit cards were targeted incurred $200 million in expenses.

The Target breach demonstrates the importance of securing building automation systems from cyber attacks.

2.10 New York Dam Attack

According to the U.S. Justice Department [56], Bowman Dam, a small dam near Rye Brook, New York was accessed by Iranian hackers in 2013. The intrusion was not sophisticated, but is believed to have been a test by the Iranians to see what systems they could access.

The Bowman Dam controls storm surges. Its SCADA system was connected to the Internet via a cellular modem. Fortunately, the SCADA system was undergoing maintenance at the time of the attack; thus, no control was possible, just status monitoring. In fact, since the dam merely functioned as a sluicegate for a small village, there was no significant threat to public safety.

Technical details of the Bowan Dam intrusion are deemed protected critical infrastructure information (PCII) and cannot be released to the public. However, it is believed that the dam was not specifically targeted. Its vulnerable Internet connection and lack of security controls were exploited by the opportunistic attackers to gain access [2]. A U.S. federal indictment disclosed that the attack was conducted by entities from ITSec Team and Mersad Company, two private computer security companies based in Iran [71]. These companies perform work for various Iranian Government organizations, including the Islamic Revolutionary Guard Corps (IRGC).

The attack on the Bowman Dam is a concern due to the country of origin of the attackers and the technical capabilities they demonstrated in directly manipulating SCADA equipment. It is possible that the Iranian attackers selected the small Bowman dam simply because it was low-hanging fruit. The important takeaway is that critical infrastructure control systems connected to the Internet are easy for potential attackers to detect and surveil, and eventually target.

2.11 Havex Malware

In 2013, a remote access Trojan named Havex (or Oldrea) that targeted industrial control systems was discovered. In 2016, the U.S. Department of Homeland Security and FBI released a report [10] that tied Havex to Russia's civilian and military intelligence services (RIS). The U.S. Government refers to this malicious activity as GRIZZLEY STEPPE; it also goes by the names Dragonfly and Energetic Bear.

Havex communicated with a command and control server that deployed modular payloads; this enabled the malware to acquire additional functionality. ICS-CERT identified and analyzed a payload that enumerated connected network resources such as computers and shared resources [23]‘. The Distributed Component Object Model (DCOM) based version of the Open Platform Communications (OPC) standard was leveraged to collect information about network resources and connected industrial control devices.

The Havex control-system-specific payload gathered server information, including CLSID, server name, program ID, OPC version, vendor information, running state, group count and server bandwidth. In addition to obtaining generic OPC server information, the Havex payload could enumerate OPC tags. However, Havex was not without flaws. It caused multiple common OPC platforms to crash intermittently; ICS-CERT has issued a warning that this could disrupt applications reliant on OPC communications [23].

The major concerns regarding Havex are its connection to Russia's civilian and military intelligence services, and the fact that it is advanced malware that targeted industrial control systems used in U.S. critical infrastructure assets. Another concern is that its command and control infrastructure enables the malware to acquire unknown enhanced capabilities.

2.12 German Steel Mill Attack

The 2014 annual report of the German Federal Office for Information Security (Bundesamt fur Sicherheit in der Informationstechnik (BSI)) mentions an attack on an unspecified German steel mill [11]. According to BSI, the attack was carried out using spear-phishing and social engineering tactics. The attackers initially gained access to the corporate network of the steel plant. From there, they worked their way into the production network. The attackers caused multiple failures of individual control systems, eventually preventing a blast furnace from shutting down in a controlled manner. This resulted in "massive damage to the plant."

The technical abilities of the attackers were described as "very advanced" [11]. Specifically, the attackers had expertise in information technology security as well as detailed knowledge of industrial control systems and the steel production process. The description in the BSI report and historical information about process plant incidents lead many to believe that the damage to the plant was intentional [42].

The German steel mill cyber attack is significant because of the physical damage that resulted and the German Government's willingness to release information about the incident. According to the BSI [11], "[t]he most significant component of this report is the demonstrated capability and willingness of an adversary to attack through traditional advanced persistent threat (APT) style methods and then advance to a cyber-physical attack with the intent to impact an operational environment."

2.13 BlackEnergy Malware

Starting in 2014, ICS-CERT published a series of alerts describing a sophisticated malware campaign that had compromised numerous industrial control systems using a variant of the BlackEnergy malware [26]. The analysis indicated this campaign had been ongoing since at least 2011. The 2016 U.S. Department of Homeland Security and FBI Joint Analysis Report [10], which identified Havex as coming from Russia's civilian and military intelligence services (RIS) group, connected BlackEnergy to the group as well.

Human-machine interface products from multiple vendors were targeted by the malware, including GE Cimplicity, Advantech/Broadwin WebAccess and

Siemens WinCC. The malware was modular and all its functionality was not necessarily used to target its victims. Typical BlackEnergy infections involved searches for network-connected file shares and removable media that could aid the malware in moving laterally in the infected environment [26].

In December 2014, the U.S. Department of Homeland Security confirmed that a BlackEnergy 3 malware variant was present in a Ukrainian energy system that was attacked to cause a power outage. ICS-CERT published a special (TLP Amber) version of an alert containing additional information about the malware, plug-ins and indicators. ICS-CERT strongly encouraged infrastructure asset owners and operators to use the indicators to look for signs of compromise in their control system environments.

In December 2014, ICS-CERT partnered with the FBI to give classified and unclassified threat briefings to critical infrastructure stakeholders across the country. Teams from ICS-CERT and the FBI traveled to fifteen cities across the United States. In total, nearly 1,600 participants involved in critical infrastructure protection across all sixteen sectors attended the briefings.

Like Havex, BlackEnergy targeted important industrial control system products. It is a major concern when adversaries target control systems used in the critical infrastructure. The BlackEnergy malware provided valuable insights into nation state actors and the tools they use to target critical infrastructure assets.

2.14 Dragonfly/Energetic Bear Campaign No. 1

On June 30, 2014, Symantec's MSS Global Threat Response described an ongoing cyber espionage campaign dubbed Dragonfly [49]. Other reports refer to the same campaign as Energetic Bear or Crouching Yeti [34]. The Dragonfly campaign primarily targeted the energy sector. The campaign focused on espionage and persistent access, with sabotage as an optional capability. The malware used the Havex (or Oldrea) malware as its favored tool and the Karagany remote access Trojan as a secondary tool. The Symantec group said that it had observed attacker activity in the United States, Turkey and Switzerland; some traces were seen in other countries as well [49].

The 2014 Dragonfly campaign was assessed to be exploratory in nature, where the attackers focused on attempting to gain access to the networks of the targeted organizations [49]. Dragonfly/Energetic Bear were later identified by the U.S. Department of Homeland Security and FBI as being connected to the GRIZZLEY STEPPE malicious activity perpetrated by Russia's civilian and military intelligence services (RIS) [10].

2.15 Ukraine Power Grid Attack No. 1

Two days before Christmas 2015, a cyber attack cut electricity to nearly a quarter-million Ukrainians. This was the first successful cyber attack on a power grid.

Reuters reported that a power company located in western Ukraine suffered a power outage, which impacted a large area that included the regional capital of Ivano-Frankivsk [55]. Attackers shut off power at 30 substations and left about 230,000 people without electricity for up to six hours. SCADA equipment was rendered inoperable and power had to be restored manually, further delaying restoration efforts [81].

Investigators discovered that attackers used the BlackEnergy malware to exploit macros in Microsoft Excel documents. The malware was planted in the company's network using spear-phishing [82]. ICS-CERT and US-CERT worked with the Ukrainian CERT and other international partners to analyze the malware, and confirmed that a BlackEnergy 3 variant was present [26]. The Ukrainian intelligence community blamed the attack on Russian actors [83]. BlackEnergy has also been publicly identified by the U.S. Department of Homeland Security and FBI as being connected to the GRIZZLEY STEPPE malicious activity perpetrated by Russia's civilian and military intelligence services (RIS) [70].

At the request of the Ukrainian Government, a U.S. interagency team comprising representatives from ICS-CERT and US-CERT, the Department of Energy, FBI and North American Electric Reliability Corporation (NERC), traveled to Ukraine to gather information about the incident and identify potential mitigations [53].

The Ukraine attack showed the world that it is possible to damage the power grid through cyber means. It was also a wake-up call to fortify the U.S. power grid against attacks. In the case of the Ukraine attack, relatively unsophisticated techniques were used to good effect. Indeed, the Ukraine power grid attack of 2015 will go down as a significant event in cyber attack history.

2.16 Kemuri Water Company Attack

In 2016, Verizon reported that an undisclosed water company experienced a cyber attack on its industrial control systems [72]. Verizon gave the water company the fictitious name "Kemuri" to protect its identity. According to Verizon, attackers accessed the water district's valve and flow control application responsible for manipulating hundreds of programmable logic controllers that managed water treatment chemical processing. The attackers then manipulated the system to alter the amount of chemicals entering the water supply. This affected water treatment and production capabilities, causing water supply recovery times to increase.

According to Verizon, a hacktivist group with ties to Syria was behind the attack. The Kemuri breach could easily have been much worse. Verizon noted that if the actors had a little more time and a little more knowledge of the industrial control systems, Kemuri and the local community could have suffered serious consequences.

A key takeaway from the Kemuri attack is that Internet-facing industrial control systems are a bad practice that can place critical infrastructure assets at serious risk. The Kemuri attack is also a reminder that malicious cyber actors are not afraid to cross the line and cause harm.

2.17 Return of Shamoon Malware

In November 2016, a second wave of attacks by the Shamoon malware was launched at selected targets in Saudi Arabia [3]. Thousands of computers in the Saudi Arabian civil aviation agency and other Gulf State organizations were wiped by Shamoon after it resurfaced some four years after attacking tens of thousands of Saudi Aramco and Qatari RasGas workstations.

Symantec discovered a strong correlation between the Timberworm cyber attack group and the Shamoon malware [63]. Timberworm appeared to have gained access to the networks of the targeted organizations weeks and, in some cases, months before the 2016 Shamoon attacks.

In December 2016, the U.S. Defense Security Service issued a security bulletin to cleared contractors warning them of the Shamoon malware [5].

The concern raised by the second Shamoon attack is the repeated use of destructive malware to target critical infrastructure assets. Critical infrastructure asset owners and operators need to be vigilant and bolster their defense postures. They must draw on the lessons learned from the Shamoon attacks to protect their assets.

2.18 Ukraine Power Grid Attack No. 2

On December 17, 2016, almost one year after Ukraine suffered a major cyber attack on its power grid, Kiev suddenly went dark. Cyber attackers had caused power grid monitoring stations to go blind. Breakers were then tripped in 30 substations, turning off electricity to approximately 225,000 customers.

To prolong the outage, the attackers launched a telephone denial-of-service attack against the utility's call center to prevent customers from reporting the outage; the same tactic was used in 2015. The intruders also rendered devices, such as serial-to-Ethernet converters, inoperable and unrecoverable to make it harder to restore electricity to customers [43]. Despite these setbacks, power was restored in three hours in most cases. However, because the attackers had sabotaged energy management systems, workers had to travel to the substations and manually close the circuit breakers that the attackers had opened remotely [81, 82].

The second Ukraine power grid attack was much more sophisticated than the first attack [43]. While the first attack used remote control software to manually trip breakers, the second attack leveraged sophisticated malware that directly manipulated SCADA systems. Lee [7], a Dragos expert, said, "[i]n my analysis, nothing about this attack looks like it's singular. The way it's built and designed and run makes it look like it was meant to be used multiple times. And not just in Ukraine."

The sophisticated malware used in the second attack is now referred to as CRASHOVERRIDE.

2.19 CRASHOVERRIDE Malware

Dragos [7], working with the Slovak anti-virus firm ESET [4], confirmed that the CRASHOVERRIDE (or Industroyer) malware was employed in the December 17, 2016 cyber attack on a Kiev, Ukraine transmission substation (Ukraine power grid attack no. 2 above).

According to the Dragos report [7], CRASHOVERRIDE was the first malware framework specifically designed and deployed to attack electric power grids. It is the fourth piece of malware tailored to target industrial control systems, with Stuxnet, BlackEnergy-2 and Havex being the first three. It is the second malware designed and deployed to disrupt industrial processes, with Stuxnet being the first [7]. The Dragos report also states that the CRASHOVERRIDE framework served no espionage purpose – its only real feature was to launch attacks that caused electric power outages.

The CRASHOVERRIDE malware framework has modules specific to industrial control protocol stacks, including IEC 101, IEC 104, IEC 61850 and OPC. It is designed to allow the inclusion of additional payloads such as DNP3, but as of this time, no such payloads have been confirmed. The malware also contains additional (non-control-system-specific) modules, such as a wiper, to delete files and disable processes on a running system in order to disrupt operations or damage equipment [7].

The CRASHOVERRIDE modules were leveraged to open circuit breakers on remote terminal units (RTUs) and force them into infinite loops in order to keep the breakers open. When power grid operators attempted to close the breakers, the substations were de-energized; thus, the breakers had to be closed manually to restore power [7].

According to the Dragos report [7], CRASHOVERRIDE could be leveraged to disrupt grid operations that would result in power outages. The power outages could last up to a few days if an attack targeted multiple sites. However, the report also mentions that there is no evidence that CRASHOVERRIDE could cause power outages to last longer than a few days. The extended outages could be achieved by targeting multiple sites simultaneously, which is entirely possible, but not easy.

On June 12, 2017, the U.S. Department of Homeland Security used the National Cyber Awareness System (NCAS) to issue a Technical Analysis Alert on June 12, 2017 that notified the U.S. critical infrastructure community about the serious threat posed by the CRASHOVERRIDE malware. The main takeaway from CRASHOVERRIDE is that a nation state actor has created an advanced reusable malware framework designed to cause power outages. This same threat actor has demonstrated on multiple occasions that it is willing and able to induce electric power outages via cyber means.

2.20 APT33 Group

In 2017, FireEye published a report detailing a cyber threat actor they named APT33 [52]. According to FireEye's analysis, APT33 is a capable group that has conducted cyber espionage operations since at least 2013. FireEye assessed that APT33 works at the behest of the Iranian Government.

APT33 has shown particular interest in aviation sector companies involved in military and commercial projects, as well as energy sector companies with ties to petrochemical production. According to FireEye, the targeting of companies involved in energy and petrochemicals mirrors previous targeting by other suspected Iranian threat groups, indicating a common interest in the sectors across Iranian actors. The targeted countries include the United States, Saudi Arabia and South Korea. FireEye also warns that APT33 may have ties to other groups with destructive capabilities.

APT33 delivered its malware by leveraging spear-phishing emails sent to employees of the targeted companies. The emails included recruitment-themed lures with links to malicious HTML application (HTA) files that contained job descriptions and links to legitimate job postings on popular employment websites. The spear-phishing emails were very relevant and appeared to be legitimate – they referenced specific job opportunities and salaries, provided links to spoofed companies' employment websites, and even included the companies' equal opportunity hiring statements.

A major concern is that the APT33 attack group has significant capabilities and ties to the destructive Shamoon malware. The group is also tied to the SHAPESHIFT malware that can wipe disks, erase volumes and delete files. FireEye believes that some of the tools used by APT33 may be shared with other Iran-based threat actors.

2.21 NotPetya Malware

Also in 2017, malware posing as the Petya ransomware surfaced in Ukraine. The earlier Petya malware targeted Microsoft Windows systems. After a system was infected, the malware encrypted the filesystem and displayed a message demanding payment in Bitcoin in order to regain access. However, while the new malware appeared to be based on and functioned like the Petya ransomware, it was different. It encrypted data on a hard drive just like Petya, but there was no way to decrypt the data. Unlike the Petya malware, the encryption was permanent; therefore, the new malware was given the name "NotPetya."

NotPetya is destructive malware. It has been enhanced to spread widely and is believed to have specifically targeted Ukraine [13]. On June 30, 2017, ICS-CERT issued an alert that warned the U.S. critical infrastructure community about the NotPetya threat [27].

In February 2018, the U.S. Government blamed the Russian military for developing and releasing NotPetya, stating that it was "reckless" and caused billions of dollars in damage [48]; it also called NotPetya the "most destructive and costly cyber attack in history" [67, 73]. The U.K. and Australian Governments also identified the Russian Government as being responsible for NotPetya [12]. The Russian Government has denied these accusations of its involvement with the malware [33, 66].

NotPetya is a significant concern because the nation state responsible for the malware – as confirmed by intelligence agencies in three countries – has demonstrated its ability and willingness to conduct destructive cyber attacks against critical infrastructure assets. A statement by the White House Press Secretary says that NotPetya has caused billions of dollars in damage across Europe, Asia and the Americas [67].

2.22 Dragonfly/Energetic Bear Campaign No. 2

In October 2017, Symantec published a report claiming that the energy sector was being targeted by a sophisticated attack group it referred to as a version of Dragonfly [58]. This group was well resourced, with a range of malware tools at its disposal and capable of launching attacks via a number of vectors. Symantec referred to this new Dragonfly activity as Dragonfly 2.0. In a vicious attack campaign, Dragonfly 2.0 compromised a number of industrial control equipment vendors, infecting their software with a remote access Trojan.

The Dragonfly 2.0 campaign shows that attackers may be entering a new phase, with new campaigns potentially providing them with access to operational systems – access that could be used for more disruptive purposes in the future. According to the Symantec report [58], this group is interested in learning how energy facilities operate as well as gaining access to operational systems. One of the report's most concerning assessments is that Dragonfly 2.0 can sabotage or gain control of industrial control systems.

On October 20, 2017, the U.S. Department of Homeland Security and FBI issued an initial alert about an advanced persistent threat that targeted government entities and organizations in the energy, nuclear, water, aviation and critical manufacturing sectors [70]. The alert described it as a multi-stage intrusion campaign that initially targeted low security and small networks, and then moved laterally to major networks and high value assets in the energy sector. Based on malware analysis and observed indicators of compromise, US-CERT has indicated with confidence that the campaign is still ongoing, and that the threat actors are actively pursuing their objectives over a long-term campaign [70].

The Dragonfly and Energetic Bear threat groups were publically identified by the U.S. Department of Homeland Security and FBI as being part of the same group they call GRIZZLEY STEPPE [70]. This information about Dragonfly reveals that the threat actor has continued its activities and that its capabilities have evolved. The Symantec Security Response Attack Investigation Team [58] states that " [the attackers] may have entered into a new phase with access to operational systems that could be used for more disruptive purposes in the future."

2.23 TRITON/Trisis/HatMan Malware

At the end of 2017, FireEye reported the existence of a new industrial control system attack framework called TRITON that was designed to disrupt critical infrastructure operations [35]. The report claims that the malware targeted industrial safety systems in the Middle East. Symantec Security Response reported this malware in late 2017, but referred to it as Trisis [64]. In December 2017, ICS-CERT also reported the same malware, but gave it a third name, HatMan [32].

The malware targeted Schneider Electric's Triconex safety instrumented system by modifying in-memory firmware to add malicious functionality. Specifically, the malware enabled the attackers to read/modify memory contents and execute custom code on demand upon receiving specially-crafted network packets from the attackers [32], as well as execute additional code that disables, inhibits or modifies the ability of a process to fail safely. The malware is especially dangerous because it targets safety systems [64].

It important to note that the TRITON malware is narrowly targeted and likely does not pose an immediate threat to other Schneider Electric customers or products. However, its capabilities, methodologies and tradecraft could be replicated by other attackers. Thus, it poses another serious threat to industrial control systems and the critical infrastructure assets they manage [8].

The most concerning aspect of TRITON is that it is the first malware to specifically target industrial safety systems that protect human lives. This capability can potentially be replicated by other attackers to cause physical damage and harm people.

3. Lessons Learned

The cyber events discussed in this chapter provide insights into the technical capabilities of key threat actors and how they have evolved. Their willingness to cause physical damage is significant.

Stuxnet was a game changer. This piece of malware demonstrated that the physical infrastructure can be significantly impacted – even destroyed – by cyber means. A key Stuxnet takeaway for critical infrastructure owners and operators is that a sophisticated and well-financed threat actor can likely attack any system it desires.

The cyber events demonstrate that several critical infrastructure assets have been attacked. The attacks are expected to increase in number and sophistication. Therefore, critical infrastructure owners and operators must develop the abilities to detect and recover from cyber attacks. Protecting all the systems in a large critical infrastructure asset from all attackers is not possible. Attacks such as Night Dragon reveal that simple techniques applied by a skillful and persistent adversary are enough to break into critical infrastructure assets, including vital energy sector assets.

The cyber events discussed above also provide visibility into the advanced techniques used in cyber attacks. The Duqu, Flame and Gauss malware demonstrate that sophisticated threat actors perform reconnaissance to collect as much information as possible to ensure success. It is especially important to understand that the first step in the cyber kill chain is reconnaissance [44].

Table 3. Most prevalent industry weaknesses (2017) [50].

Weakness Area	Rank	Risk
Boundary Protection	1	* Undetected unauthorized activity in critical systems. * Weak boundaries between industrial control networks and enterprise networks.
Identification and Authentication (Organizational Users)	2	* Lack of accountability and traceability for user actions when accounts are compromised. * Increased difficulty in managing accounts when users leave an organization, especially especially users with administrative access.
Allocation of Resources	3	* No backup or alternate personnel to fill a position if the primary is unable to work. * Loss of critical control systems knowledge.
Physical Access Control	4	* Unauthorized physical access to field equipment and locations provides increased opportunities to: – Maliciously modify, delete or copy device programs and firmware. – Access the industrial control network. – Steal or vandalize cyber assets. – Add rogue devices to capture and retransmit network traffic.
Account Management	5	* Compromise of unsecured password communications. * Password compromise could enable unauthorized access to critical systems.
Least Functionality	6	* Increased vectors for malicious party access to critical systems. * Rogue internal access.

Information-stealing malware – as exemplified by Duqu, Flame and Gauss – is how sophisticated attackers begin the cyber kill chain.

The Target breach demonstrates that one of the weakest links may be the security of building automation systems. More than half-a-billion dollars in costs were incurred as a result of poor building automation security.

The most important lesson is that nation states are actively developing capabilities to attack critical infrastructure assets. The two Ukraine attacks demonstrate that cyber attacks can disrupt and damage an electric power grid. GRIZZLEY STEPPE malicious activity perpetrated by Russia's civilian and military intelligence services (RIS) shows that nation states have the resources to develop and deploy sophisticated attacks on the critical infrastructure. Just

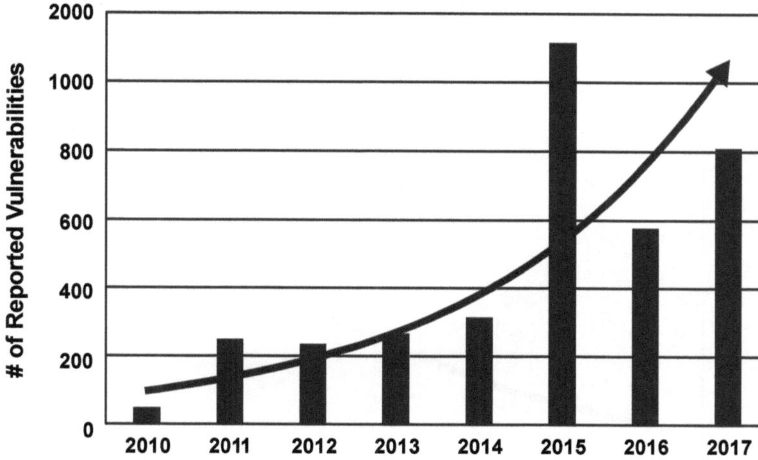

Figure 1. Reported industrial control system vulnerabilities [51].

as important is the fact that attackers are willing and able to launch destructive attacks.

The U.S. Department of Homeland Security conducted more than 130 industrial control system security assessments in 2017. Table 3 lists the top six areas of weakness. Boundary protection was ranked as the most prevalent weakness and has been the top weakness since 2014. The risks from boundary protection vulnerabilities are: (i) undetected unauthorized activity in critical systems; and (ii) weak boundaries between industrial control networks and enterprise networks.

Critical infrastructure asset owners and operators can undertake basic cyber hygiene actions to mitigate the risks [25]. However, more research and development efforts are needed to strengthen boundary protection for industrial control systems.

Figure 1 shows the number of industrial control system vulnerabilities reported annually to the U.S. Department of Homeland Security. Although not all vulnerabilities are reported to the U.S. Department of Homeland Security, the presented data is a good proxy for demonstrating the growth of industrial control system vulnerabilities. The massive increase in reported vulnerabilities from approximately 48 in 2010 to 806 in 2017 underscores the need to focus on protection as well as mitigation of the negative impacts of attacks.

Figure 2 highlights the trends in cyber attacks on industrial control systems. The notional graphic illustrates that the number and complexity of cyber attacks on industrial control systems are increasing. The increased complexity of cyber attacks makes them more difficult to detect and mitigate.

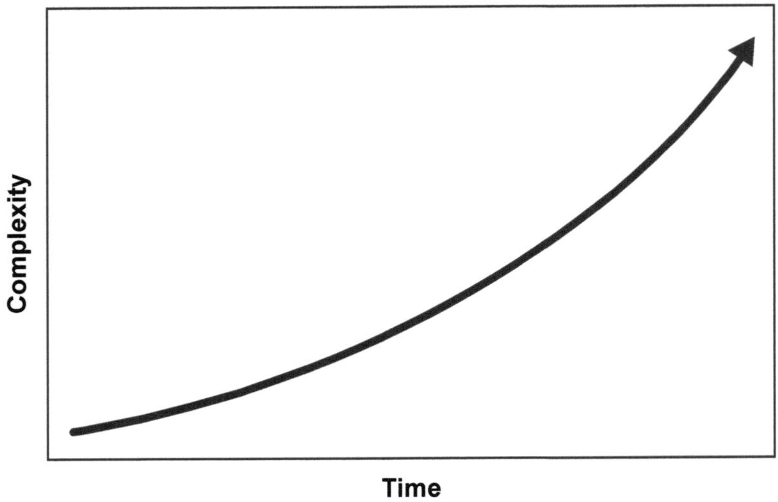

Figure 2. Trends in industrial control system cyber attacks.

4. Conclusions

The analysis of publicly-reported cyber incidents involving critical infrastructure assets provides valuable insights into industrial control system threats and vulnerabilities. Also, it highlights the changing landscape and growing threats to industrial control systems and, by extension, the critical infrastructure. The skill level of sophisticated threat actors is also increasing as is the frequency of attacks targeting critical infrastructures and the systems that control them.

Cyber threats are very real and appropriate investments in cyber security should be made by critical infrastructure asset owners and operators. Many of the threat actors targeting industrial control systems are well resourced and have advanced skills and knowledge. The defenders of these systems must have adequate resources as well as advanced skills and knowledge to prepare for and respond to cyber attacks.

Critical infrastructure protection, already an urgent problem in our time, will be compounded as the Internet of Things increases its penetration. The Internet of Things comprises ubiquitous networked devices – sensors and actuators – that support novel and innovative capabilities. These devices have become entrenched in our daily lives from the devices we wear to the vehicles we drive to the devices that manage critical infrastructure assets. Because Internet of Things systems are extensions of industrial control systems, cyber security will become more complex and require even greater attention to protect the critical infrastructure. This chapter is a call to arms to the critical infrastructure community to prepare for and respond to cyber attacks now and in the future.

References

[1] B. Bencsath, Duqu, Flame, Gauss: Followers of Stuxnet, presented at the *RSA Conference Europe*, 2012.

[2] J. Berger, A dam, small and unsung, is caught up in an Iranian hacking case, *The New York Times*, March 25, 2016.

[3] S. Chan, Cyberattacks strike Saudi Arabia, harming aviation agency, *The New York Times*, December 1, 2016.

[4] A. Cherepanov, WIN32/INDUSTROYER: A New Threat for Industrial Control Systems, Version 2017-06-12, ESET, Bratislava, Slovakia (`www.welivesecurity.com/wp-content/uploads/2017/06/Win32_Industroyer.pdf`), 2017.

[5] J. Cox, Department of Defense warns contractors about Iran-linked malware, *Motherboard*, December 16, 2016.

[6] M. Crawford, Utility hack led to security overhaul, *Computerworld*, February 16, 2006.

[7] Dragos, CRASHOVERRIDE: Analysis of the threat to electric grid operations, Hanover, Maryland (`www.dragos.com/blog/crashoverride/CrashOverride-01.pdf`), 2017.

[8] Dragos, TRISIS malware: Analysis of safety system targeted malware, Hanover, Maryland (`www.dragos.com/blog/trisis/TRISIS-01.pdf`), 2017.

[9] Department of Homeland Security (DHS), Backoff: New Point of Sale Malware, Washington, DC, 2014.

[10] Department of Homeland Security (DHS) and Federal Bureau of Investigation (FBI), Joint Analysis Report (JAR-16-20296A): GRIZZLY STEPPE – Russian Malicious Cyber Activity, December 29, 2016.

[11] Federal Office for Information Security, The State of IT Security in Germany 2014, BSI-LB15503e, Bonn, Germany, 2014.

[12] Foreign and Commonwealth Office, National Cyber Security Centre and Lord Ahmad of Wimbledon, Foreign Office minister condemns Russia for NotPetya attacks, News Story, London, United Kingdom, February 15, 2018.

[13] J. Fruhlinger, Petya ransomware and NotPetya malware: What you need to know now, *CSO*, October 17, 2017.

[14] B. Gourley, Most violent cyber attack noted to date: 2008 pipeline explosion caused by remote hacking, *CTOvision.com*, December 13, 2014.

[15] HazardEx, Russian hackers now thought to have caused 2008 Turkish oil pipeline explosion, Tonbridge, United Kingdom, December 21, 2014.

[16] Homeland Security News Wire, 2008 Turkish oil pipeline explosion may have been Stuxnet precursor, Washington, DC, December 17, 2014.

[17] S. Hong, *Wireless: From Marconi's Black-Box to the Audion*, MIT Press, Cambridge, Massachusetts, 2001.

[18] G. Hughes, The cyberspace invaders, *The Age*, June 22, 2003.

[19] Industrial Control Systems Cyber Emergency Response Team (ICS-CERT), Gas pipeline cyber intrusion campaign, *ICS-CERT Monthly Monitor*, p. 1, April 2012.

[20] Industrial Control Systems Cyber Emergency Response Team (ICS-CERT), Gas pipeline cyber intrusion campaign – Update, *ICS-CERT Monthly Monitor*, p. 1, June-July 2012.

[21] Industrial Control Systems Cyber Emergency Response Team (ICS-CERT), Advisory (ICSA-11-041-01A), McAfee Night Dragon Report (Update A), Idaho Falls, Idaho, January 2, 2014.

[22] Industrial Control Systems Cyber Emergency Response Team (ICS-CERT), Advisory (ICSA-100-238-01B), Stuxnet Malware Mitigation (Update B), Idaho Falls, Idaho, January 8, 2014.

[23] Industrial Control Systems Cyber Emergency Response Team (ICS-CERT), Advisory (ICSA-14-178-01), ICS Focused Malware, Idaho Falls, Idaho, July 1, 2014.

[24] Industrial Control Systems Cyber Emergency Response Team (ICS-CERT), Best Practices for Continuity of Operations (Handling Destructive Malware), Idaho Falls, Idaho, January 22, 2015.

[25] Industrial Control Systems Cyber Emergency Response Team (ICS-CERT), Recommended Practice: Improving Industrial Control System Cybersecurity with Defense-in-Depth Strategies, Idaho Falls, Idaho, 2016.

[26] Industrial Control Systems Cyber Emergency Response Team (ICS-CERT), Alert (ICS-ALERT-14-281-01E), Ongoing Sophisticated Malware Campaign Compromising ICS (Update E), Idaho Falls, Idaho, December 9, 2016.

[27] Industrial Control Systems Cyber Emergency Response Team (ICS-CERT), Alert (ICS-ALERT-17-181-01C), Petya Malware Variant (Update C), Idaho Falls, Idaho, July 10, 2017.

[28] Industrial Control Systems Cyber Emergency Response Team (ICS-CERT), Joint Security Awareness Report (JSAR-12-151-01A), sKyWIper/Flame Information-Stealing Malware (Update A), Idaho Falls, Idaho, April 18, 2017.

[29] Industrial Control Systems Cyber Emergency Response Team (ICS-CERT), Joint Security Awareness Report (JSAR-12-222-01), Gauss Information-Stealing Malware, Idaho Falls, Idaho, April 18, 2017.

[30] Industrial Control Systems Cyber Emergency Response Team (ICS-CERT), Joint Security Awareness Report (JSAR-12-241-01B), Shamoon/DistTrack Malware (Update B), Idaho Falls, Idaho, April 18, 2017.

[31] Industrial Control Systems Cyber Emergency Response Team (ICS-CERT), Joint Security Awareness Report (JSAR-11-312-01): W32.Duqu-Malware, Idaho Falls, Idaho, April 18, 2017.

[32] Industrial Control Systems Cyber Emergency Response Team (ICS-CERT), Analysis Report, Malware Analysis, MAR-17-352-01 HatMan – Safety System Targeted Malware (Update A), Idaho Falls, Idaho, April 10, 2018.

[33] P. Ivanova, Kremlin rejects U.S. accusation that Russia is behind cyber attack, *Reuters*, February 16, 2018.

[34] K. Jackson Higgins, "Energetic" Bear under the microscope, *Dark Reading*, July 31, 2014.

[35] B. Johnson, D. Caban, M. Krotofil, D. Scali, N. Brubaker and C. Glyer, Attackers deploy new ICS attack framework "TRITON" and cause operational disruption to critical infrastructure, *Threat Research Blog*, FireEye, Milipitas, California, December 14, 2017.

[36] Kaspersky Lab, Resource 207: Kaspersky Lab research proves that Stuxnet and Flame developers are connected, Press Release, Woburn, Massachusetts, June 11, 2012.

[37] Kaspersky Lab, Kaspersky Lab discovers "Gauss" – A new complex cyber-threat designed to monitor online banking accounts, Press Release, Woburn, Massachusetts, August 9, 2012.

[38] G. Keizer, Is Stuxnet the "best" malware ever? *Computerworld*, September 16, 2010.

[39] B. Krebs, Target hackers broke in via HVAC company, *Krebs on Security*, February 14, 2014.

[40] R. Langner, To Kill a Centrifuge: A Technical Analysis of What Stuxnet's Creators Tried to Achieve, The Langner Group, Arlington, Virginia, 2013.

[41] R. Lee, Closing the case on the reported 2008 Russian cyber attack on the BTC pipeline, *SANS Industrial Control Systems Security Blog*, SANS Institute, Bethesda, Maryland, June 19, 2015.

[42] R. Lee, M. Assante and T. Conway, German Steel Mill Cyber Attack, ICS Defense Use Case (DUC), SANS Industrial Control Systems, SANS Institute, Bethesda, Maryland, 2014.

[43] R. Lee, M. Assante and T. Conway, Analysis of the Cyber Attack on the Ukrainian Power Grid, Electricity Information Sharing and Analysis Center (E-ISAC), Washington, DC, 2016.

[44] Lockheed Martin, The cyber kill chain, Bethesda, Maryland (`www.lockheedmartin.com/us/what-we-do/aerospace-defense/cyber/cyber-kill-chain.html`), 2018.

[45] V. Lynch, Cost of 2013 Target data breach nears $300 million, *Hashed Out*, The SSl Store, St. Petersburg, Florida (`www.thesslstore.com/blog/2013-target-data-breach-settled`), May 26, 2017.

[46] N. Maskelyne, Electrical syntony and wireless telegraphy, *The Electrician*, vol. 51, pp. 357–360, 1903.

[47] McAfee, Global Energy Cyberattacks: Night Dragon, Version 1.4, White Paper, Santa Clara, California, 2011.

[48] A. McLean, Australia also points finger at Russia for NotPetya, *ZDNet*, February 15, 2018.

[49] MSS Global Threat Response, Emerging threat: Dragonfly/Energetic Bear – APT Group, *Symantec Official Blog*, Symantec, Mountain View, California, June 30, 2014.

[50] National Cybersecurity and Communications Integration Center (NCCIC), Fiscal year 2017 ICS assessment summary, *ICS-CERT Monitor*, pp. 3–5, November-December 2017.

[51] National Cybersecurity and Communications Integration Center (NCCIC) and Industrial Control Systems Cyber Emergency Response Team (ICS-CERT), 2017 ICS-CERT Annual Vulnerability Coordination Report, Department of Homeland Security, Washington, DC, 2017.

[52] J. O'Leary, J. Kimble, K. Vanderlee and N. Fraser, Insights into Iranian cyber espionage: APT33 targets aerospace and energy sectors and has ties to destructive malware, *Threat Research Blog*, FireEye, Milipitas, California, September 20, 2017.

[53] A. Ozment and G. Touhill, DHS works with critical infrastructure owners and operators to raise awareness of cyber threats, Public Statement, Department of Homeland Security, Washington, DC, March 7, 2016.

[54] N. Perlroth, In cyberattack on Saudi firm, U.S. sees Iran firing back, *The New York Times*, October 23, 2012.

[55] P. Polityuk, Ukraine to probe suspected Russian cyber attack on grid, *Reuters*, December 31, 2015.

[56] S. Prokupecz, T. Kopan and S. Moghe, Former official: Iranians hacked into New York dam, *CNN*, December 22, 2015.

[57] J. Robertson and J. Riley, Mysterious '08 Turkey pipeline blast opened new cyberwar, *Bloomberg*, December 10, 2014.

[58] Security Response Attack Investigation Team, Dragonfly: Western energy sector targeted by sophisticated attack group, *Threat Intelligence Blog*, Symantec, Mountain View, California, October 20, 2017.

[59] J. Slay and M. Miller, Lessons learned from the Maroochy Water breach, in *Critical Infrastructure Protection*, E. Goetz and S. Shenoi (Eds.), Boston, Massachusetts, pp. 73–82, 2007.

[60] T. Smith, Hacker jailed for revenge sewage attacks, *The Register*, October 31, 2001.

[61] Symantec Security Response, W32.Duqu: The Precursor to the Next Stuxnet, Version 1.4, Symantec, Mountain View, California, 2011.

[62] Symantec Security Response, The Shamoon attacks, *Symantec Official Blog*, Symantec, Mountain View, California, August 16, 2012.

[63] Symantec Security Response, Shamoon: Multi-staged destructive attacks limited to specific targets, *Symantec Official Blog*, Symantec, Mountain View, California, February 27, 2017.

[64] Symantec Security Response, Triton: New malware threatens industrial safety systems, *Threat Intelligence Blog*, Symantec, Mountain View, California, December 14, 2017.

[65] H. Tanriverdi, Die Tatwaffe fehlt (The murder weapon is missing), *Sueddeutsche Zeitung*, June 19, 2015.

[66] TASS, Kremlin slams "Russophobic" allegations that pin NotPetya cyber attack on Russia, February 15, 2018.

[67] The White House, Statement from the Press Secretary, Washington, DC (`www.whitehouse.gov/briefings-statements/statement-press-secretary-25`), February 15, 2018.

[68] D. Turner, Prepared Testimony and Statement for the Record of Dean Turner, Director, Global Intelligence Network, Symantec Security Response, Symantec Corporation, Hearing on Securing Critical Infrastructure in the Age of Stuxnet, Committee on Homeland Security and Governmental Affairs, United States Senate, Washington, DC (`www.hsgac.senate.gov/download/2010-11-17-turner-testimony-revised2`), November 17, 2010.

[69] United States Computer Emergency Readiness Team (US-CERT), Alert (TA17-163A), CrashOverride Malware, Washington, DC, July 27, 2017.

[70] United States Computer Emergency Readiness Team (US-CERT), Alert (TA17-293A), Advanced Persistent Threat Targeting Energy and Other Critical Infrastructure Sectors, Washington, DC, March 15, 2018.

[71] United States District Court, Southern District of New York, Sealed Indictment: United States of America v. Ahmad Fathi et al., New York (`justice.gov/opa/file/834996/download`), 2016.

[72] Verizon, Data Breach Digest: Scenarios from the Field, New York, 2016.

[73] D. Volz and S. Young, White House blames Russia for "reckless" NotPetya cyber attack, *Reuters*, February 15, 2018.

[74] S. Ward, ModPoS: Highly-sophisticated, stealthy malware targeting U.S. PoS systems with high likelihood of broader campaigns, *Threat Research Blog*, FireEye, Milipitas, California, November 24, 2015.

[75] D. Yadron and P. Ziobro, Before Target, they hacked the heating guy, *The Wall Street Journal*, February 5, 2014.

[76] K. Zetter, How digital detectives deciphered Stuxnet, the most menacing malware in history, *Wired*, July 11, 2011.

[77] K. Zetter, Qatari gas company hit with virus in wave of attacks on energy companies, *Wired*, August 30, 2012.

[78] K. Zetter, The malware that duped Target has been found, *Wired*, January 16, 2014.

[79] K. Zetter, An unprecedented look at Stuxnet, the world's first digital weapon, *Wired*, November 3, 2014.

[80] K. Zetter, Attackers stole certificate from Foxconn to hack Kaspersky with Duqu 2.0, *Wired*, June 15, 2015.

[81] K. Zetter, Everything we know about Ukraine's power plant hack, *Wired*, January 20, 2016.

[82] K. Zetter, Inside the cunning, unprecedented hack of Ukraine's power grid, *Wired*, March 3, 2016.

[83] N. Zinets, Ukraine hit by 6,500 hack attacks, sees Russian "cyberwar," December 29, 2016.

Chapter 13

AN INTEGRATED CONTROL AND INTRUSION DETECTION SYSTEM FOR SMART GRID SECURITY

Eniye Tebekaemi, Duminda Wijesekera and Paulo Costa

Abstract Several control architectures have been proposed for smart grids based on centralized, decentralized or hybrid models. This chapter describes the Secure Overlay Communications and Control Model, a peer-to-peer, decentralized control and communications model with its own communications protocols and intrusion detection mechanisms that integrate a physical power system and its communications and control systems. This chapter also demonstrates how the model can help mitigate cyber attacks on a power system.

Keywords: Smart grid, communications, control, security, intrusion detection

1. Introduction

A networked control system uses a feedback control loop that requires control and feedback signals to be exchanged between its components over a communications network. The feedback signals contain periodic sensor measurements of the system that may vary during each iteration. The central controllers use the signals to estimate the current state of the system and, when necessary, the controllers send signals to actuators to adjust the behavior of the system. Traditionally, the communications network of a control system has been isolated from the Internet, with all the components (sensors, actuators and controllers) residing in the same physical location. However, the components of a smart grid are not co-located and the communications network is not isolated, making the resulting cyber-physical system highly vulnerable to cyber and physical attacks.

A modern power grid is centrally managed using the communications and control architecture shown in Figure 1. The central controller obtains telemetry data from all the locations and attempts to estimate the current state of the distributed system. The control and automation functions make control decisions

Published by Springer Nature Switzerland AG 2018.
J. Staggs and S. Shenoi (Eds.): Critical Infrastructure Protection XII, IFIP AICT 542, pp. 243–264, 2018.
https://doi.org/10.1007/978-3-030-04537-1_13

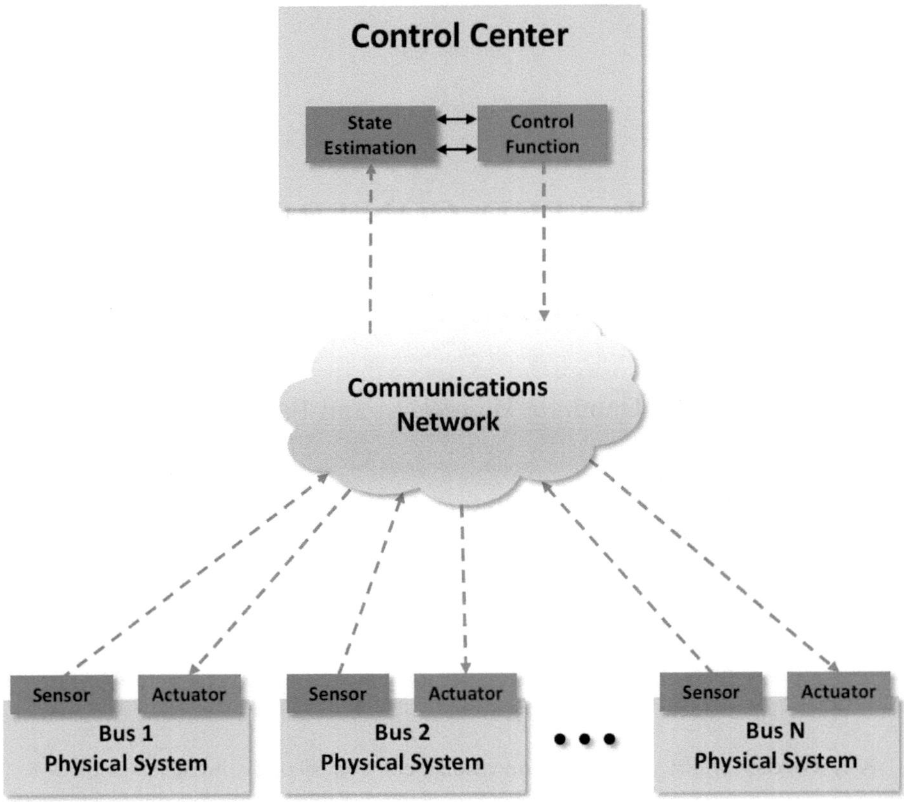

Figure 1. Cyber-physical system.

based on the estimated system state. Grid components use remote terminal units to send telemetry data and receive control commands from the central controller housed in a supervisory control and data acquisition (SCADA) system. The remote terminal units may be equipped with cryptographic tools and intrusion detection systems that validate messages purportedly sent by the central controller.

1.1 State Estimation

The objective of state estimation is to compute (with sufficient accuracy) the operational state of the power grid from measurements (bus voltage magnitudes and angles, branch currents and branch real and reactive power values) taken by sensors and communicated over the distributed communications network. The state estimator, which is an important component of the control center, computes the system state from sensor measurements. The relationship between the measurements and the system state is given by:

$$z = h(s) + e \tag{1}$$

where $z = (z_i, z_2, \cdots, z_k)$ is the measurement data vector; $s = (s_1, s_2, \cdots, s_n)$ is the true state vector; e is the measurement error (usually white Gaussian noise); and h is a non-linear scalar function that relates the measured data z to the state variables in s. The equation is typically solved using the weighted least squares method as described in [12] to obtain the estimated state vector $\hat{s}$.

1.2 Control Function

The objective of power system control functions is to constrain system behavior (by controlling power regulation transformers, capacitor banks, circuit breakers, etc.) to meet objectives such as optimal power flow, voltage regulation, power quality and/or economic dispatch. In the case of a smart grid, the objective functions are automation functions such as self-healing and restoration [8, 9, 22, 26], dynamic volt/var optimization [1, 25, 27] and priority load management [2–4, 18], among others. The control functions rely primarily on the state estimator to obtain the current state of the power system in order to determine the optimal control vector that constrains the behavior of the power system.

1.3 Communications

Two communications protocols – distributed network protocol (DNP3) [6] and IEC Power Utility Automation Standard (IEC 61850) [7] – are predominantly used in power systems communications and control. DNP3 is a centralized master/slave protocol used by most SCADA systems to control field devices at remote locations. Each location is polled by the master (SCADA central controller) and the information obtained is used to make control decisions that are enforced by actuators at remote locations. IEC 61850 is a layered standard that defines three protocols: (i) manufacturing messaging service (MMS) protocol; (ii) generic object-oriented substation event (GOOSE) protocol; and (iii) sampled value (SV) protocol. Manufacturing messaging service is a centralized connection-oriented client/server protocol used by a central controller to control lower-level devices in SCADA-based substations. GOOSE and sampled value are multicast subscriber/publisher protocols used to interact with and control field devices such as sensors and circuit breakers. The GOOSE and sampled value protocols are inherently insecure and used only for communications that originate and terminate in the same physical location.

1.4 False Data Injection Attacks

The smart grid attacks considered in this work fall broadly in the false data injection attack category. False data injection attacks seek to corrupt system state estimation by injecting false data in the messages sent from sensors in remote locations to the central controller or directly controlling actuators by injecting false commands from the control controller to actuators in remote locations.

The architecture in Figure 1 has the following attack entry points:

- **Communications Channel:** An attacker with access to the network communications channel may be able to observe and inject data into the communications stream between the central controller and buses. An attacker located at the control center side could gain global visibility of the network and attack any remote bus.

- **Remote Bus:** An attacker with physical access to remote bus sensors could physically alter the sensors to produce incorrect measurements that result in erroneous system states computed by the state estimator [23].

- **Control Center:** The control center houses the management network that is connected to the Internet. This makes the control center vulnerable to traditional cyber attacks that could be leveraged to gain access to the power grid communications network and perform false data injection attacks. The 2015 attack on the Ukrainian power grid [11] exemplifies the use of a traditional cyber attack on a centrally-managed power system followed by false data injection attacks.

1.5 Research Objective

Cyber security controls for computer networks seek to meet some or all of the traditional goals of confidentiality, integrity and availability. While confidentiality retains its original meaning, the concepts of integrity and availability are defined a little differently for the smart grid. In this context, integrity means that the data does not violate the operational constraints of the physical system and availability means that the physical system operates predictably and reliably in an optimal manner even when data is compromised.

Integrity and availability together define the resilience of a cyber-physical system. The most important cyber security objective for the smart grid is resilience. This is because it is impossible to provide absolute guarantees about defeating all cyber attack activity. Therefore, the resilience goal is to ensure that the smart grid operates reliably and predictably under cyber attacks, even when portions of the grid are already compromised.

This work focuses on mitigating cyber attacks using a resilient communications and control architecture. Specifically, it employs the Secure Overlay Communications and Control Model (SOCOM), a novel peer-to-peer, decentralized control and communications model with its own communications protocols and intrusion detection mechanisms that integrate a physical power system and its communications and control systems. A power grid intrusion detection system (SOCOM-IDS) is designed specifically for SOCOM. SOCOM-IDS integrates the coupling characteristics of the smart grid – physical system properties, automation/control function behavioral properties and communications network properties.

2. Related Work

Three aspects should be considered when designing intrusion detection and prevention systems for decentralized cyber-physical systems such as the smart grid: (i) data integrity; (ii) state integrity; and (iii) process integrity. Data integrity, which ensures that data has not been tampered with when it transits from node to node; is usually enforced using cryptographic solutions. The global system state is estimated using data obtained from various points (buses) in the system, and it is imperative that the integrity of system state estimation is maintained for the automation functions to work correctly. Each automation function makes a control decision based on its perception of the global system state relative to the local states and is governed by a process. The process involves a series of actions and interactions between the physical system, nodes (controllers and intelligent electronic devices) and the communications network required to implement the automation function. Most research in this area focuses on one or two of these three aspects.

Yang et al. [24] have designed an encryption-based system that detects false data injection in smart grids during data aggregation (state estimation). Hong and Lin [5] have presented a collaborative intrusion detection system that detects false data injection in sampled values and GOOSE messages based on the semantic anomalies in the sampled value and GOOSE packet header information. Li et al. [10] have designed a rule-based collaborative false data detection method, where the nodes share and compare measured data collected from sensors. Talha and Ray [19] have proposed a framework for MAC-layer wireless intrusion detection and response for smart grid applications; in their system, nodes collaboratively detect flooding attacks at the MAC layer that may result in denial of service and switch the wireless transmission channel as a countermeasure. Zhang et al. [28] have proposed a distributed intrusion detection system that engages intelligent multilayer analysis modules positioned at each network level of the grid to detect and classify malicious data and possible cyber attacks.

Lin et al. [13] have proposed a method for detecting and mitigating control-related attacks on power grids using runtime semantic security analysis of control messages sent over the communications network. Mashima et al. [14] have designed a concrete command mediation scheme called autonomous command-delaying to enhance grid resilience. They introduce artificial delays between intelligent electronic devices and the control center to provide the control center with a time buffer to detect attacks and subsequently cancel malicious commands. Sakis Meliopoulos et al. [17] have developed a cyber-physical co-model for detecting data and control-related attacks. They created a distributed dynamic state estimator that decentralizes the state estimation process, thereby reducing the cyber attack points and the processing overhead at the control center.

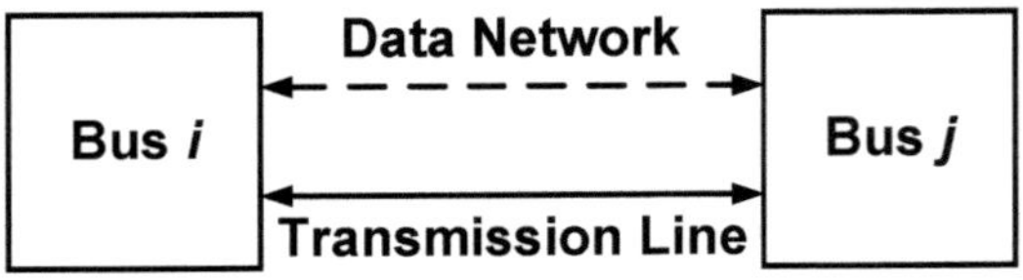

Figure 2. Double coupling property.

3. Proposed Model

A smart grid incorporates automation functions that coordinate the widely distributed components of the grid to ensure reliable, efficient and safe delivery of power. Attacks on the smart grid target the correct operation of automation functions by: (i) corrupting data exchanged over the communications network; and/or (ii) attacking physical equipment so that it is unable to operate correctly. The objective of SOCOM-IDS is to detect and mitigate the cyber and physical attacks on automation functions and their corresponding processes. In order for SOCOM-IDS to adequately protect the automation functions that control physical power distribution, it has to understand the physical distribution system, control system and network system behavior that define the automation/control process.

The power grid communications/control architecture discussed in Section 1 has an obvious flaw – an attacker can maximize the attack impact by focusing on the control center; if the control center, is compromised then the whole system may be compromised. To address this flaw, several architectures have been proposed that employ a decentralized communications/control model or a hybrid centralized-decentralized model. These new architectures often fall short for the following reasons:

- They focus mainly on control and rely on an existing centralized communications model.
- They do not incorporate cyber security as a major factor in their models and designs.
- They rely on high-level decentralized communications protocols (e.g., JADE [21]) that cannot be readily implemented on low-level field devices.

3.1 SOCOM Overview

The physical power system is inherently decentralized. Power transmission lines provide point-to-point connections between the distributed components and power flows only between directly connected terminals. SOCOM has been designed to mirror the natural behavior of power systems.

Consider the configuration in Figure 2 where buses i and j are also directly connected by a power transmission line modeled as:

$$\begin{bmatrix} V_{i,j} \\ I_{i,j} \end{bmatrix} = \begin{bmatrix} A_{j,i} & B_{j,i} \\ C_{j,i} & D_{j,i} \end{bmatrix} \begin{bmatrix} V_{j,i} \\ I_{j,i} \end{bmatrix} \quad (2)$$

where the matrix $\begin{bmatrix} A_{j,i} & B_{j,i} \\ C_{j,i} & D_{j,i} \end{bmatrix}$ is the characteristic impedance or power transfer characteristics of the transmission line; $A = V_S/V_R$ is the voltage ratio; $B = V_S/I_R$ is the short circuit resistance; $C = I_S/V_R$ is the open circuit conductance; and $D = I_S/I_R$ is the current ratio. Buses i and j are directly connected by a data network and exchange state information.

Consider two neighboring buses i and j where $x_{i,j} = \begin{bmatrix} A_{i,j} & B_{i,j} \\ C_{i,j} & D_{i,j} \end{bmatrix}$ denotes the power transfer characteristics from bus i to bus j; $s_{i,j} = \begin{bmatrix} V_{i,j} \\ I_{i,j} \end{bmatrix}$ is the state of the line $\{i,j\}$ at bus i; $Z^*_{RVI_{i,j}} = x_{i,j} \cdot Z_{LVI_{i,j}} = x_{i,j} \cdot (h(s_{i,j}) + e_i)$ is the line state measurement of bus j estimated at bus i; and $Z_{RVI_{i,j}} = Z_{LVI_{j,i}} = h(s_{j,i}) + e_j$ is the line state measurement sent over the network from bus j to bus i.

Under normal operating conditions:

$$\begin{aligned} Z^*_{RVI_{i,j}} &\stackrel{?}{=} Z_{RVI_{i,j}} \\ x_{i,j} \cdot h(s_{i,j}) - h(s_{j,i}) &= e_j - x_{i,j} \cdot e_i \end{aligned} \quad (3)$$

where $e_j - x_{i,j} \cdot e_i$ is the estimation error. Therefore:

$$|e_j - x_{i,j} \cdot e_i| = |Z^*_{RVI_{i,j}} - Z_{RVI_{i,j}}| < \zeta \quad (4)$$

where ζ is the error detection threshold or estimation error threshold.

SOCOM uses the characteristic impedance of power transmission lines to model the physical power grid system as a sparse matrix of pairs of directly connected nodes. Each node holds a subset of the system state information matrix that is used to estimate the system state and make control decisions.

SOCOM Architecture. Decentralized autonomous functions for smart grid can benefit from using decentralized communication protocols. However, a major challenge is the reluctance of utility providers to make the necessary investments because they already have older but functional technology. SOCOM runs as middleware on the existing TCP/IP communications infrastructure employed by utilities. This creates a logically decentralized network for the efficient operation of decentralized automation functions.

SOCOM offers the following advantages:

- **Administration:** Engineers are reluctant to cede control of power systems to autonomous intelligent electronic devices (IEDs). Because the SOCOM overlay model is only logical, utility managers can still have direct access to the underlying communications network and retain the ability to observe and intercede in administering the power system.

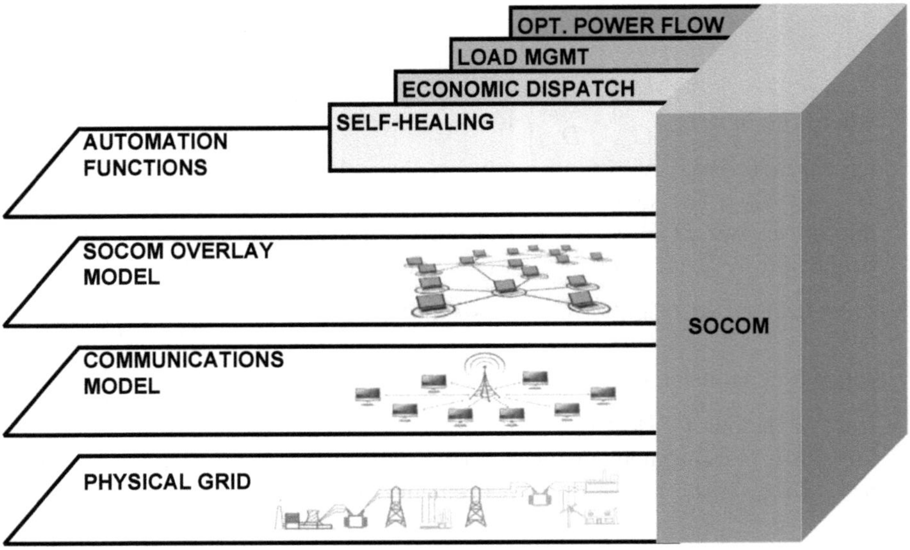

Figure 3. SOCOM architecture.

- **Cost:** No structural modifications to the existing communications infrastructure are required. The overlay middleware is implemented between the automation functions and physical communications network as shown in Figure 3.

- **Portability:** The overlay model executes in the network, Internet, transport or application layers of the TCP/IP network. The implementation would, of course, depend on user objectives and requirements.

- **Ease of Use:** Automation functions are oblivious to the physical communications layer and vice versa. Consequently, regardless of the communications protocols, automation functions can be adapted to run on the overlay model.

- **Implementation:** The overlay is lightweight and suitable for direct hardware implementation on field electronic devices and field programmable gate array (FPGA) based controllers.

SOCOM Protocol. The SOCOM protocol is a lightweight asynchronous messaging platform designed for decentralized automation and control in cyber-physical systems [20]. The SOCOM protocol (Figure 4) executes as middleware (overlay network) between the smart grid automation functions and the physical communications network as shown in Figure 3. The overlay network layer is structured to mirror the physical system layer (bus network), where each node

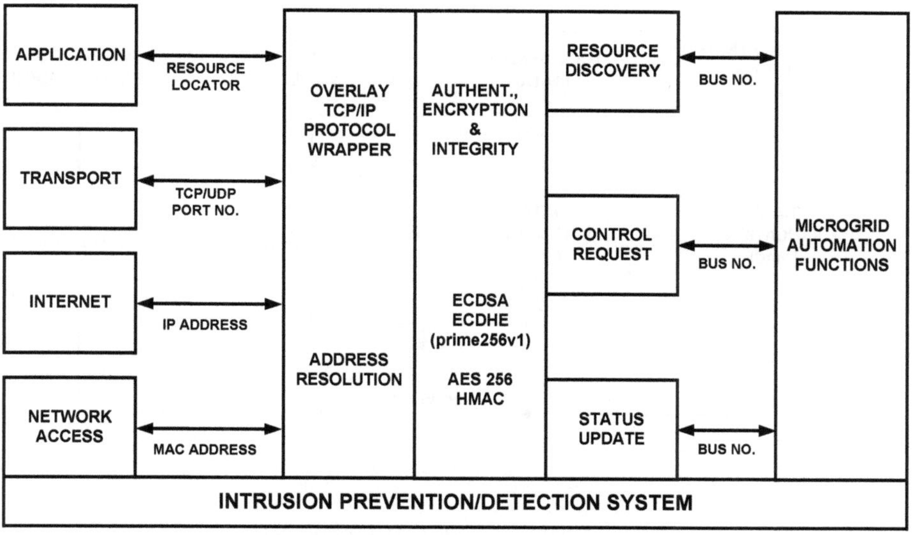

Figure 4. SOCOM protocol.

represents a local (bus) controller that can communicate only with its physically connected peers.

SOCOM uses three major protocols: (i) resource discovery protocol; (ii) control request protocol; and (iii) status update protocol. SOCOM has a security layer that provides communications confidentiality, integrity and authentication if needed, and a TCP/IP wrapper for address resolution. Using these protocols, local controllers in the smart grid can locate resources, update their status and initiate control operations in response to optimization objectives in a secure and logically decentralized manner.

The SOCOM protocols have various features:

- **Resource Discovery Protocol (RDP):** This gossip-like protocol is used to locate resources in the smart grid. A resource may be an energy source, storage component, electric load or any other device that may provide, transform or consume energy in the smart grid.

- **Control Request Protocol (CRP):** This request/response protocol remotely executes control actions on resources that are directly connected to peer buses. For example, a bus controller can request a peer bus controller to connect or disconnect a power line to alter the flow of power, possibly in response to a disturbance in order to recover from line faults.

- **Status Update Protocol (SUP):** This point-to-point protocol sends and receives bus information to and from directly connected buses. Each bus sends bus status messages at set time intervals or immediately when specific bus information changes.

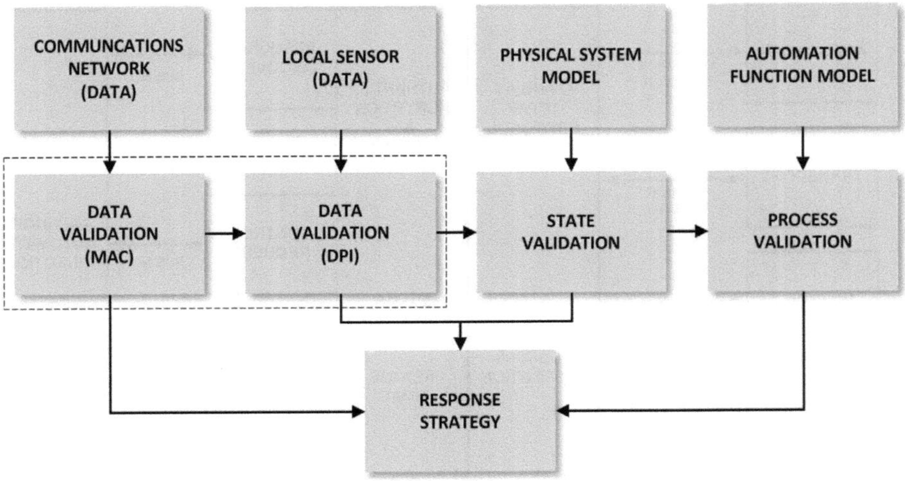

Figure 5. SOCOM-IDS model.

3.2 SOCOM-IDS Model

The SOCOM-IDS model employs a modular strategy for attack detection and response in a microgrid (i.e., part of a smart grid). It incorporates three detection modules, each of which is compartmentalized to run independently of the other modules. Figure 5 shows the structural layout of SOCOM-IDS.

Data Validation Module. The goal of the data validation module is to detect false data injection attacks on the nodes in a microgrid. The module has two components. The first component, data validation (MAC), uses cryptographic controls to validate network data received from neighboring nodes. Each bus controller has a hard-coded (permanent) private/public key pair that initiates the ephemeral elliptic curve Diffie-Hellman (ECDHE) key exchange process with other peer bus controllers to generate session keys. After the session keys are generated, a symmetric algorithm (AES) is used for encryption and the keyed hash message authentication code (HMAC) is used to ensure message integrity.

The second component, data validation (DPI), uses deep packet inspection to check for voltage and current values that exceed predetermined values. The detection problem is formulated as a binary decision:

$$\begin{aligned} FALSE &: |Z^*_{RVI_{i,j}} - Z_{RVI_{i,j}}| \leq \zeta \\ TRUE &: |Z^*_{RVI_{i,j}} - Z_{RVI_{i,j}}| > \zeta \end{aligned} \quad (5)$$

where the claim that the data has been modified is verified when the equation evaluates to true. The data validation module estimates the neighbor node bus voltage magnitudes and phase angles, branch currents and direct and reactive power values from local sensor measurements. These values are compared against the neighbor node state measurements obtained over the network. Potential bad data is detected when the variation exceeds the bad data detection threshold.

State Validation Module. Each node estimates the state of the microgrid using information obtained from SOCOM messages exchanged with neighboring nodes. The estimated state is evaluated against the constraints and guarding conditions of the modeled physical system. The constraints are obtained from the physical laws that govern electric power systems.

The state validation module is based on three laws of electricity:

- Let $Z^{I\leftarrow in}_{RVI_i} = \left[Z^{I\rightarrow out}_{RVI_{i,j}} : \{j \in J \subset M_i\}\right]_{J\times 1}$ denote the current measurements from all the neighboring buses from which bus i draws current. Let $Z^{I\rightarrow out}_{RVI_i} = \left[Z^{I\rightarrow out}_{RVI_{i,k}} : \{k \in K \subset M_i\}\right]_{K\times 1}$ denote the current measurements from all the neighboring buses that draw current from bus i. Then, the sum of currents flowing into a node is equal to the sum of currents flowing out:

$$\sum_{j=1}^{J} Z^{I\leftarrow in}_{RVI_{i,j}} \stackrel{?}{=} \sum_{k=1}^{K} Z^{I\rightarrow out}_{RVI_{i,k}} \tag{6}$$

- The voltage $Z^{V}_{RVI_{i,j}}$ and current $Z^{I}_{RVI_{i,j}}$ measurements received from bus j should be equal to the estimated branch power $x_{i,j}\cdot Z^{V}_{LVI_{i,j}} * x_{i,j}\cdot Z^{I}_{LVI_{i,j}}$ measured locally at bus i for line $\{i,j\}$:

$$x_{i,j}\cdot Z^{V}_{LVI_{i,j}} * x_{i,j}\cdot Z^{I}_{LVI_{i,j}} \stackrel{?}{=} Z^{V}_{RVI_{i,j}} * Z^{I}_{RVI_{i,j}} \tag{7}$$

- Let LD_u be the consumer load that is directly connected to bus u and let GEN_v be the power generator that is directly connected to bus v. In a closed system, the total power used by the load is equal to the total power drawn from the power source. Each node estimates the total power used by loads in the microgrid and the total power drawn from all the sources using resource discovery protocol message exchanges:

$$\sum_{q=1}^{u} LD_q + \varpi = \sum_{r=1}^{v} GEN^{used}_r \tag{8}$$

where $\sum_{q=1}^{u} LD_q$ is the sum of the bus loads in the grid; $\sum_{r=1}^{v} GEN^{used}_r$ is the total power generated by all the sources in the grid; bus u and bus

v are the load bus and source bus, respectively; and ϖ is the estimated maximum power loss in the grid.

Process Validation Module. The process validation module is unique to each automation function. A process is a series of actions and interactions between physical system components, intelligent controllers (or intelligent electronic devices) and communications network devices that are needed to implement an automation function under normal operating conditions. Each automation function has distinct process behavior that is useful in designing security solutions that are tailored to meet its unique requirements.

For example, consider the self-healing automation function described by the state diagram in Figure 6. The goal of the healing function is to ensure that a failed bus i can independently generate a new grid configuration, which restores power to satisfy the following constraints:

- The load on bus i, which has to be restored, must be less than the sum of the available capacity of all the available power generation sources in the power grid.
- The bus voltage at bus i after power restoration must not violate the bus voltage constraints.
- The load on transmission lines must not be less than its maximum capacity.
- The switching overhead must be minimized. For example, the least number of number of switchgear device configuration changes should be performed to restore power.

The self-healing automation function is described by the state diagram in Figure 6. The goal of the healing function is to ensure that the failed bus i can independently generate local configuration changes to restore power in a manner that satisfies the constraints listed above. The new configuration, which is generated by the failed bus, is sent to neighboring buses.

The self-healing process has four states: (i) NORMAL; (ii) FAIL; (iii) RECOVER; and (iv) BAD:

- **NORMAL:** During the normal operating state, the bus continuously monitors its voltage state (using local sensors) and the voltage states of its neighboring nodes.
- **FAIL:** Power lines incorporate relays that detect faults and trigger circuit breakers in response to faults. The triggering of these protective relays may result in power failures that affect one or more buses in the microgrid.
- **RECOVER:** If a failure occurs and the self-healing function is enabled, then the affected bus i independently generates a new configuration to control local and neighboring switchgear devices in order to restore power based on the self-healing algorithm.

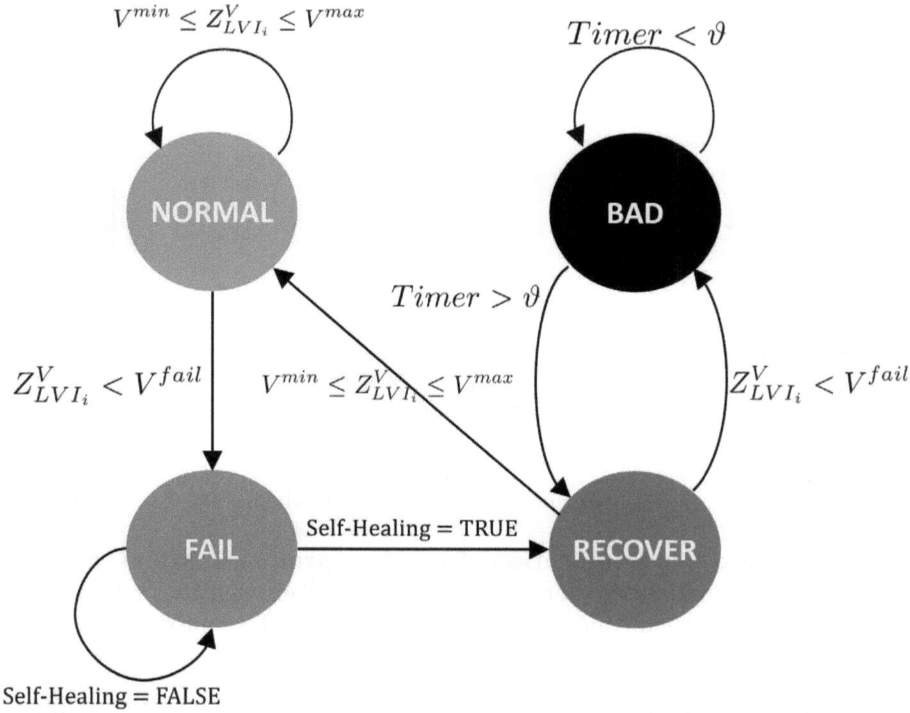

Figure 6. Self-healing state diagram.

- **BAD:** The bus enters the bad state when no configuration solution is found that restores power in a manner that satisfies the self-healing function constraints.

The self-healing process follows the following sequence of messages from a failure to service restoration:

$$SUP_{NORMAL} \rightarrow SUP_{FAIL} \rightarrow RDP \rightarrow CRP_{HEAL} \rightarrow SUP_{NORMAL} \quad (9)$$

Response Strategy. Upon detecting an intrusion, SOCOM-IDS attempts to stop the attack by performing the following tasks in order:

- **Change the Implementation Layer:** SOCOM can run on the MAC layer, network layer, transport layer (UDP) or application layer. When an intrusion is detected by a node, a change layer message is sent by the detecting node to all its neighboring nodes.

- **Change Cryptographic Keys:** If the intrusion persists, then the node generates new cryptographic keys and initiates a key exchange procedure.

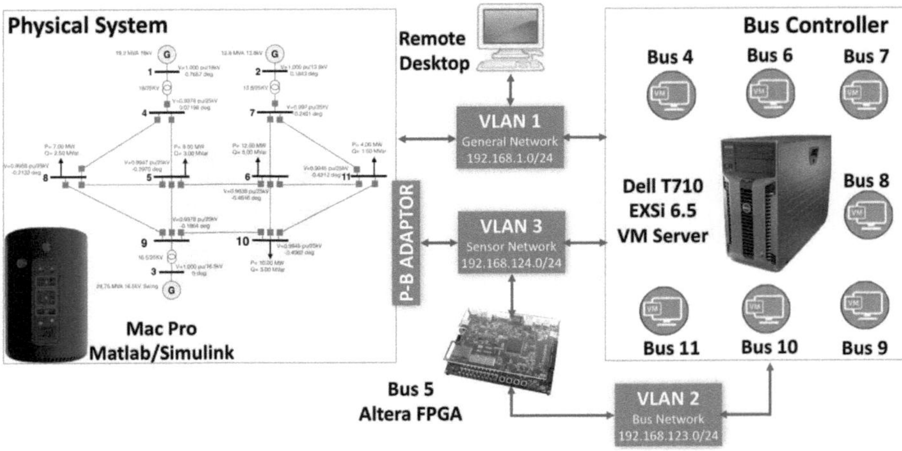

Figure 7. Experimental testbed.

- **Discard Communications from Compromised Node(s):** If the intrusion continues to persist, then it is most likely that the originating node may have been compromised. Messages from the compromised node are discarded.

- **Disable Secondary Control Functions:** Discarding network messages may have an adverse effect on secondary control functions. If more than a predetermined number of neighboring nodes are compromised or the secondary control function is unable to run effectively, then the secondary control function is disabled.

4. Implementation and Results

Figure 7 shows the experimental testbed. The physical power grid was simulated using Matlab/Simulink, Simscape Power Systems [15] and Simulink Real-Time [16] applications. Simscape Power Systems provided component libraries and analysis tools for modeling and simulating electrical power systems. Simulink Real-Time enabled the creation of real-time applications from Simulink models. The applications supported the implementation and execution of an eleven-bus physical power grid in real-time on a Mac Pro server (3 GHz 8-Core Intel Xeon E5, 64 GB RAM). The physical power grid comprised three power generator sources, three transformers (one for each source), five load buses and current/voltage sensors and switchgear devices.

Eight bus controllers were developed based on the SOCOM communications and control protocol. Seven of the eight buses were implemented as virtual machines and the remaining bus was implemented on an FPGA device. The seven virtual machines ran on a Dell T710 server (2.66 GHz 6-Core x2 Intel Xeon X5650, 64 GB RAM). Each bus controller received sensor measurements and sent control messages to the corresponding physical bus over UDP messages

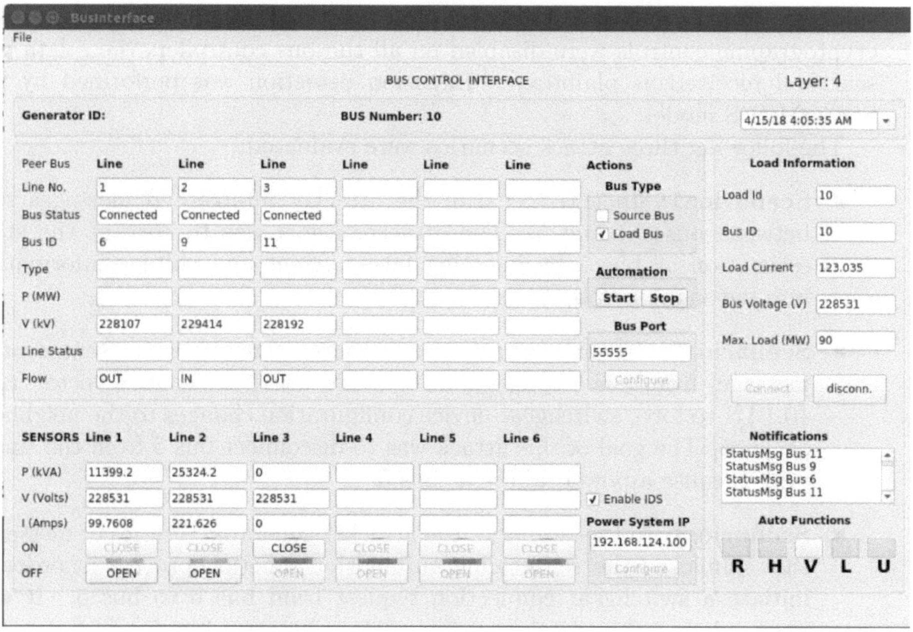

Figure 8. SOCOM bus interface.

through the physical-bus (P-B) controller adaptor. The adaptor routed UDP packets from the physical buses to the corresponding bus controller, and from the bus controllers to the corresponding physical buses. Figure 8 shows the bus control and configuration interface.

4.1 FPGA Implementation

The implementation employed a Cyclone IV-E EP4CE115F29C7 FPGA and an Altera DE2-115 Development and Education Board. The model comprised a Nios II processor that executed application programs, a JTAG UART component for supporting communications between the processor and host computer, a Triple-Speed Ethernet IP Core for implementing the MAC sublayer and a partial physical layer, a synchronous dynamic random-access memory (SDRAM) for program code and data, and two scatter-gather direct memory access (SGDMA) controllers for data transmission and receiving functions to and from the MAC sublayer. The model also incorporated flash memory for storing MAC and IP addresses, input/output peripherals used as output indicators and control inputs for the bus controller.

4.2 Attack Scenarios

Three attack scenarios were developed to evaluate the performance of the SOCOM-IDS in protecting a smart grid The scenarios involved disruptions of smart grid operations and its automation functions. In the attack scenarios, cryptographic controls were disabled on all the bus controllers (i.e., data was sent and received as plaintext). Intrusion detection was performed by the SOCOM-IDS model.

The following three attack scenarios were evaluated:

- **Scenario 1:** In this scenario, the attacker intercepted messages sent between buses 4 and 5. The attacker's goal was to corrupt the state estimation at bus 5 by injecting false current and voltage information into messages sent by bus 4.

- **Scenario 2:** In this scenario, the attacker generated and sent control messages from bus 5 to neighboring buses using the control vector $a_4 = \{0, 0\}_i^M$ to force switchgear device configuration changes to the neighbors of bus 5. The goal of this attack was to disconnect bus 5 from the smart grid to cause a power failure at bus 5.

- **Scenario 3:** In this scenario, the attacker generated a series of messages that mimicked the self-healing automation function process in order to initiate a switchgear connection request from bus 6 to bus 5. It was assumed that the switchgear device state between bus 5 and 6 was not connected and that the attacker understood how the self-healing process worked. The goal of the attacker was to force a disruption in the power flow of the smart grid by routing power in an unauthorized manner.

Attackers have varying knowledge about power systems, SOCOM operational behavior and physical access. This impacts their ability to disrupt the smart grid. The experiments assumed the following three categories of attackers:

- **Category 1:** Attackers in this category have limited knowledge about smart grid network protocols. They can sniff and modify network traffic, but have no understanding of how power systems and automation functions work. The attackers are basically script-kiddies who launch random attacks without clear objectives.

- **Category 2:** Attackers in this category have basic knowledge of smart grid network protocols and can sniff and modify network traffic. They have a basic understanding of power systems, but no understanding of the automation functions. The attackers can craft valid messages in order to deceive state estimators in the smart grid and trigger switchgear devices.

- **Category 3:** Attacker in this category have complete understanding of smart grid network protocols and detailed knowledge of power system functionality. The attackers also have an expert understanding of smart grid automation functions and the underlying processes and network behavior. The attackers in this category can craft sequences of messages to manipulate automation functions.

Scenario 2 assumed the presence of a Category 2 attacker. The attacker spoofed bus 5 and sent valid control request protocol messages to buses 4, 6, 8 and 9 to disconnect their switchgear device connections to bus 5. The malicious messages were detected by the SOCOM-IDS process validation module. The data validation module discovered that the malicious messages did not belong to an automation function process running on the smart grid and, therefore, flagged them as false messages.

4.3 Results

SOCOM-IDS was tested against attacks in Scenario 1. The attacker, who was assumed to be in Category 1, generated random status messages. One hundred status update protocol messages were generated with random voltages and currents in the ranges 24 kV to 25 kV and 300 A to 400 A, respectively. Figure 9 shows the random measurement values sent every five seconds by the attacker (who spoofed bus 4) to bus 5 compared against the expected measurements at bus 5. The SOCOM-IDS data validation module detected all the false messages with no false alarms or missed detections.

An attacker in Scenario 3 would generally be in Category 3. The corresponding attack was detected by the SOCOM-IDS state validation module. Figure 10 shows the sequence of messages received by bus 5 during a self-healing process initiated by bus 6. As discussed above, buses 4, 6, 8 and 9 sent duplicated resource discovery protocol messages to bus 5 reflecting the same changes in the source and load information. These messages were used in Equation (8) to verify if a failure actually occurred. A significant drop in total power drawn from source buses (bus failure causing load disconnection) indicated that a power failure had occurred.

The data validation module and the process validation module are designed for on-line operation. Figure 11 shows the runtime performance of the SOCOM-IDS data validation and process validation modules.

5. Conclusions

This chapter has presented the Secure Overlay Communications and Control Model Intrusion Detection System (SOCOM-IDS) for smart grid security. SOCOM-IDS provides an extra layer of security over traditional network security controls by integrating the physical and behavioral properties of a power system. Its primary objective is to ensure the resilient operation of a smart grid under cyber attacks. The intrusion detection modules in SOCOM-IDS constantly validate the communications between buses in a smart grid to ensure that operational constraints are not violated. The modules were evaluated using a self-healing automation function developed for smart grids and the results demonstrate that SOCOM-IDS is able to detect a variety of control-related and state-estimation cyber attacks on a simulated smart grid.

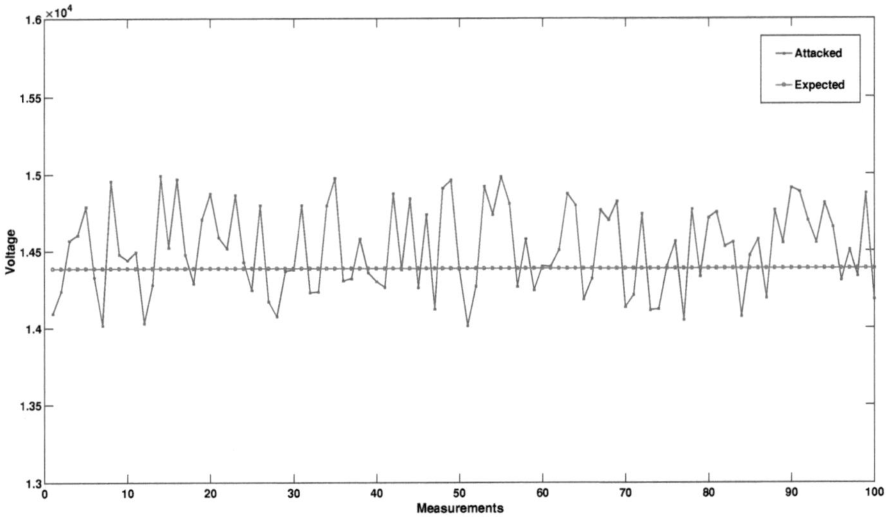

(a) Bus 4 voltage measurements.

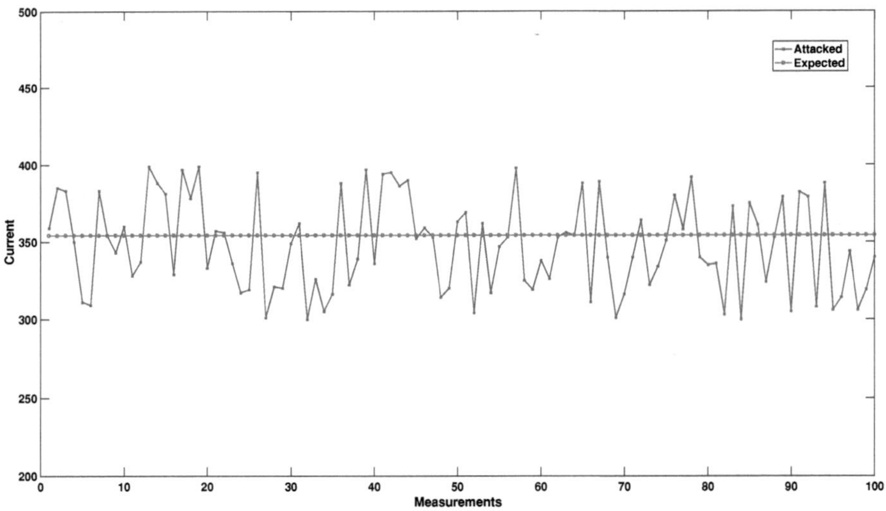

(b) Bus 4–5 line current measurements.

Figure 9. Random attack values vs. expected state measurement values.

Figure 10. Self-healing message sequence.

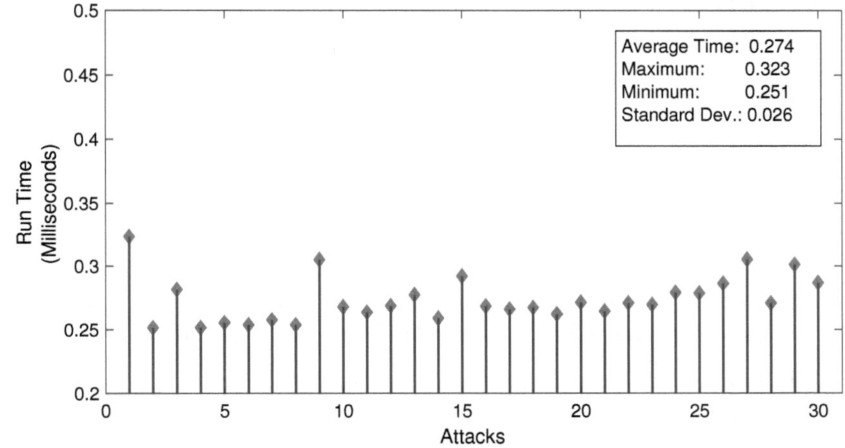

(a) Data validation module performance.

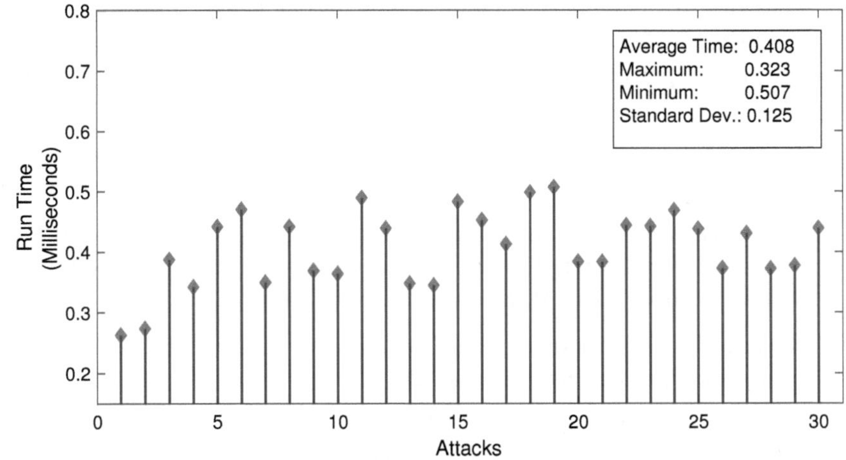

(b) Process validation module performance.

Figure 11. Runtime performance of the SOCOM-IDS validation modules.

This chapter also demonstrates the importance of communications/control architectures for implementing cyber security in cyber-physical systems such as a power grid. The SOCOM architecture provides a framework that integrates physical system properties and behavior into cyber security controls in an intuitively-appealing manner. The SOCOM framework is extensible and its application extends beyond power systems. Indeed, it is easily adapted to any cyber-physical system for which secure decentralized automation is a requirement.

References

[1] H. Cho, T. Hai, I. Chung, J. Cho and J. Kim, Distributed and autonomous control system for voltage regulation in low-voltage DC distribution systems, *Proceedings of the International Conference on Condition Monitoring and Diagnosis*, pp. 806–810, 2016.

[2] A. Clausen, A. Umair, Z. Ma and B. Norregaard Jorgensen, Demand response integration through agent-based coordination of consumers in virtual power plants, in *PRIMA 2016: Principles and Practice of Multi-Agent Systems*, M. Baldoni, A. Chopra, T. Son, K. Hirayama and P. Torroni (Eds.), Springer, Cham, Switzerland, pp. 313–322, 2016.

[3] L. Gomes, P. Faria, H. Morais, Z. Vale and C. Ramos, Distributed, agent-based intelligent system for demand response program simulation in smart grids, *IEEE Intelligent Systems*, vol. 29(1), pp. 56–65, 2014.

[4] L. Hernandez, C. Baladron, J. Aguiar, B. Carro, A. Sanchez-Esguevillas, J. Lloret, D. Chinarro, J. Gomez-Sanz and D. Cook, A multi-agent system architecture for smart grid management and forecasting of energy demand in virtual power plants, *IEEE Communications*, vol. 51(1), pp. 106–113, 2013.

[5] J. Hong and C. Liu, Intelligent electronic devices with collaborative intrusion detection systems, to appear in *IEEE Transactions on Smart Grid*.

[6] Institute of Electrical and Electronics Engineers, IEEE 1815-2012 – IEEE Standard for Electric Power Systems Communications – Distributed Network Protocol (DNP3), Piscataway, New Jersey, 2012.

[7] International Electrotechnical Commission, IEC 61850 Power Utility Automation, Geneva, Switzerland, 2013.

[8] X. Ji, J. Liu, X. Yan and H. Wang, Research on self-healing technology of smart distribution network based on multi-agent system, *Proceedings of the Chinese Control and Decision Conference*, pp. 6132–6137, 2016.

[9] Y. Kumar and R. Bhimasingu, Enabling self-healing microgrids by the improvement of resiliency using closed loop virtual DC motor and induction generator control scheme, *Proceedings of the IEEE Power and Energy Society General Meeting*, 2016.

[10] B. Li, R. Lu, W. Wang and K. Choo, Distributed host-based collaborative detection for false data injection attacks in smart grid cyber-physical systems, *Journal of Parallel and Distributed Computing*, vol. 103, pp. 32–41, 2017.

[11] G. Liang, S. Weller, J. Zhao, F. Luo and Z. Dong, The 2015 Ukraine Blackout: Implications for false data injection attacks, *IEEE Transactions on Power Systems*, vol. 32(4), pp. 3317–3318, 2017.

[12] G. Liang, J. Zhao, F. Luo, S. Weller and Z. Dong, A review of false data injection attacks against modern power systems, *IEEE Transactions on Smart Grid*, vol. 8(4), pp. 1630–1638, 2017.

[13] H. Lin, A. Slagell, Z. Kalbarczyk, P. Sauer and R. Iyer, Runtime semantic security analysis to detect and mitigate control-related attacks in power grids, *IEEE Transactions on Smart Grid*, vol. 9(1), pp. 163–178, 2018.

[14] D. Mashima, P. Gunathilaka and B. Chen, Artificial command delaying for secure substation remote control: Design and implementation, to appear in *IEEE Transactions on Smart Grid*.

[15] MathWorks, Simscape Power Systems, Natick, Massachusetts (`www.mathworks.com/products/simpower.html`), 2018.

[16] MathWorks, Simulink Real-Time, Natick, Massachusetts (`www.mathworks.com/products/simulink-real-time.html`), 2018.

[17] A. Sakis Meliopoulos, G. Cokkinides, R. Fan and L. Sun, Data attack detection and command authentication via cyber-physical co-modeling, *IEEE Design and Test*, vol. 34(4), pp. 34–43, 2017.

[18] V. Singh, N. Kishor and P. Samuel, Distributed multi-agent-system-based load frequency control for multi-area power systems in smart grids, *IEEE Transactions on Industrial Electronics*, vol. 64(6), pp. 5151–5160, 2017.

[19] B. Talha and A. Ray, A framework for MAC layer wireless intrusion detection and response for smart grid applications, *Proceedings of the Fourteenth International Conference on Industrial Informatics*, pp. 598–605, 2016.

[20] E. Tebekaemi and D. Wijesekera, A communications model for decentralized autonomous control of the power grid, *IEEE International Conference on Communications*, 2018.

[21] Telecom Italia Lab, Java Agent Development Framework (JADE), Telecom Italia Group, Turin, Italy (`jade.tilab.com`), 2018.

[22] Z. Wang, B. Chen, J. Wang and C. Chen, Networked microgrids for self-healing power systems, *IEEE Transactions on Smart Grid*, vol. 7(1), pp. 310–319, 2016.

[23] K. Weaver, Smart meter deployments result in a cyber attack surface of "unprecedented scale," *Smart Grid Awareness*, SkyVision Solutions, Naperville, Illinois (`smartgridawareness.org/2017/01/07/cyber-attack-surface-of-unprecedented-scale`), January 7, 2017.

[24] L. Yang and F. Li, Detecting false data injection in smart grid in-network aggregation, *Proceedings of the Fourth IEEE International Conference on Smart Grid Communications*, pp. 408–413, 2013.

[25] N. Yorino, Y. Zoka, M. Watanabe and T. Kurushima, An optimal autonomous decentralized control method for voltage control devices using a multi-agent system, *IEEE Transactions on Power Systems*, vol. 30(5), pp. 2225–2233, 2015.

[26] M. Zaki El-Sharafy and H. Farag, Self-healing restoration of smart microgrids in islanded mode of operation, in *Smart City* 360°, A. Leon-Garcia, R. Lenort, D. Holman, D. Stas, V. Krutilova, P. Wicher, D. Caganova, D. Spirkova, J. Golej and K. Nguyen (Eds.), Springer, Cham, Switzerland, pp. 395–407, 2016.

[27] X. Zhang, A. Flueck and C. Nguyen, Agent-based distributed volt/var control with distributed power flow solver in smart grid, *IEEE Transactions on Smart Grid*, vol. 7(2), pp. 600–607, 2016.

[28] Y. Zhang, L. Wang, W. Sun, R. Green II and M. Alam, Distributed intrusion detection system in a multi-layer network architecture of smart grids, *IEEE Transactions on Smart Grid*, vol. 2(4), pp. 796–808, 2011.

Chapter 14

GENERATING ABNORMAL INDUSTRIAL CONTROL NETWORK TRAFFIC FOR INTRUSION DETECTION SYSTEM TESTING

Joo-Yeop Song, Woomyo Lee, Jeong-Han Yun, Hyunjae Park, Sin-Kyu Kim and Young-June Choi

Abstract Industrial control systems are widely used across the critical infrastructure sectors. Anomaly-based intrusion detection is an attractive approach for identifying potential attacks that leverage industrial control systems to target critical infrastructure assets. In order to analyze the performance of an anomaly-based intrusion detection system, extensive testing should be performed by considering variations of specific cyber threat scenarios, including victims, attack timing, traffic volume and transmitted contents. However, due to security concerns and the potential impact on operations, it is very difficult, if not impossible, to collect abnormal network traffic from real-world industrial control systems. This chapter addresses the problem by proposing a method for automatically generating a variety of anomalous test traffic based on cyber threat scenarios related to industrial control systems.

Keywords: Industrial control systems, anomaly detection, traffic generation

1. Introduction

Industrial control systems are used in a variety of critical infrastructure assets such as power plants, waterworks, railways and transportation systems. The security of industrial control systems in the critical infrastructure is a grave concern due to the increased risk of external attacks and the potentially serious impact on operations [7, 23]. Therefore, it is important to develop sophisticated systems that can rapidly and accurately detect anomalous industrial control network behavior due to potential attacks.

Published by Springer Nature Switzerland AG 2018.
J. Staggs and S. Shenoi (Eds.): Critical Infrastructure Protection XII, IFIP AICT 542, pp. 265–281, 2018.
https://doi.org/10.1007/978-3-030-04537-1_14

Intrusion detection systems, which have been used for decades to detect and respond to abnormal operations in information technology systems and networks, are increasingly used in operational technology infrastructures such as industrial control networks. Intrusion detection systems are classified as misuse detection systems and anomaly detection systems [5]. Misuse detection relies on attack signatures – patterns and characteristics – to identify attacks. Therefore, misuse detection is ineffective against zero-day attacks and clever variants of known attacks. In addition, the massive network flows, diversity of attacks and increasing numbers of new attacks make it difficult for modern misuse detection systems to keep up with the threats.

Anomaly detection relies on deviations from normal usage patterns that are specified or learned. The approach is attractive for use in industrial control networks because of their stable structure, predictable traffic and relatively low traffic volumes [1, 20]. An anomaly-based intrusion detection system learns a statistical model of normal activities, which it compares against data pertaining to current activities in order to detect behavioral abnormalities, including those caused by undetected or zero-day attacks.

The same cyber attack can be executed on different targets at different times and with variations in its content. Depending on the environment, an anomaly-based intrusion detection system may or may not detect the same attack. Therefore, to evaluate the performance of an anomaly-based intrusion detection system, extensive testing has to be conducted using variations of each cyber threat scenario, including the targets, attack timing, traffic characteristics and transmitted content. Unfortunately, due to security concerns and the potential operational impact, it is very difficult, if not impossible, to evaluate cyber threat scenarios on real-world industrial control systems.

A solution to this problem is to use a testbed that models a real industrial control network and the physical infrastructure. The testbed can then be employed to collect normal and abnormal traffic. However, a high-fidelity testbed is expensive to implement and operate; in any case, it would never completely model the actual assets. Additionally, it is infeasible to create and analyze a large number of cyber attack scenarios, especially when each scenario can have numerous variations.

Efforts have been made to collect real-world traffic using honeypots [19], but such traffic does not adequately model real industrial control environments. A possible solution is to generate abnormal industrial control network traffic by modifying normal traffic to model cyber threat scenarios while maintaining the characteristics of the normal traffic to the extent possible. For each cyber threat scenario, the nature of anomalous network traffic varies. Therefore, the characteristics of abnormal traffic could be modified based on the specific points of time, target sessions and characteristics of the cyber threat scenarios, and a number of cases could be generated to perform accurate performance analysis. However, depending on the specific scenario, it may be difficult to manually modify normal traffic based on variants of the cyber threat scenario.

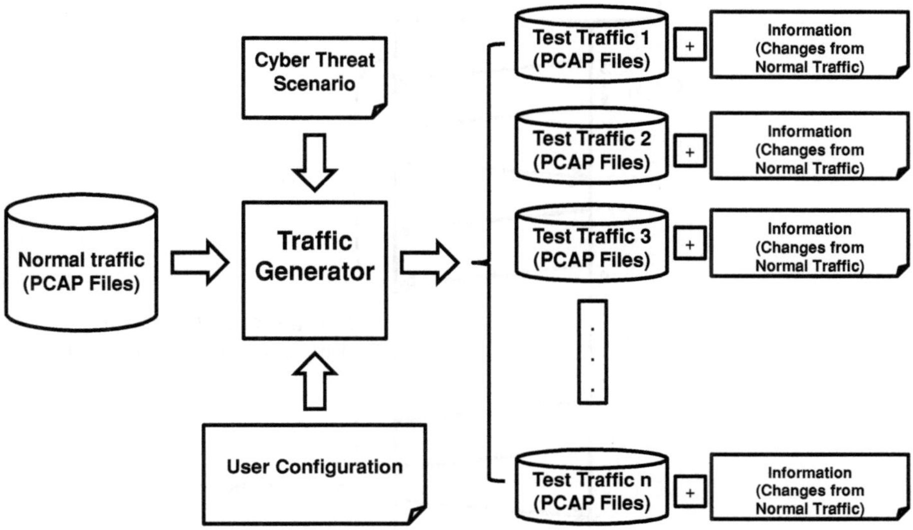

Figure 1. Abnormal test traffic generator.

To address these problems, this chapter proposes a method for automatically generating a variety of anomalous test traffic based on cyber threat scenarios related to industrial control networks. Figure 1 presents a schematic diagram of the abnormal test traffic generator. The automated generation of abnormal traffic requires a method that clearly describes the cyber threat scenarios to be tested. The method involves the specification of "actions" on industrial control network traffic. The characteristics of the point of occurrence, target and abnormal traffic are accordingly adjusted. This creates a number of abnormal scenario cases and abnormal traffic is generated by modifying normal traffic according to each case. Test data can also be generated by combining multiple scenarios.

Packets are the basic communications units of network traffic. In the case of TCP networks, the transmitted data is split into packets, and it is difficult to describe the traffic characteristics by considering individual packets in isolation. On the other hand, in industrial control networks, it is difficult to distinguish transactions since the protocols are often proprietary in nature. Yun et al. [22] have proposed a method for distinguishing transactions in industrial control network traffic. The method, which is shown in Figure 2, distinguishes transactions when the inter-packet arrival time is larger than a predefined threshold. Thus, although the transmitted data is divided into multiple packets, the test traffic is generated in units of transactions that model abnormal traffic more effectively.

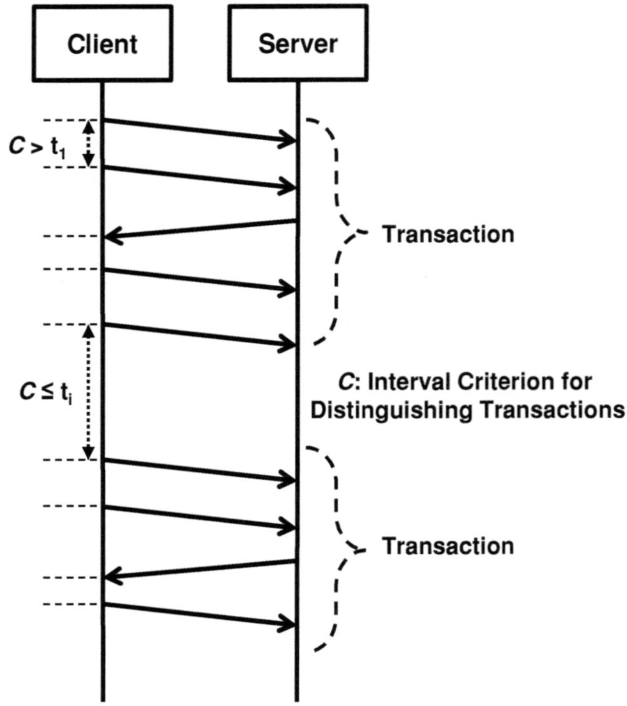

Figure 2. Distinguishing transactions by inter-packet arrival time.

2. Related Work

Some intrusion detection system testing tools generate network traffic that corresponds to known cyber attacks or Snort rules [10, 16, 17]. These tools are useful for measuring the detection rates of intrusion detection systems that rely on attack signatures. However, in order to use these tools for performance analyses of anomaly-based intrusion detection systems, it is necessary to properly mix the generated attack traffic and normal traffic.

Industrial control system testbeds can be used to analyze vulnerabilities, threats and the impacts of attacks. A testbed may be developed using real systems, simulators or a combination of real and simulated systems. Popular simulation tools include Simulink, Stateflow and dSPACE [2, 8, 11]. The tools support automatic code generation, task scheduling and fault management applications for modeling, simulation and testing. Some researchers have used programmable logic controllers and control protocol emulators for constructing honeypots that provide anomalous traffic [3].

SCADA system testbeds have been developed at the national level for security research and analysis. One example is the National SCADA Testbed (NSTB) developed by the U.S. Department of Energy [2]. Other SCADA testbeds have been evaluated by the U.S. National Institute of Standards and Technology (NIST) and the British Columbia Institute of Technology (BCIT) in Canada. In Europe, testbeds are operational in Grenoble, France; CERN in Geneva, Switzerland; and at the European Joint Research Centre in Ispra, Italy [14]. Christiansson and Luiijf [4] discuss the development of a European SCADA security testbed. Japan also uses an industrial control system testbed for various purposes, including vulnerability analysis [9].

The National SCADA Testbed [2] combines state-of-the-art facilities at national laboratories with expert research, development, analysis and training to identify and address security vulnerabilities and threats in the energy sector. The test and research facilities include field-scale control systems, and advanced visualization and modeling tools.

Other SCADA testbeds have been developed to support similar activities as the National SCADA Testbed. However, they are large and expensive, and are only available to selected researchers. The complexity and scale of a testbed can be reduced, but the results obtained do not adequately model real-world systems. The absence of high-fidelity testbeds that provide open access to researchers has made it very difficult to independently evaluate the research results published in the SCADA systems security literature.

Two other test methods are possible. The first relies on data gathered from real-world systems. In this case, it is possible to perform practical analyses of real traffic. However, it is difficult to conduct evaluations because attack scenarios involve traffic that often does not exist in the captured traffic, requiring attack traffic to be generated artificially.

The second method is to use publicly-available test data provided by organizations that operate testbeds. In the field of industrial control systems, some datasets have been made available, including for secure water treatment [6], S7Comm [18] and Modbus [13]. These datasets enable researchers to quantitatively evaluate the performance of different security techniques and tools. However, when testing anomaly detection systems, it is necessary to experiment with many variations of each abnormal situation. Unfortunately, publicly-available datasets do not maintain adequate amounts of such data.

3. Abnormal Traffic Generation

Given normal traffic and an attack scenario, the test traffic generator (TG) automatically generates a variety of abnormal traffic by changing: (i) target packets (i.e., packets selected to represent anomalies in normal traffic); (ii) generation times (i.e., specific times during which attacks occur repeatedly or regularly in normal traffic); and (iii) applied IP addresses (i.e., changes to the IP source address and/or IP destination address of packets to specific IP addresses corresponding to attack scenarios).

First, the traffic generator selects normal traffic for a certain condition that forms the basis of the scenario. Next, it modifies the selected normal traffic according to the scenario. The basic traffic generation process is as follows:

- **Preprocessing:** The traffic generator receives PCAP-type normal traffic and selects the traffic related to the specific communications section (IP, edge, session and service) specified by the user and uses it to generate abnormal traffic. The selected traffic is referred to as "base traffic."

- **Target Traffic Extraction:** The traffic generator receives the number and length of the target traffic from the user. It then extracts the target traffic by randomly selecting the start time from the base traffic. The target traffic is part of the base traffic and is used to generate abnormal traffic. A variety of abnormal traffic is generated for a single attack scenario by extracting target traffic at various points in time from the base traffic and using it to generate abnormal traffic.

- **Target Traffic Modification:** The traffic generator modifies the target traffic packets according to the attack scenario to generate abnormal traffic. The traffic is transformed by performing an "action" on target traffic. An action involves modifying, adding or deleting some packets or transactions. This creates test traffic corresponding to cyber attacks. The characteristics of the abnormal traffic expected according to the attack are defined as "actions."

  In a real industrial control system, it is highly likely that various types of cyber attacks are performed periodically on multiple devices. To simulate this, n cyber attacks as expressed as n actions, and multiple actions are performed in parallel or sequentially on abnormal traffic. The user selects the number of actions according to the attack scenario and creates a scenario file by selecting elements such as the attack time and frequency, target packet selection criterion and packet transformation method for each action.

  A variety of cases can be created by changing the details of an action based on some condition without fixing it to specific values. In other words, the various test traffic corresponding to an attack scenario can be automatically generated and used for performance evaluation, improving the reliability of the results.

  The traffic generator implements packet-based and transaction-based traffic modifications. A packet is the basic unit of network communications. However, when data is transmitted in the network, it is broken up into multiple packets. For example, when using the TCP protocol, data is divided into several packets and the receiver sends a response to each packet. It is difficult to express the characteristics of such traffic by examining individual packets. An attack is more likely to manifest itself in a transaction than in an individual packet.

3.1 Time and Periodicity of Actions

Multiple actions on target traffic can be performed simultaneously according to each attack cycle. The user has to select the number of actions based on

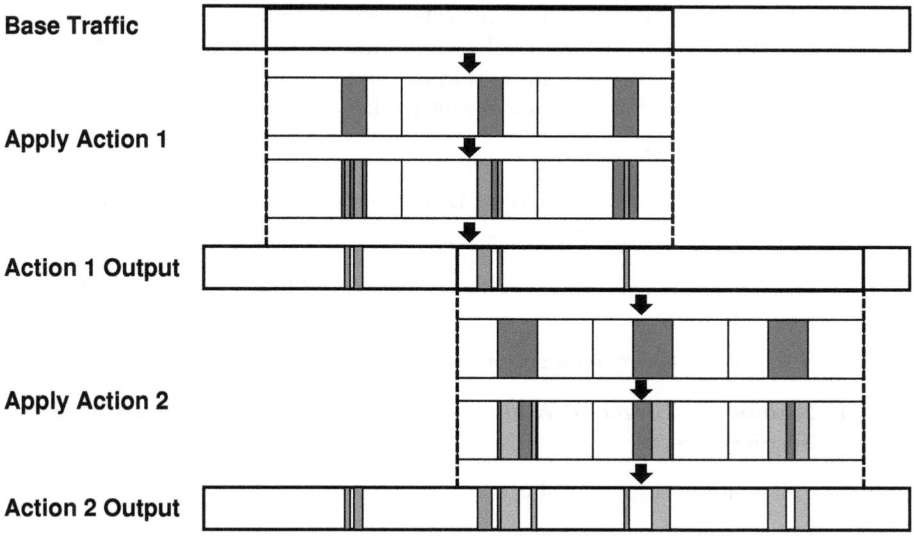

Figure 3. Combination of two actions.

the attack scenario, and then set the attack cycle, attack start time and attack end time for each action.

Figure 3 shows that, when Action 1 is applied, the test traffic, which is the output of Action 1, is generated from the base traffic. When Action 2 is applied to the test traffic transformed by Action 1, test traffic on which Action 1 and Action 2 are simultaneously applied is generated. This procedure generates test traffic corresponding to multiple combined attacks. Since multiple attacks can occur at the same time in a real environment, it is possible to express this situation via multiple action definitions. This can also be used to evaluate whether or not a specific attack type is classified correctly under multiple attacks. If scenarios that define actions are shared and reused, then experimental results and intrusion detection performance can be compared using common actions in normal traffic corresponding to each user. A user can create and test individual traffic with shared actions or traffic with multiple actions in combination with other actions. This produces a variety of anomalous traffic for testing purposes.

3.2 Target Packets of an Action

In order to transform traffic, the target packets used for transformation should be selected for each target data. The target data is part of the target traffic and is segmented at a specific time. A cyber attack on an actual industrial control system involves a specific target IP address, edge (IP address to IP address), session and service. Therefore, the user should specify the criteria for selecting target packets in a scenario.

The traffic generator provides packet-based and transaction-based transformations. When transforming traffic on a transaction-by-transaction basis, the user should select a target transaction instead of a target packet.

The user options for selecting target packets are summarized as follows:

I. Target Types:

1. Attack occurs at a specific IP address.
2. Attack occurs at a specific edge (IP address, IP address).
3. Attack occurs on a specific session (IP address, port, protocol, port, IP address).
4. Attack occurs on a specific service (protocol, port).

II. Number of Target IP Addresses (N_{TI}), Edges (N_{TE}), Sessions (N_{TSS}), Services (N_{TS}):

1. Enter a constant value.
2. Enter an occurrence rate ($x1\%$).

– $N_{Tx1} = x1\%$ of the number of IP addresses/edges/sessions/services used in target traffic.

III. Target IP Address Selection:

1. Input a target IP address/edge/session/service and use it in all the target data.
2. Select a target IP address/edge/session/service randomly for each target data. If a smaller number of IP addresses is used for specific target data, then all the IP addresses in the target data become target IP addresses. The same is true for edge and session.
3. Randomly select target traffic and use it all the target data.

IV. Number of Target Packets (N_{TP}):

1. Enter a constant value.
2. Enter an occurrence rate ($x2\%$).

– N_{TP} = total number of packets in target traffic $\times$ $x2\%$/number of target data.

Based on the four options listed above, the traffic generator selects N_{TP} packets in the target traffic.

3.3 Traffic Modification by an Action

The traffic generator supports four operations for directly modifying target packets or transactions:

- **Payload Change:** Change the payload of the target packet based on the byte section provided by the user.

- **Packet Transmission Rate Change:** Change the packet transmission rate by modifying the packet number of normal traffic by adding or deleting a packet to normal traffic or changing the byte length of the target packet.
- **Packet Replacement/Addition:** Replace the target packet or add an attack packet to the target packet.
- **Header Change:** Change the transmission time and IP address of a target packet. For example, to represent a replay attack, the headers containing the transmission time of the target packet and IP address information should be changed and added to the normal traffic. In order to represent a packet forgery in an intermediate attack on a specific (IP address, IP address) interval, a target packet is selected in the interval and the payload of the selected target packet is modified and added to the normal traffic.

The user options for target traffic transformation are stored in the scenario file. The options are summarized as follows:

I. Payload Change:

1. Change confirmation
   (a) Make a change. When changing to T_R, the same option applies to all the packets in T_R.
   (b) Do not make a change. The remaining options (2 and 3) are not input.
2. Enter a payload change interval (byte).
3. How to change the payload.
   (a) Enter the change value.
   (b) Change to a random value.

II. Packet Transmission Rate Change:

1. Change confirmation.
   (a) Make a change. When changing to T_R, the same option applies to all the packets in T_R.
   (b) Do not make a change. The remaining options (2 and 3) are not input.
2. Count change (increase, decrease or maintain the number of packets).
   (a) Increase: Number of test traffic packets is greater than the number of target traffic packets. Increase the amount of test traffic by copying the target packet based on the packet growth rate ($x3\%$) selected by the user.
   (b) Reduction: Number of test traffic packets is smaller than the number of target traffic packets. Reduce the number of test traffic by deleting part of the target packet based on the packet reduction rate ($x3\%$) selected by the user. $N_{AP} = N_{TP} \times x3\%$.

   (c) No change: Number of test traffic packets and number of target traffic packets do not change.

3. Count change (increase, decrease or maintain the number of packets).
   (a) Increase: Change the byte length of the target packet by appending a random value to the end of the target packet.
   (b) Reduction: Remove the payload part of the target packet to change the byte length of the target packet.
   (c) No change.

III. Packet Replacement/Addition:

1. Delete the target packet and replace it with an attack packet.
2. Add an attack packet to the target packet.

IV. Header Change:

1. Change confirmation.
   (a) Make a change. When changing to T_R, the same option applies to all the packets in T_R.
   (b) Do not make a change. The remaining options (2, 3 and 4) are not input.
2. Transmission time change.
   (a) Sequential offset: Transmission time of the target packet is shifted by an offset time provided by the user and employed as the transmission time of the attack packet (at this time, the packet leaving the action period is discarded).
   (b) Sequential random: Keep only the transmission order of the target packet and randomly transmit the generated attack packet in the action period.
   (c) Random: Randomly transmit the generated attack packet in the action period.
   (d) No change.
3. IP address change.
   (a) Randomly change the target packet IP address to an IP address in the base traffic and use it as the IP address of the attack packet.
   (b) Change the target packet IP address to a user-specified IP address.
   (c) No change.
4. Session change.
   (a) Change within the target session.
   (b) Change the session associated with the transmission/reception of the target packet.
   (c) Randomly select one of the (IP address, port) values in the target traffic.
   (d) Change the session to a user-specified session on a transaction-by-transaction basis.
   (e) No change.

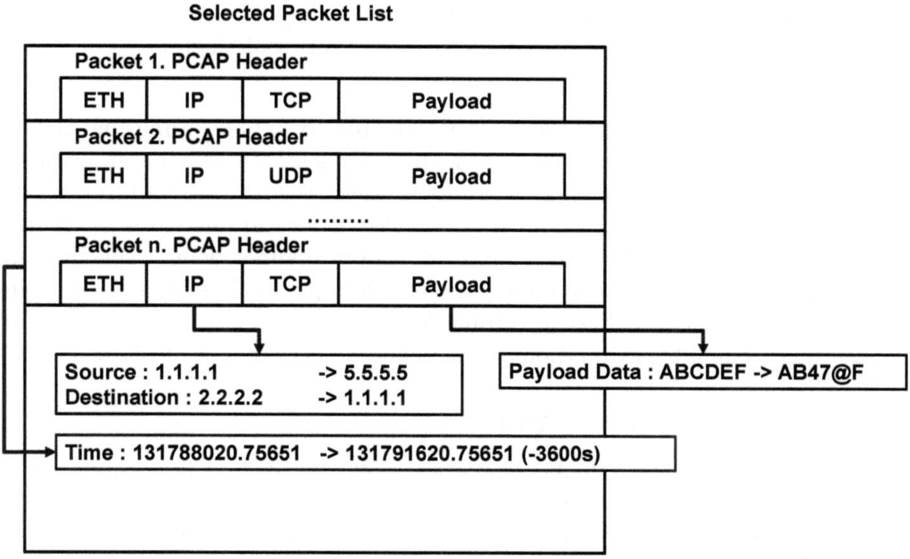

Figure 4. Example of packet modification.

Figure 4 shows an example of packet modification. Packets in the selected list are modified based on the input condition values. The packets are then combined with normal traffic to generate anomalous traffic.

A user may define traffic modifications based on specific cyber threat scenarios by listing the actions that yield the following effects:

- By specifying a protocol, it is possible to express abnormal behavior using a protocol vulnerability or to select abnormal behavior that occurs at a specific point (IP address). By changing the IP addresses in common packets, it is possible to represent a distributed denial-of-service attack that transmits packets from various IP addresses to specific IP addresses, or a man-in-the-middle attack that intercepts packets from certain IP addresses and sends them to other IP addresses.

- Network attacks can occur simultaneously or repeatedly at various time intervals. It is possible to represent attacks that occur at specific times and an attack that occurs repeatedly.

- Increasing the amount of traffic can represent abnormal behavior corresponding to a denial-of-service attack. Reducing the amount of traffic can represent abnormal behavior corresponding to intentional packet drops. Since this method increases or decreases the amount of traffic at several levels, the denial-of-service criterion can be determined by considering the general packet volume and throughput in the network environment. If throughput information is not available, it is possible to predict a denial-of-service attack by specifying an acceptable scale factor.

- By changing IP addresses, it is possible to express abnormal behavior such as communications at unexpected locations or communications at abnormal times at various locations. Details such as IP addresses, times and transmission content can be created for various cases by changing these configurations in a fixed or random manner or in a specific range. In other words, it is possible to create a large number of cases for a single scenario. The resulting automatically-generated test traffic can support highly-reliable evaluations of intrusion detection system performance.

4. Implementation

In the experiments, PCAP traffic was collected from a real industrial control network and passed to the traffic generator. The traffic collection was accomplished using an application programming interface (API) – `libpcap` for Unix/Linux systems and WinPcap for Windows systems. Since the PCAP traffic was collected in an industrial control network, it contained information about the real environment.

The traffic generator created anomalous PCAP traffic from the collected PCAP traffic, which was added to the original PCAP traffic to create the test PCAP traffic. Since the real environment was reflected in the original traffic, the test traffic captured normal operations as well as attacks. After creating the test traffic, it may be sent to a network, machine learning system or a security device (intrusion detection system or firewall) for learning and testing.

The traffic generator was written in Python 2.7. Wireshark was employed to leverage its PCAP splitting and merging functions (`editcap` and `mergecap`). The `scapy` library was used for PCAP `read` and `write` functions and the `multiprocessing` library was used for speed up. The performance was increased by dividing a large-capacity PCAP file into 1,000 units using `editcap` and then reading it with `multiprocessing`. Note that the selection of 1,000 units was arbitrary and a user may increase or decrease the number of units based on memory availability.

4.1 Preprocessing

The traffic generator receives PCAP-type normal traffic from the collected network traffic and generates base traffic by selecting only the traffic related to specific IP addresses/edges/sessions/services designated by a user. The user inputs a CSV file with preprocessing options to the traffic generator as shown in Table 1. Note that "–" means any and "$r(n)$" means select the number n randomly. If multiple rules (preprocessing conditions) are provided as in Table 1, then the packets that satisfy at least one rule are included in the base traffic.

The following options are included in Table 1:

Table 1. Preprocessing options.

Option	IP_{src}	$Port_{src}$	Protocol	IP_{dest}	$Port_{dest}$	Bidirectional
1	IP1	–	–	–	–	No
2	IP2	–	–	IP3	–	No
3	IP4	Port1	–	IP5	Port2	Yes
4	–	–	Proto1	–	–	No
5	IP6	Port3	Proto2	IP7	Port4	Yes
6	$r(50)$	–	–	–	–	No

- **Option 1:** All the packets sent from and received at IP1 (IP address selection).
- **Option 2:** All the packets sent between IP2 and IP3 (edge selection).
- **Option 3:** All the packets sent from IP4-Port1 to IP5-Port2 (session selection).
- **Option 4:** All the packets using Proto1 (service selection).
- **Option 5:** All the packets sent from IP6-Port3 to IP7-Port4 using Proto2.
- **Option 6:** Fifty randomly-selected IP addresses from among the IP addresses in the input data, and all the packets transmitted from and received at the 50 IP addresses.

The traffic generator can also provide information about the IP addresses/edges/sessions/services for traffic that a user can employ to create an attack model. Each file provides a list of IP addresses/edges/sessions/services used by the traffic. If base traffic is already available, the traffic generator can proceed directly to the target traffic generation step without any preliminary work.

4.2 User Configuration File

The traffic generator modifies normal traffic according to the characteristics of an attack scenario to create abnormal traffic. A user inputs a scenario (discussed in Sections 3.2 and 3.3 and Table 1) in the form of a CSV file that embodies the characteristics of the test method and attack scenario. The traffic generator then creates: (i) target traffic according to the options listed in the scenario file; (ii) divides the target traffic into target data representing attack periods; and (iii) generates abnormal traffic by performing actions on the target data. The abnormal traffic that is generated is also in the PCAP format and has the same size as the target traffic.

Table 2 shows a scenario file that simulates a query injection attack by changing the payloads of randomly-selected target packets in target traffic. Since only the payload is changed, not the header, it corresponds to a man-in-the-middle (MiTM) attack. The target traffic is divided into five pieces of

Table 2. Generation of abnormal traffic for a query injection attack.

Target Traffic Generation	- Number of target traffic (N_T): 100 - Length of target traffic(L_T): 5 min		
Target Data Generation	- Number of actions (N_{Act}): 1 - Attack period (P_A): 1 min - Starting point (t_{A1}): 0 min - Ending point (t_{A2}): 5 min (Five target data of one minute in length are generated)		
Action	**Target Packet Type**	- Action period ($t_{A1} \sim t_{A2}$): $0 \sim 60$ s Total period and action period set to same value - Target type: 2. Attack occurs at a specific edge (IP address, IP address) - Number of target IP addresses/edges/sessions/ services: 2. Enter the occurrence rate (x1%) N_{TE} = 1% of number of edges in target traffic - How to specify target edge: 3. Randomly select target traffic and use the same in all target data	
	Target Packet Selection	- Number of target packets (N_{TP}): 2. Enter occurrence rate = 0.01% N_{TP} = Packets in target traffic $\times$ 0.0001/5 - Select target packets: 1. Randomly select N_{TP} target packets from packets using target edge in target data	
	Target Traffic Transformation	**Payload**	1. Change confirmation: (a) Perform the change 2. Enter payload change interval: 1$\sim$5 bytes 3. How to change payload: (c) Change to random value
		Traffic Volume	1. Change confirmation: (b) No change
		Replacement/ Addition	1. Delete target packet and replace it with attack packet
		Header	1. Change confirmation: (b) No change

target data of one minute each to perform an action. If the action period and total period are the same, then the target data length would be meaningless because the attack does not have any periodicity.

When it is executed, the traffic generator produces the target packet list, modified packet list, test traffic and test traffic log information. By comparing the target packet list against the modified packet list, it is possible to con-

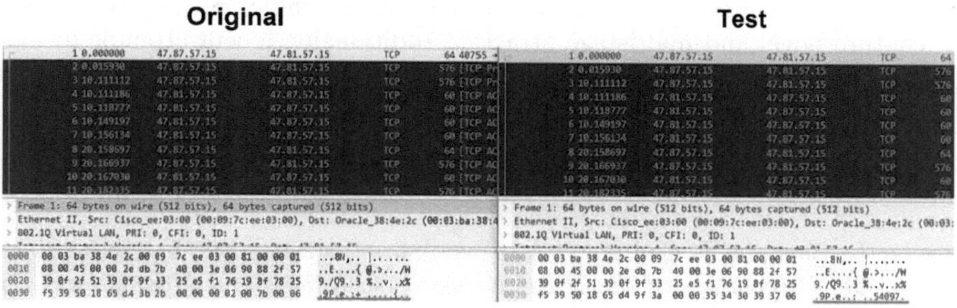

Figure 5. Traffic generation for a query injection attack.

firm whether or not traffic modifications were performed based on the attack scenario. The test traffic log stores the options pertaining to the scenario file and information about traffic generation. After the test traffic is generated using the scenario file and the collected industrial control network traffic, the resulting target traffic and modified packets are shown in Figure 5.

The proposed approach can change protocol commands as desired (e.g., to DNP3, IEC61850 or Modbus). Injection is modeled by specifying the byte portion that contains the command and changing it to another command desired by the user. This method handles bytes; therefore, if the structure of the protocol is known, the desired protocol commands can be generated.

5. Conclusions

The principal challenge in conducting research on securing industrial control networks from cyber attacks is the lack of availability of real-world network traffic that reflects normal and anomalous operations. Although it is possible to collect traffic under normal operating conditions, due to security concerns and the potential impact on operations, it is very difficult, if not impossible, to collect abnormal network traffic from real-world industrial control systems. While testbeds can overcome this limitation, they are expensive to implement and operate; moreover, they will never completely model their real counterparts. Additionally, it is infeasible to create and analyze a large number of cyber attack scenarios, especially when each scenario can have numerous variations.

This chapter has addressed the problem by proposing a method for automatically generating a variety of anomalous test traffic based on cyber threat scenarios related to industrial control systems. The proposed method starts with normal traffic that is collected from a real industrial control network. Leveraging abnormal scenarios provided by users, the method automatically generates anomalous (attack) traffic based on target connections, time, traffic amounts and transmission content that satisfy the scenarios. The anomalous traffic is added to the original traffic to create the test traffic for developing and evaluating intrusion detection systems.

Future research will enhance the automated traffic generation process to capture novel and multistage attacks. Additionally, it will attempt to model the potential impacts of traffic with manipulated packets and/or transactions on real industrial control devices.

References

[1] R. Barbosa, R. Sadre and A. Pras, A first look into SCADA network traffic, *Proceedings of the IEEE Network Operations and Management Symposium*, pp. 518–521, 2012.

[2] H. Christiansson and E. Luiijf, Creating a European SCADA security testbed, in *Critical Infrastructure Protection*, E. Goetz and S. Shenoi (Eds.), Springer, Boston, Massachusetts, pp. 237–247, 2007.

[3] Conpot Development Team, CONPOT: ICS/SCADA Honeypot (`conpot.org`), 2018.

[4] C. Davis, J. Tate, H. Okhravi, C. Grier, T. Overbye and D. Nicol, SCADA cyber security testbed development, *Proceedings of the Thirty-Eighth North American Power Symposium*, pp. 483–488, 2006.

[5] O. Depren, M. Topallar, E. Anarim and M. Ciliz, An intelligent intrusion detection system (IDS) for anomaly and misuse detection in computer networks, *Expert Systems with Applications*, vol. 29(4), pp. 713–722, 2005.

[6] J. Goh, S. Adepu, K. Junejo and A. Mathur, A dataset to support research in the design of secure water treatment systems, *Proceedings of the Eleventh International Conference on Critical Information Infrastructures Security*, pp. 88–99, 2016.

[7] A. Hahn and M. Govindarasu, Cyber attack exposure evaluation framework for the smart grid, *IEEE Transactions on Smart Grid*, vol. 2(4), pp. 835–843, 2011.

[8] A. Hahn, B. Kregel, M. Govindarasu, J. Fitzpatrick, R. Adnan, S. Sridhar and M. Higdon, Development of the PowerCyber SCADA security testbed, *Proceedings of the Sixth Annual Workshop on Cyber Security and Information Intelligence Research*, article no. 21, 2010.

[9] Information-Technology Promotion Agency, About IPA, Tokyo, Japan (`www.ipa.go.jp`), 2018.

[10] Ixia, Test Architecture, Calabasas, California (`www.ixiacom.com/solutions/test-architecture`), 2018.

[11] M. Knauff, J. McLaughlin, C. Dafis, D. Niebur, P. Singh, H. Kwatny and C. Nwankpa, Simulink model of a lithium-ion battery for the hybrid power system testbed, *Proceedings of the ASNE Intelligent Ships Symposium*, 2007.

[12] A. Lazarevic, L. Ertoz, V. Kumar, A. Ozgur and J. Srivastava, A comparative study of anomaly detection schemes in network intrusion detection, *Proceedings of the SIAM International Conference on Data Mining*, pp. 25–36, 2003.

[13] A. Lemay and J. Fernandez, Providing SCADA network datasets for intrusion detection research, *Proceedings of the Ninth USENIX Workshop on Cyber Security Experimentation and Test*, 2016.

[14] S. Luders, Control systems under attack? *Proceedings of the Tenth International Conference on Accelerator and Large Experimental Physics Control Systems*, 2005.

[15] S. Mukkamala, G. Janoski and A. Sung, Intrusion detection using neural networks and support vector machines, *Proceedings of the International Joint Conference on Neural Networks*, vol. 2, pp. 1702–1707, 2002.

[16] pevma, rule2alert (`github.com/pevma/rule2alert`), 2014.

[17] pytbull, What is pytbull? (`pytbull.sourceforge.net`), 2018.

[18] N. Rodofile, T. Schmidt, S. Sherry, C. Djamaludin, K. Radke and E. Foo, Process control cyber attacks and labeled datasets on S7Comm critical infrastructure, *Proceedings of the Twenty-Second Australasian Conference on Information Security and Privacy*, Part II, pp. 452–459, 2017.

[19] J. Song, H. Takakura, Y. Okabe, M. Eto, D. Inoue and K. Nakao, Statistical analysis of honeypot data and building of Kyoto 2006+ dataset for NIDS evaluation, *Proceedings of the First Workshop on Building Analysis Datasets and Gathering Experience Returns for Security*, pp. 29–36, 2011.

[20] K. Stouffer, V. Pillitteri, S. Lightman, M. Abrams and A. Hahn, Guide to Industrial Control Systems (ICS) Security, NIST Special Publication 800-82, Revision 2, National Institute of Standards and Technology, Gaithersburg, Maryland, 2015.

[21] J. Wan, A. Canedo and M. Al Faruque, Security-aware functional modeling of cyber-physical systems, *Proceedings of the Twentieth IEEE Conference on Emerging Technologies and Factory Automation*, 2015.

[22] J. Yun, S. Jeon, K. Kim and W. Kim, Burst-based anomaly detection for the DNP3 protocol, *International Journal of Control and Automation*, vol. 6(2), pp. 313–324, 2013.

[23] B. Zhu, A. Joseph and S. Sastry, A taxonomy of cyber attacks on SCADA systems, *Proceedings of the International Conference on Internet of Things and Fourth IEEE International Conference on Cyber, Physical and Social Computing*, pp. 380–388, 2011.

Chapter 15

VARIABLE SPEED SIMULATION FOR ACCELERATED INDUSTRIAL CONTROL SYSTEM CYBER TRAINING

Luke Bradford, Barry Mullins, Stephen Dunlap and Timothy Lacey

Abstract Industrial control systems employ a variety of hardware, software and network protocols to control physical processes that are critical to societal functions. It is vital that industrial control system operators receive quality training to defend against cyber attacks. Hands-on training exercises with real-world control systems enable operators to learn defensive techniques and understand the real-world impacts of their control decisions. However, cyber attacks and operator actions have unforeseen effects that can take a significant amount of time to manifest and potentially cause physical harm to systems, making high-fidelity training exercises costly and time-consuming.

This chapter presents a methodology for accelerating training exercises by simulating and predicting the effects of cyber events in partially-simulated control systems. A hardware-in-the-loop simulation comprising a software-modeled water tank and a commercially-available programmable logic controller are used to demonstrate the feasibility of the methodology. The experimental results demonstrate that the effects of cyber events can be accurately simulated at speeds faster than real time, significantly enhancing operator training regimens.

Keywords: Industrial control systems, training, hardware-in-the-loop simulation

1. Introduction

Most critical infrastructure operators lack the training to prevent and properly respond to sophisticated cyber attacks against industrial control systems [1]. Additionally, information security personnel lack an understanding of industrial control systems and cannot predict the impacts of changes to control systems and networks. Adversaries know that cyber attacks launched against industrial control systems have the potential to cause significant physical harm to critical infrastructure assets. The ability to cause physical damage and the lack

J. Staggs and S. Shenoi (Eds.): Critical Infrastructure Protection XII, IFIP AICT 542, pp. 283–306, 2018.
https://doi.org/10.1007/978-3-030-04537-1_15

of well-trained personnel render industrial control systems attractive targets for adversaries. Exacerbating the lack of cyber-capable control system operators is the fact that most training programs provide instruction at the basic or intermediate knowledge levels [7].

The absence of thorough, advanced training regimens for industrial control system security is primarily due to the lack of robust facilities that provide experience with real-world control system components, physical systems and processes. Consequently, there is an urgent need for training environments that blend real control system components with physical processes to enable trainees to understand the effects of cyber attacks and defensive actions. Since cyber effects often take significant amounts of time to manifest themselves, replicating their impacts in a learning environment becomes infeasible. Additionally, unforeseen consequences of cyber events have the potential to cause catastrophic damage to the equipment used for training. The potential damage to equipment makes high-fidelity training exercises prohibitively expensive. Thus, a critical component of any solution is the ability to quickly model the effects of cyber attacks so that more time can be devoted to analysis and evaluation, which correspond to higher levels of learning [4].

An ideal training environment is a full-scale, real-world facility with several interconnected processes [7]. However, using such an environment for training is expensive and time-consuming. To address these challenges, this chapter proposes a methodology for augmenting industrial control system security training environments by enabling exercise coordinators to rapidly model and predict the effects of cyber events. The methodology engages a hardware-in-the-loop (HiL) simulation and commercially-available programmable logic controllers (PLCs) to increase the speed of a physical process while enabling the programmable logic controllers to operate as intended. The speed-up feature enables trainees to gain valuable expertise and to understand the consequences of their actions while limiting the possibility of physical damage to equipment.

2. Background

This section provides background information about industrial control systems, cyber training environments and hardware-in-the-loop simulation, and discusses related work.

2.1 Industrial Control Systems

The United States describes the critical infrastructure as systems and assets, both physical and virtual, so crucial to the country that their destruction or incapacity would threaten U.S. national security, the economy, power supply, public health, public safety and other areas [5]. Presidential Policy Directive 21 [6] lists sixteen U.S. critical infrastructure sectors, including energy, transportation, water and communications. Assets in all the critical infrastructure sectors engage industrial control systems in order to achieve increased automation, efficiency and maintainability.

Industrial control systems monitor and control industrial processes with the help of sensors, actuators, control units and networks [9]. In an industrial control system, control units receive data from sensors. Based on the data received from sensors, the control units direct actuators to control the processes being monitored by the sensors in order to produce the desired outputs [9]. An industrial control system can be implemented locally, such as within the confines of a factory, or across numerous devices and pieces of equipment over a large geographical area, such as a power grid. Industrial control systems are typically systems of systems that control multiple interconnected, mutually dependent processes that function together to achieve industrial objectives.

2.2 Cyber Training Environments

Plumley et al. [7] have specified various levels of cognitive complexity for control system training environments. Their goal was to create an industrial control system educational framework that could offer training regimens for a range of organizational budgets and needs. The primary determiner of the level of a training environment is the realism it provides in the context of real industrial control systems. The complexity of the training scenarios that can be provided by a training environment depends on the amount of realism provided. The level of cognitive complexity increases as the training environment realism increases [7]. With this in mind, Plumley and colleagues created and mapped four levels of industrial control system training environments to Bloom's Taxonomy [4] in order to create a comprehensive industrial control system training framework. Training environments capable of administering exercises at higher levels of thinking map to higher levels of the taxonomy while providing access to all the lower levels of the taxonomy.

Bloom's Taxonomy, which was created by Benjamin Bloom (1913-1999), classifies educational objectives based on cognitive complexity [4]. The taxonomy is widely used by educators to structure courses that enable students to learn, apply knowledge, think critically and create new ideas. Bloom's Taxonomy was revised in 2001 to comprise six categories of educational goals [7]. The taxonomy progresses from the lowest cognitive level of basic understanding to the highest cognitive level, which is the creation of original ideas. Bloom's Taxonomy offers a means for aligning educational tools to specific levels of cognitive complexity. Bloom emphasized the acquisition of concrete knowledge before increasing the complexity of a training regimen. In other words, it is imperative that trainees master their current levels in the taxonomy before proceeding to higher levels. This is why, in many high-risk or critical fields, training involves several levels of simulation, where the complexity of each level progressively increases until trainees are proficient enough to attempt real tasks using real equipment.

A Level 1 training environment is entirely software defined. It uses software to simulate an industrial controller or control system. Level 1 environments encompass the lowest two levels of Bloom's Taxonomy – remembering and understanding [7].

A Level 2 training environment includes an automated process that creates real physical effects. However, instead of employing the same hardware and software used in industry, it incorporates simple embedded systems (e.g., Arduino and Raspberry Pi devices) that are programmed to monitor and control physical processes using common programming languages (e.g., C and Python). Level 2 environments cover the applying and analyzing levels of Bloom's Taxonomy [7].

A Level 3 environment uses real hardware and software. The hardware and software control a partial industrial control system [7]. For example, a Level 3 environment for training prison guards would enable them to control the locks on a block of prison cell doors. These environments are not full-scale industrial control systems, but they enable trainees to familiarize themselves with real-world equipment, industrial networks and process systems. Since they are not full-scale systems, they cannot provide an understanding of real-world systems, where a malfunction in one process can affect other processes due to system interdependencies. Level 3 environments reach the evaluating level of Bloom's Taxonomy. Trainees can compare observations against standard operational criteria and data. Realistic data enables them to make realistic evaluations that transfer to real-world systems. To maximize realism and minimize cost and space, Level 3 environments employ hardware-in-the-loop simulations of industrial processes that are controlled by real hardware. Hardware-in-the-loop simulations eliminate the need to incorporate large and expensive physical equipment in training environments by replacing them with accurate simulations that interface with real-world industrial control hardware.

A Level 4 environment is a real-world, full-scale industrial control system training facility. The training facility is essentially identical to its real-world counterpart [7]. Level 4 environments reach the highest level of thinking in Bloom's Taxonomy – creating. In the context of industrial control system training, this type of cognitive complexity cannot be achieved without the real system. Trainees have the ability to view and manipulate every component in the actual industrial environment. New methods and solutions can be devised, tested and applied to the real system, and their effects can be observed.

Cyber attacks are often slow moving and are, therefore, difficult to detect. Stuxnet, the most famous malware to target industrial control systems, took months to conduct its attacks [11]. Since cyber attacks often employ the low-and-slow attack paradigm, it can take a long time to witness their effects. Thus, a key component of a realistic industrial control system training environment is a means for speeding up the progress of slow moving attacks and their impacts during training activities.

2.3 Hardware-in-the-Loop Simulation

The industrial control system training environment discussed in this chapter is a Level 3 system that can speed up simulations of physical processes, enabling trainees to quickly detect and observe the progression of cyber events, and predict their effects. It employs a hardware-in-the-loop simulation of an industrial process that is controlled by an actual programmable logic controller. Hardware-in-the-loop simulation is a common technique used in industry to develop and evaluate complex, real-time, embedded systems [8]. It yields a simulated process under control that creates a realistic test environment for embedded systems. During testing, the embedded system interacts with the process simulation. A key component of hardware-in-the-loop simulation is the electrical emulation of sensors and actuators, which serves as the interface between the simulation and embedded system under test. The process simulation determines the values of the electrically-emulated sensors. These values are read by the embedded system under test as feedback. The control algorithm running on the embedded system under test outputs actuator control signals based on the feedback received. The output control signals produce changes to the variable values in the simulation, including the values measured by the sensors. Thus, hardware-in-the-loop simulation incorporates a complete control loop.

Hardware-in-the-loop simulation is commonly used to test embedded systems because, in many cases, it is more efficient (and safer) than connecting the embedded system directly to a real process. For example, the simulation can enhance the quality of testing by increasing the scope of test scenarios and overcoming the testing limitations imposed by a real process. Using a real process also prevents the embedded system from being tested under failure conditions. Furthermore, the simulation assists in developing embedded systems under tight schedules that do not allow testing to be delayed until a process prototype becomes available. Finally, it is more economical to conduct testing using a high-fidelity, real-time hardware-in-the-loop simulation instead of a real process.

2.4 Related Work

Saco et al. [8] noted that the ideal learning environment for industrial control system operators is a real-world plant, but they acknowledged that such environments would be large, expensive and potentially dangerous to be used by novices. Recognizing the need for high-fidelity simulation tools that could provide effective training regimens, Saco and colleagues proposed a hardware-in-the-loop, real-time simulation system. MATLAB Simulink, a software tool for modeling dynamic systems, was used to model the control algorithm and plant (water tank with an inflow pump and pneumatic drainage valve). The control algorithm was converted to a C program using the Simulink Real Time Workshop software. The C code was downloaded to real-time prototyping hardware (dSpace 1102 floating-point controller board) and executed independently of MATLAB. After the C code was downloaded to the board and executed, the controller hardware was able to control the water level in the tank simulation. The functionality provided by Simulink enabled trainees to implement and refine their own (similar) hardware-in-the-loop systems to gain better understanding of the physical system and control principles.

Thornton and Morris [10] state that the ideal environment for studying cyber attacks against industrial control systems is a real system with real hardware, software and communications technologies. Since such environments are prohibitively expensive, Thornton and Morris developed a virtual laboratory testbed that was mobile, sharable and expandable. The testbed incorporated a Simulink gas pipeline simulation with sensors and actuators, a virtual programmable logic controller simulated with Python, and a human-machine interface (HMI). The Simulink gas pipeline and the virtual logic controller communicated via JavaScript Object Notation (JSON) attribute-value pairs contained in user datagram protocol (UDP) packets. The virtual logic controller communicated with other devices such as the human-machine interface and physical programmable logic controllers via Modbus/TCP, a standard industrial control system protocol. Communications between the virtual logic controller and physical logic controllers enabled the virtual laboratory testbed to produce accurate control system network traffic for analysis.

Unfortunately, the two environments described above do not incorporate real-world control system hardware. Incorporating industrial control system hardware brings simulated environments much closer to real industrial systems, especially when attempting to understand normal operations and attack scenarios. Moreover, the two environments cannot speed up operations to quickly model the effects of attacks and provide more time for operator training and analysis. For example, without a speedup capability, low and slow attacks such as Stuxnet cannot be replicated quickly enough to observe their effects and learn from them in a reasonable timeframe.

3. Methodology

This section describes the proposed methodology, including the test systems and experimental design.

3.1 Test Systems

The test environment incorporated two systems that implemented a water tank control loop. A Lab-Volt 3531 training system was used as the baseline, real-world control system [3]. A simulated water tank system replicated the operation of the Lab-Volt training system. The Lab-Volt system and the simulated water tank incorporated the control loop shown in Figure 1. The programmable logic controller in each control loop maintained the water level in the tank at a user-defined set point. It polled the water level sensor to obtain the current water level in the tank. Based on the current water level, the controller sent a command to the drainage valve to increase or decrease the outflow rate.

Lab-Volt 3531 Training System. Figure 2 shows the Lab-Volt 3531 system components. An Allen-Bradley ControlLogix 1756-L55 programmable logic controller that was programmed using RSLogix 5000 from Rockwell Au-

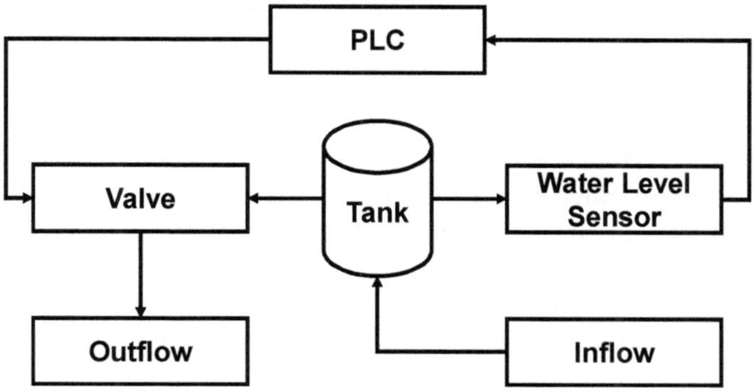

Figure 1. Test system control loop.

Figure 2. Lab-Volt 3531 training system.

tomation served as the primary control unit. The inflow rate of water into the cylindrical tank (8" diameter and 36" height) was controlled by an alternating current pump driven by an Allen-Bradley PowerFlex 40 variable frequency drive (VFD). A differential pressure transmitter (DPT) continuously monitored the inflow rate using a Venturi tube, which created a differential pressure proportional to the rate of flow through the tube. Taps at the high pressure and low pressure portions of the tube were connected to the differential pressure transmitter, which measured the pressure difference and computed the flow rate. The differential pressure transmitter sent the flow rate value to the programmable logic controller as a 4-20 mA analog signal. The programmable logic controller employed a proportional-integral-derivative (PID) control strategy to determine the appropriate pump speed in order to ensure that the water inflow rate remained constant at the set point. The computed pump speed was transmitted to the variable frequency drive as an analog value. An Allen-Bradley PanelView Plus 600 human-machine interface was used to monitor and configure the system.

A globe type valve located at the bottom of the tank enabled water to exit the tank and drain into the holding basin. The valve closed partially or completely based on the analog signal it received from the programmable logic controller. The valve was pneumatically operated and equipped with a spring-and-diaphragm actuator. A plug in the valve restricted the flow of water into the tank outlet. The plug was designed to have a fixed linear relationship between the distance traveled by the valve stem and the amount of flow allowed through the valve. When the valve received an analog signal from the logic controller, the current-to-pressure converter of the valve linearly transformed the analog signal to a pneumatic pressure, which was applied to the surface of the valve diaphragm, producing a force that overcame the spring force and moved the plug up or down. The plug restricted the flow of water through the valve from 0% to 100%. The percentage of flow allowed through the valve is referred to as the valve position.

A second differential pressure transmitter measured the pressure in the bottom tank to compute its water level. The water level value was transmitted to the programmable logic controller as a 4-20 mA analog signal. The programmable logic controller used a PID control strategy to determine the valve position based on the water level value supplied by the differential pressure transmitter. The controller transmitted the desired valve position to the valve as a 4-20 mA analog signal. The PID control strategy maintained the water level in the tank with minimal overshoot, undershoot and set point deviation.

Simulated Water Tank. The process simulator experiment employed a hardware-in-the-loop simulation of the physical process and a ControlLogix 1756-L55 programmable logic controller. Figure 3 shows the simulated and real components of the test system. The hardware-in-the-loop simulation enabled the system to use real industrial control system hardware without having to incorporate physical equipment such as pumps and tanks. The simulated physical process managed by the programmable logic controller mirrored the Lab-Volt water tank system. A pump filled the tank with water at a constant inflow rate and a drainage valve at the bottom of the tank allowed water to exit. The simulation accurately replicated the behavior of the Lab-Volt system.

The simulated water tank was implemented as a MATLAB Simulink model. Simulink is a graphical programming environment that can model a variety of systems by selecting blocks from various block libraries and connecting them via input/output (I/O) arrows. Each block includes customizable features; some blocks enable users to write custom code. Simulink simplified the modeling process because it eliminated the need to write code for complex mathematical functions. Its graphical environment assisted in visualizing the system.

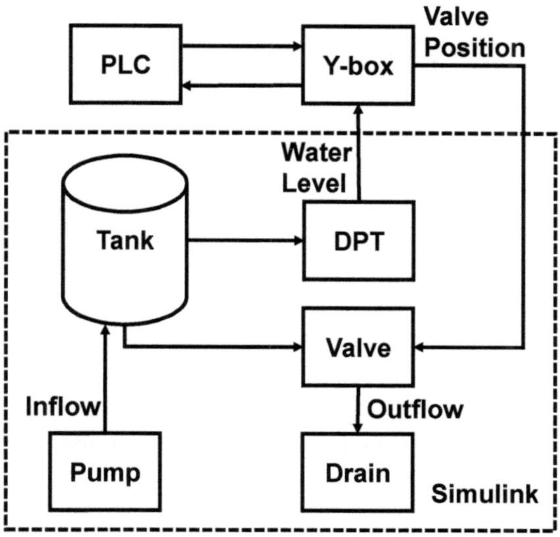

Figure 3. HiL simulation of the Lab-Volt 3531 system.

A MATLAB function block was developed to capture the water tank dynamics. The function block implemented Equations (1) through (5) below that define the dynamics and physical characteristics of the water tank.

The cross-sectional area of the water tank A (in^2) is computed as:

$$A = \pi r^2 \tag{1}$$

where r is the radius of the tank in inches.

The height (level) of water in the tank H in inches is computed as:

$$H = \frac{Vol}{A} \tag{2}$$

where Vol is the volume of water in the tank (in^3).

The inflow rate Q_{in} (in^3/min) is computed as:

$$Q_{in} = 588.9 \tag{3}$$

Note that Q_{in} is a user-provided value that could be changed at any point in time during the simulation; it was set to 588.9 in^3/min.

The outflow rate Q_{out} (in^3/min) is computed as:

$$Q_{out} = V_P \cdot V_C \cdot \sqrt{H\ (\text{psi})} \tag{4}$$

where V_P is the position of the drainage valve that ranges from zero to one ($V_P = 0.5$ corresponds to the valve being half closed); and V_C is the dimensionless valve constant/flow coefficient ($V_C = 6$). Note that Q_{out} increases as the

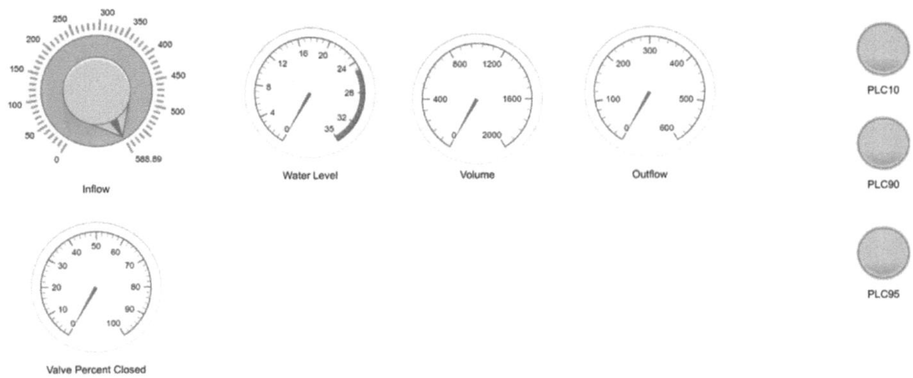

Figure 4. Simulink water tank graphical user interface.

height (level) of water in the tank increases. This reflects the effect of water pressure on the outflow rate.

The rate of change of water volume ΔVol (in^3/min) in the tank with respect to time is computed as:

$$\Delta Vol = Q_{in} - Q_{out} \tag{5}$$

The signal generated by the ΔVol output was fed to an integrator block, which calculated the integral of the derivative with respect to time in order to compute the volume Vol; initially $Vol = 0$, meaning that the tank was empty. This was reflected as an initial condition in the configuration options of the integrator block.

The Simulink Dashboard library supports the rapid development of graphical user interfaces (GUIs) for monitoring and controlling simulated processes. Figure 4 shows the graphical user interface for the simulated water tank. The interface includes a knob block for setting the inflow rate and gauge blocks for monitoring the water level, volume, current drainage valve configuration and outflow rate. The interface also has three light blocks that indicate if the current water level in the tank is at a critical level. By default, the lights shine green. When the water level reaches a critical height, the corresponding light shines red. The three lights correspond to whether the tank is less than 10% full, greater than 90% full and greater than 95% full, respectively.

Figure 5 shows a speedup block from the Real-Time Pacer library, which enabled the simulation speed to be increased or decreased as desired. The speedup block was configured by entering a number that represented the speedup factor for the simulation. For example, a speedup factor of two doubled the speed of the simulation.

A MATLAB function block reported the current water level in the tank to the programmable logic controller every 0.1 seconds. In order to maintain this reporting rate, the sample time t_S of the MATLAB function block was adjusted according to the simulation speedup factor f. When the simulation

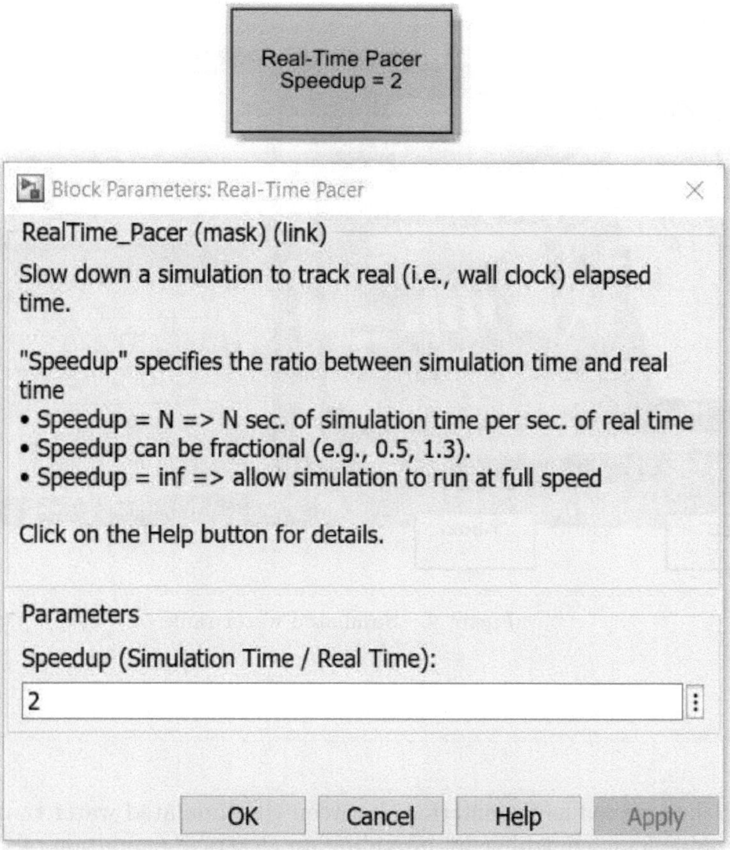

Figure 5. Simulink speedup block.

Figure 6. Setting the sample time.

was executed in real time, the sample time was set to 0.1 seconds as shown in Figure 6. A speedup factor of two required the sample time to be set to 0.2 seconds and a speedup factor of ten required the sample time to be set to 1 second. In general, the sample time t_S is computed as:

$$t_S = \frac{f}{10} \tag{6}$$

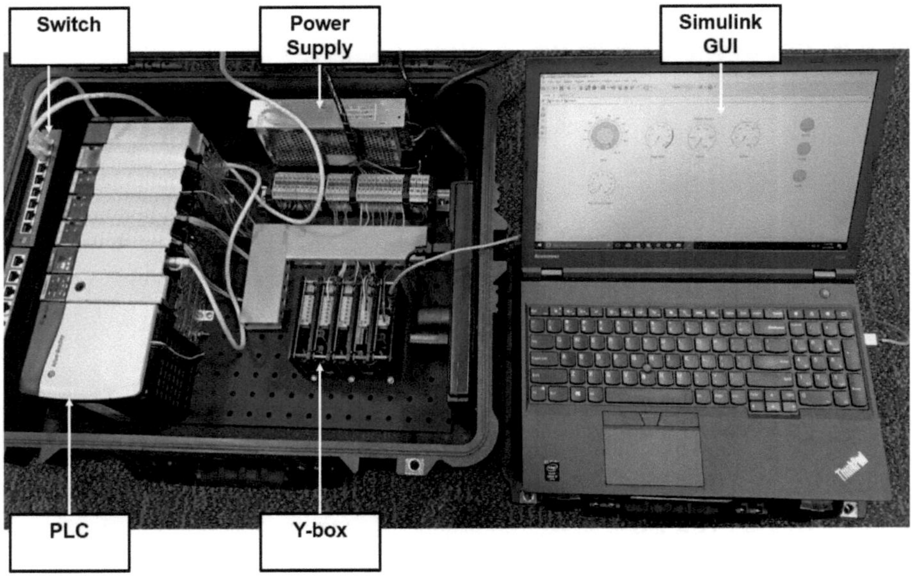

Figure 7. Simulated water tank.

A Y-box served as the interface between the simulated water tank and programmable logic controller by providing an electrical emulation of the sensors and actuators. The Y-box, which was created to support the development of hardware-in-the-loop simulations, received currents and voltages as inputs and generated currents and voltages as outputs based on commands received over a universal serial bus (USB) port [2]. The inputs and outputs enabled the Y-box to interface with a programmable logic controller in the same manner as regular sensors and actuators. The simulation and the Y-box communicated via a serial link. The test system utilized the same model programmable logic controller and ladder logic used to control the Lab-Volt system. Note that the ladder logic used for the simulation did not include the PID controller that controlled the inflow rate.

In the Simulink model, the inflow rate was set as a constant parameter. Minimal changes were made to the ladder logic to enable the programmable logic controller to operate in the simulated environment. The project path was updated with the IP address of the programmable logic controller used in the simulated system. Also, the module numbers were updated to match the hardware present in the programmable logic controller used to control the simulated system. The code section containing the PID controller that adjusted the inflow rate was not configured to activate because it was not used by the Simulink model. Figure 7 shows the setup of the simulated system.

Table 1. Run names.

Name	Meaning
LVTn	Lab-Volt 3531 Trial n
Simx1Tn	HiL System Real-Time Trial n
Simx2Tn	HiL System 2 × Real-Time Trial n
Simx10Tn	HiL System 10 × Real-Time Trial n

3.2 Experimental Design

Validation testing was performed to verify the accuracy and consistency of the simulation. The experiments compared the operation of the simulation against the operation of an actual water tank, specifically, against the Lab-Volt training system. The experiments sought to demonstrate that the simulation reflected the normal operation of a real water tank when executed at real time and at faster simulation rates. Previous pilot studies showed that the Lab-Volt system was consistent from run to run. Specifically, the standard deviation of the average differences between Lab-Volt runs was less than 0.01%. Thus, water level recordings from only three Lab-Volt runs were used in the experiments.

The water tank simulation was also executed three times at each speed specified in Table 1 for a total of nine runs; Table 1 shows the names of the runs executed in the experiments. The water level recordings from the three Lab-Volt runs were compared against the water level recordings from the nine simulated water tank runs. In order to ensure accurate comparisons, all the Lab-Volt and simulation runs followed the procedure shown in Figure 8, which depicts a typical Lab-Volt run in the experiments.

Each run in the experiments comprised the following steps:

- The pump was started, which caused the water level in the tank to rise.

- A Python script was executed to read and write tags in the programmable logic controller ladder logic. The script set the water level set point to 30% full and began to record the current water level and the time at half-second intervals starting at zero seconds. Water levels for the simulation and Lab-Volt runs were measured as percentages, where 0% corresponded to an empty tank and 100% corresponded to a full tank.

- As the water level in the tank approached 30%, the Python script monitored the rising water level to detect the water level steady state at 30%. In the experiments, steady state was considered to be reached when the water level reached the set point and remained at the set point with minimal deviations (less than ±0.5% from the set point). A well-tuned PID controller ensured that the process variable reached the set point with minimal overshoot (less than 3%). The PID controller then corrected the overshoot and the process variable remained close to the set point

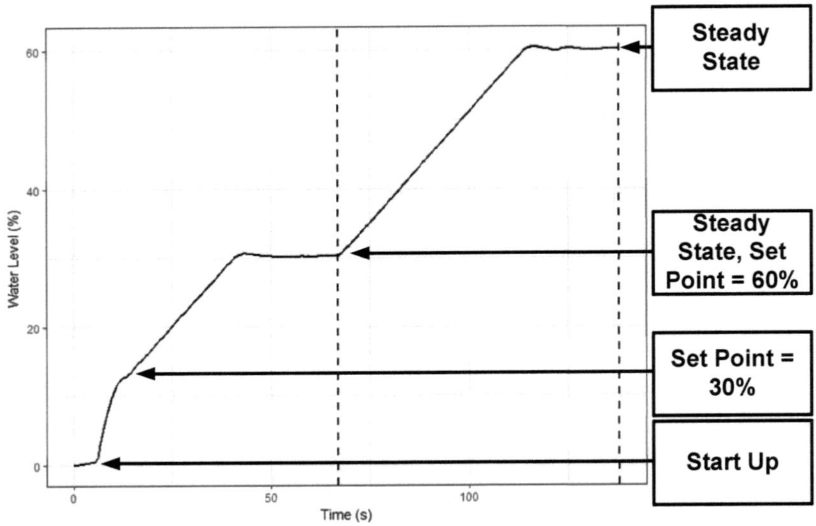

Figure 8. Experimental procedure.

with minimal deviations. The Python script detected the steady state by checking if the current water level was within ±0.1% of the set point. After meeting this threshold, the Python script waited for 25 seconds to enable the PID controller to correct any initial overshoot. After this time, the Python script sent a confirmation that the steady state was reached. Pilot studies demonstrated that a wait time of 25 seconds was adequate for the PID controllers in the simulation and Lab-Volt system to correct initial overshoots and achieve the water level steady state.

- After the script detected the steady state at 30%, it changed the water level set point to 60%. The water level in the tank began to rise and the script waited to detect when the water level was steady at 60%.

- After the script detected that the water level had reached steady state at 60%, it stopped collecting data and ended the run.

Each run in the experiments yielded a curve with water level on the y-axis and time on the x-axis. In order to ensure consistent analysis, the experiments considered an interval of time in which all three Lab-Volt runs achieved steady state at 30%, rose to 60% and achieved steady state at 60%.

After the three Lab-Volt runs were completed and the raw data was collected, another Python script adjusted the three curves so that they reached 45% at the same time (designated as zero seconds). In the first Lab-Volt curve, the Python script used linear interpolation to compute the time at which the curve reached 45% and subtracted this time from all other times in the curve. The script repeated this process for the other two Lab-Volt curves. These adjustments shifted the curves so that that they centered at zero seconds with a height of 45%, the midpoint between the 30% and 60% set points.

After the three curves were adjusted, an 80-second time interval was selected. During this time interval, the three Lab-Volt runs completed the required behavior of achieving steady state at 30%, rising to 60% and achieving steady state at 60%. Before the curves from different runs were compared, all the curves were adjusted so that they reached 45% at zero seconds. Subsequent analysis compared only the portions of the curves within the 80-second time interval to ensure consistent comparisons. The consistency enabled accurate comparisons of each run against every other run in the experiments.

Table 2 shows the matrix used to compare all the runs. Note that the simulation run names are shortened in the matrix. Simx1Tn is denoted as Sx1Tn, Simx2Tn is denoted as Sx2Tn and Simx10Tn is denoted as Sx10Tn.

The experiments compared the curves from the trials by considering the average difference between them. The average difference corresponded to the average distance between the water level for the first curve and the water level for the second curve at each point in time. In order to compute the average difference between two curves, a Python script first adjusted the timestamps for each curve to account for speedup. For example, when the Sx1T1 run was being compared against the Sx10T1 run, all the timestamps in the Sx10T1 run were multiplied by ten to account for the speedup in the Sx10T1 run. The matrix cell represented by this comparison has row Sx10T1 and column Sx1T1.

Next, the Python script adjusted both curves so that they reached 45% at the same time, which was designated as zero seconds. In the case of the Sx1T1 curve, the Python script used linear interpolation to compute the time when the curve reached 45% and subtracted this time from all the other times in the curve. The script repeated this process for the Sx10T1 curve. These adjustments shifted both curves so that that they centered at zero seconds for a water level of 45%, the midpoint between the 30% and 60% set points.

At this point, the script removed all the portions from the two curves that were not included in the 80-second time interval. Then, the script iterated through the remaining timestamps for the Sx1T1 curve and employed linear interpolation to determine the corresponding heights in the Sx10T1 curve. At this point, the Python script had three curves – the Sx1T1 curve, the Sx10T1 curve and the new Sx10T1 curve containing only the timestamps from the Sx1T1 curve and their associated water levels computed via linear interpolation. The matching timestamps enabled the Sx1T1 and Sx10T1 runs to be compared in a straightforward manner.

Finally, the Python script iterated through the Sx1T1 curve and the new Sx10T1 curve. For each timestamp, the script calculated the absolute value of the difference between the water level in the Sx1T1 curve and the water level in the new Sx10T1 curve. After iterating through both curves, the Python script added all the absolute values, divided the sum by the number of timestamps and returned the quotient as the average difference between the two runs.

Table 2. Test matrix.

	LVT1	LVT2	LVT3	Sx1T1	Sx1T2	Sx1T3	Sx2T1	Sx2T2	Sx2T3	Sx10T1	Sx10T2	Sx10T3
LVT1												
LVT2	X											
LVT3	X	X										
Sx1T1	X	X	X									
Sx1T2	X	X	X	X								
Sx1T3	X	X	X	X	X							
Sx2T1	X	X	X	X	X	X						
Sx2T2	X	X	X	X	X	X	X					
Sx2T3	X	X	X	X	X	X	X	X				
Sx10T1	X	X	X	X	X	X	X	X	X			
Sx10T2	X	X	X	X	X	X	X	X	X	X		
Sx10T3	X	X	X	X	X	X	X	X	X	X	X	

Two evaluation metrics were employed in the experiments. The first metric measured whether or not the simulation completed the required behavior of achieving the water level steady state at 30%, raising the water level to 60% and achieving the water level steady state at 60% within the 80-second time interval. If a simulation run completed the required behavior within the time interval, then the run passed the first metric; otherwise, it failed the metric.

The second metric considered the average difference between two runs. The experiments used the mean of the average differences between the three Lab-Volt runs as the threshold for the second evaluation metric. A high-fidelity simulation was expected to have a similar average difference when compared to the Lab-Volt system. If the average difference between two runs was less than or equal to the mean of the average differences between the Lab-Volt runs, then the runs passed the second metric; otherwise, they failed the metric.

4. Experimental Results

This section describes the experimental results.

4.1 Metric 1 (Required Behavior)

All the simulation runs completed the required behavior within the 80-second time interval. Regardless of the simulation speed, all the simulation runs passed the first evaluation metric. In order to verify that each run passed the first metric, all the runs were graphed. For each graph, the portion of the graph containing the 80-second time interval was visually inspected to ensure that the simulation run achieved the required behavior.

Figures 9, 10 and 11 demonstrate the accuracy of the simulation in real-time, two times the speed and ten times the speed, respectively. Note that the time scales for the Simx2 and Simx10 graphs were multiplied by their respective speedup factors to match real time. Each of the three graphs represents a typical comparison of the simulation against the Lab-Volt system. The slopes of the Lab-Volt run and the Simx1, Simx2 and Simx10 runs are almost identical. The slight differences in the slopes is most likely due to variations of the inflow rate in the Lab-Volt system. As mentioned above, the PID controller in the Lab-Volt system that maintained a constant inflow rate produced minor fluctuations in the inflow rate whenever it was forced to adjust the speed of the pump. Another possible cause for the slight differences in the slopes is the trial-and-error method used to tune the PID controller in the simulation. The main differences between the graphs occur as they approach 60% steady state. The differences are due to a control delay in the Lab-Volt system, corresponding to the amount of time between the programmable logic controller sending a command to the valve and the valve adjusting the plug to the correct position. The simulation did not model this control delay and was, therefore, not affected by the delay.

Figure 12 shows how well the simulation runs with speedup factors of two and ten match the simulation run executed at real time. Note that the time

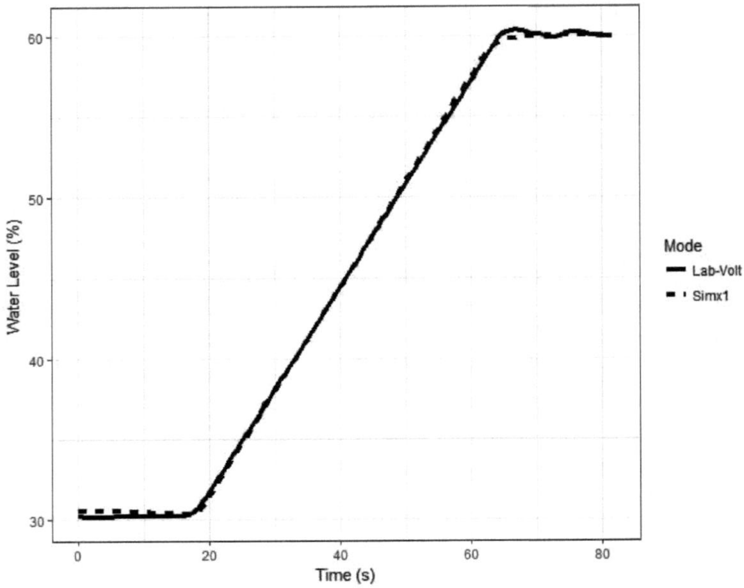

Figure 9. Lab-Volt vs. Simx1.

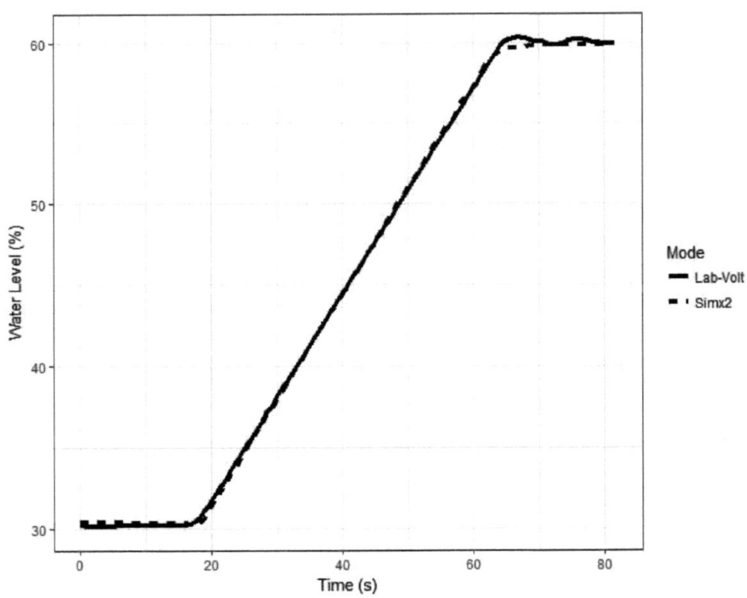

Figure 10. Lab-Volt vs. Simx2.

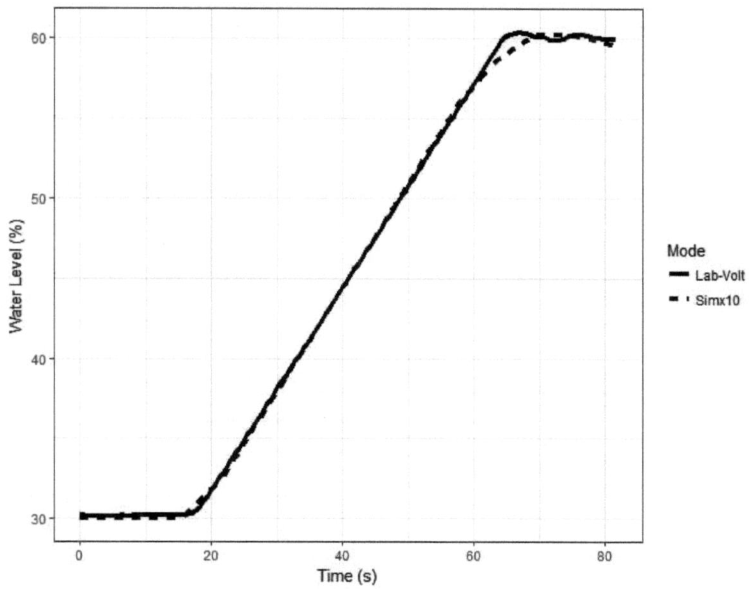

Figure 11. Lab-Volt vs. Simx10.

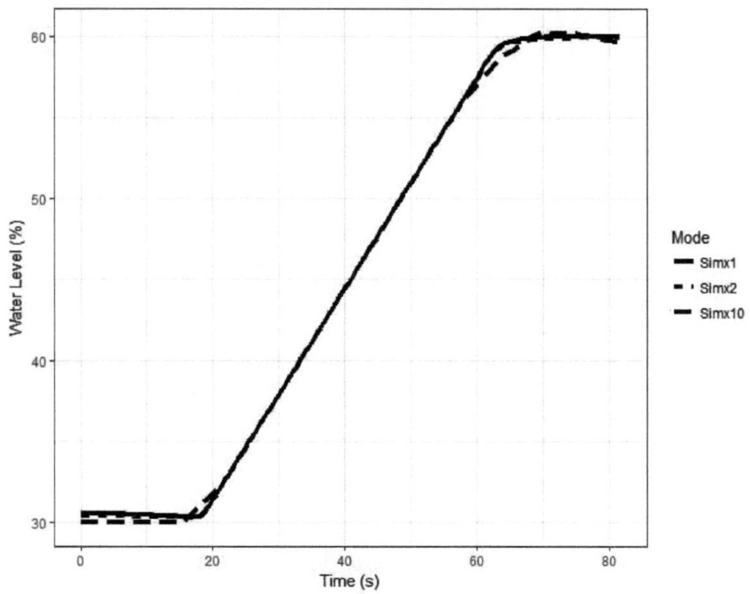

Figure 12. Simx1 vs. Simx2 vs. Simx10.

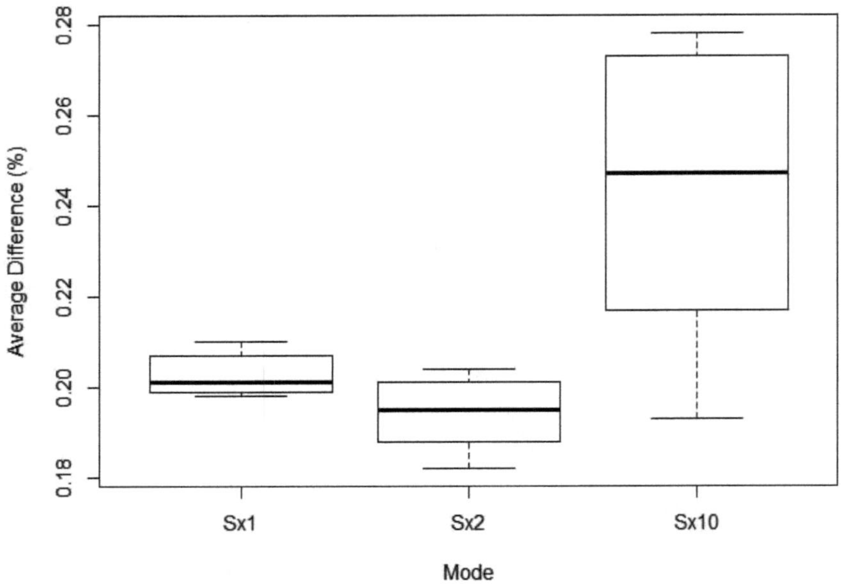

Figure 13. Lab-Volt vs. simulation.

scales for the Simx2 and Simx10 graphs were multiplied by their respective speedup factors to match real time. The primary differences between the Simx1, Simx2 and Simx10 graphs occur as they approach the 60% steady state. The differences most likely result from the trial-and-error method used to tune the PID controller in the simulation.

Table 3. Simulation accuracy – average differences (%).

Comparison	Mean	Min	Max	StDev
Lab-Volt vs. Lab-Volt	0.055	0.045	0.061	0.009
Lab-Volt vs. Simx1	0.203	0.198	0.210	0.005
Lab-Volt vs. Simx2	0.194	0.182	0.204	0.007
Lab-Volt vs. Simx10	0.245	0.193	0.278	0.031

4.2 Metric 2 (Average Difference)

Figure 13 and Table 3 summarize the average differences between the Lab-Volt system and the simulation.

The average difference when comparing the Lab-Volt runs is well below 0.1%. When executed at real time, the average difference between the Lab-Volt and simulation runs ranges from 0.198% to 0.21%. When executed at two times

Table 4. Run consistency – average differences (%).

	Mean	Min	Max	StDev
Lab-Volt	0.055	0.045	0.061	0.009
Simx1	0.021	0.014	0.026	0.006
Simx2	0.037	0.031	0.048	0.009
Simx10	0.102	0.045	0.131	0.049

faster than real time, the average difference between the Lab-Volt and simulation runs ranges from 0.182% to 0.204%. When executed at ten times faster than real time, the average differences between the Lab-Volt and simulation runs ranges from 0.193% to 0.278%. The mean of the average differences for all the comparisons between the simulation and the real water tank is 0.214%, which is well above 0.055%, the mean of the average differences between the Lab-Volt runs. These results demonstrate that the average difference between the simulation and Lab-Volt runs is significantly greater than the average difference between the Lab-Volt runs regardless of the simulation speed. Thus, the proposed simulation does not pass the second evaluation metric.

Although the simulation does not pass the second evaluation metric, all the simulation results are relatively consistent as shown in Figure 13. In fact, all the simulation runs produce relatively low average differences ranging from 0.182% to 0.278%. The mean of the average differences for the Simx2 runs is slightly lower than the mean for the Simx1 runs.

A permutation test comparing the average differences from the Simx1 and Simx2 runs produced a p-value of 0.01354, which is greater than the 0.01 threshold for the 99% confidence level. Thus, the permutation test showed that no significant difference exists between the mean of the average differences for the Simx2 runs and the mean of the average differences for the Simx1 runs at the 99% confidence level. The null hypothesis that the two means are equal cannot be rejected because the p-value is greater than 0.01. This implies that the simulation accuracy is the same when the simulation is run at real time and at two times real time. Overall, the experimental results demonstrate that the proposed simulation consistently models the Lab-Volt system with less than $\pm 0.28\%$ error on average at any point in time. The average differences between the simulation and the real water tank are consistently low with a standard deviation of 0.028%, even when the simulations were executed at speeds much faster than real time.

4.3 Consistency

Table 4 demonstrates the consistency of multiple runs of the Lab-Volt system and the simulation. The first row of the table represents the consistency of the Lab-Volt system from run to run. The mean of the average differences between Lab-Volt runs is 0.055%, the minimum average difference is 0.045% and the

Table 5. Effect of speedup – average differences (%).

Comparison	Mean	Min	Max	StDev
Simx1 vs. Simx1	0.021	0.014	0.026	0.006
Simx1 vs. Lab-Volt	0.203	0.198	0.210	0.005
Simx1 vs. Simx2	0.070	0.056	0.081	0.008
Simx1 vs. Simx10	0.256	0.220	0.280	0.025

maximum average difference is 0.061%. The standard deviation for the Lab-Volt runs is 0.009%. The other rows in the table demonstrate the consistency of the simulation from run to run at each speed. When executed at real time, the average differences between the simulation runs are more consistent than the average differences between the Lab-Volt runs. The consistency in the average differences decreases as the simulation speed increases. The table shows that there is much less consistency from run to run when the simulation was executed with a speedup factor of ten. This decrease in consistency may be due to the reduced precision of the linear interpolation technique used to compute the curves for the Simx10 runs. Since the simulations generated fewer points when they were executed at higher speeds, there were fewer reference points for the linear interpolation computations. Consequently, the Simx10 curves have higher variability.

4.4 Simulation Speedup

Table 5 summarizes the average differences between: (i) simulation runs executed at real time; (ii) simulation runs executed at real time versus the Lab-Volt system; (iii) simulation runs executed at real time versus simulation runs executed with a speedup factor of two; and (iv) simulation runs executed at real time versus simulation runs executed with a speedup factor of ten.

The first row in Table 5 shows that the simulation yields consistent results from run to run when executed at real time. The second and third rows demonstrate that, when executed at real time, the simulation is consistent with the Lab-Volt system and the simulation executed with a speedup factor of two, respectively. The fourth row shows that the average difference between the simulation executed at real-time is consistent with the simulation executed with a speedup factor of ten. The average differences between the simulation executed at real time and the simulation executed with a speedup factor of two are much lower than the average differences between the simulation executed at real time and the simulation executed with a speedup factor of ten. As mentioned above, this difference is due to the imprecise tuning of the PID controller in the simulation, which has a greater effect when the simulation is executed with a speedup factor of ten.

5. Conclusions

The proposed methodology has been designed to accelerate the pace of industrial control system training exercises. The methodology incorporates commercial programmable logic controllers to enable trainees to work with real control system components. A hardware-in-the-loop simulation is employed to speed up the simulated physical process, providing trainees with an understanding of the impacts of their control actions during normal operations, attacks and other cyber events. While consistency and accuracy are lost at high simulation speeds, the experimental results reveal that the physical process responds appropriately. In each test case, when a trainee changed the set point, the programmable logic controller responded to the change and adjusted the water level in the tank. Simulation runs were performed at real time, twice as fast as real time and ten times as fast as real time. As expected, the fidelity of the accelerated simulation runs depends on the accuracy and fidelity of the physical process model. Nevertheless, the environment incorporating a simulated system in conjunction with a full-scale industrial control system enables trainees to gain operational expertise efficiently, safely and at reduced cost.

While this research has demonstrated the feasibility of the simulation speedup methodology, additional work is required to decrease the negative impacts induced by increasing the simulation speed. The simulation should also be tested to determine how well it copes with diverse operating conditions and models the effects of attacks and other cyber events. Additionally, refinements should be made to improve the consistency and accuracy of the simulation. Other future improvements include: (i) devising a universal approach for tuning PID controller parameters; (ii) incorporating additional interconnected processes and components; (iii) testing the upper bound of the speedup factor to determine how fast the proposed methodology can accurately and consistently accelerate cyber events; and (iv) formulating an approach for seamlessly and concurrently transferring cyber events from full-scale, real-world testbeds such as Level 4 environments to the simulation system.

Note that the views expressed in this chapter are those of the authors and do not reflect the official policy or position of the U.S. Air Force, U.S. Department of Defense or U.S. Government.

Acknowledgement

This research was partially supported by the U.S. Department of Homeland Security Industrial Control Systems Cyber Emergency Response Team (ICS-CERT).

References

[1] J. Butts and M. Glover, How industrial control system security training is falling short, in *Critical Infrastructure Protection IX*, M. Rice and S. Shenoi (Eds.), Springer, Cham, Switzerland, pp. 135–149, 2015.

[2] A. Chaves, M. Rice, S. Dunlap and J. Pecarina, Improving the cyber resilience of industrial control systems, *International Journal of Critical Infrastructure Protection*, vol. 17, pp. 30–48, 2017.

[3] Festo Didactic, Familiarization with the Training System: Pressure, Flow and Level, User Guide 86004-E0, Quebec, Canada, 2016.

[4] M. Forehand, Bloom's Taxonomy, in *Emerging Perspectives on Learning, Teaching and Technology*, M. Orey (Ed.), Global Text Project, University of Georgia, Athens, Georgia (`textbookequity.org/Textbooks/Orey_Emergin_Perspectives_Learning.pdf`), pp. 41–47, 2010.

[5] J. Moteff and P. Parfomak, Critical Infrastructure and Key Assets: Definition and Identification, CRS Report for Congress, RL32631, Congressional Research Service, Washington, DC, 2004.

[6] B. Obama, Presidential Policy Directive 21: Critical Infrastructure Security and Resilience, The White House, Washington, DC, 2013.

[7] E. Plumley, M. Rice, S. Dunlap and J. Pecarina, Categorization of cyber training environments for industrial control systems, in *Critical Infrastructure Protection XI*, M. Rice and S. Shenoi (Eds.), Springer, Cham, Switzerland, pp. 243–271, 2017.

[8] R. Saco, E. Pires and C. Godfrid, Real-time controlled laboratory plant for control education, in *Proceedings of the Thirty-Second Annual Conference on Frontiers in Education*, pp. T2D-12–T2D-16, 2002.

[9] K. Stouffer, J. Falco and K. Scarfone, Guide to Industrial Control Systems (ICS) Security, NIST Special Publication 800-82, National Institute of Standards and Technology, Gaithersburg, Maryland, 2011.

[10] Z. Thornton and T. Morris, Enhancing a virtual SCADA laboratory using Simulink, in *Critical Infrastructure Protection IX*, M. Rice and S. Shenoi (Eds.), Springer, Cham, Switzerland, pp. 119–133, 2015.

[11] K. Zetter, An unprecedented look at Stuxnet, the world's first digital weapon, *Wired*, November 3, 2014.

MIX
Papier aus verantwortungsvollen Quellen
Paper from responsible sources
FSC® C105338

If you have any concerns about our products,
you can contact us on
ProductSafety@springernature.com

In case Publisher is established outside the EU,
the EU authorized representative is:
Springer Nature Customer Service Center GmbH
Europaplatz 3, 69115 Heidelberg, Germany

Printed by Libri Plureos GmbH
in Hamburg, Germany